Repetitive DNA

AF778110

Genome Dynamics

Vol. 7

Series Editor

Michael Schmid Würzburg

Repetitive DNA

Volume Editor

Manuel A. Garrido-Ramos Granada

26 figures, 11 in color, and 1 table, 2012

KARGER

Basel · Freiburg · Paris · London · New York · New Delhi · Bangkok · Beijing · Tokyo · Kuala Lumpur · Singapore · Sydney

Dr. Manuel A. Garrido-Ramos
Departamento de Genética
Facultad de Ciencias
Universidad de Granada
Avda. Fuentenueva s/n
18071 Granada (Spain)

Library of Congress Cataloging-in-Publication Data

Repetitive DNA / volume editor, Manuel A. Garrido-Ramos.
p. ; cm. -- (Genome dynamics, ISSN 1660-9263 ; v. 7)
Includes bibliographical references and index.
ISBN 978-3-318-02149-3 (hard cover : alk. paper) -- ISBN 978-3-318-02150-9 (e-ISBN)
I. Garrido-Ramos, Manuel A. II. Series: Genome dynamics ; v. 7. 1660-9263.
[DNLM: 1. DNA--genetics. 2. Repetitive Sequences, Nucleic Acid. 3. Genomics--methods. W1 GE336DK v.7 2012 / QU 58.5]

614.5'81--dc23

2012014216

Bibliographic Indices. This publication is listed in bibliographic services, including Current Contents®.

Disclaimer. The statements, opinions and data contained in this publication are solely those of the individual authors and contributors and not of the publisher and the editor(s). The appearance of advertisements in the book is not a warranty, endorsement, or approval of the products or services advertised or of their effectiveness, quality or safety. The publisher and the editor(s) disclaim responsibility for any injury to persons or property resulting from any ideas, methods, instructions or products referred to in the content or advertisements.

Drug Dosage. The authors and the publisher have exerted every effort to ensure that drug selection and dosage set forth in this text are in accord with current recommendations and practice at the time of publication. However, in view of ongoing research, changes in government regulations, and the constant flow of information relating to drug therapy and drug reactions, the reader is urged to check the package insert for each drug for any change in indications and dosage and for added warnings and precautions. This is particularly important when the recommended agent is a new and/or infrequently employed drug.

All rights reserved. No part of this publication may be translated into other languages, reproduced or utilized in any form or by any means electronic or mechanical, including photocopying, recording, microcopying, or by any information storage and retrieval system, without permission in writing from the publisher.

© Copyright 2012 by S. Karger AG, P.O. Box, CH–4009 Basel (Switzerland)
www.karger.com
Printed in Germany on acid-free and non-aging paper (ISO 9706) by Bosch Druck GmbH, Ergolding
ISSN 1660–9263
e-ISSN 1662–3797
ISBN 978–3–318–02149–3
e-ISBN 978–3–318–02150–9

Contents

Editorial

As has been clearly stated by the former Series Editor of *Genome Dynamics*, Jean-Nicolas Volff, this book series aims to provide readers with an up-to-date overview on genome structure and diversity. Therefore, it is a great pleasure to introduce volume 7 entitled 'Repetitive DNA'. The existence of repetitive DNAs in the genomes of eukaryotes was first recognized in 1961 by Kit [1] and Sueoka [2] by virtue of their unique buoyant density in DNA density gradient centrifugation using caesium chloride or caesium sulphate. During the following 50 years, molecular biology revealed an astonishing richness of diverse reiterated DNA classes, such as transposon-derived sequences, inactive retroposed copies of cellular genes, simple sequence repeats, segmental duplications, and large blocks of tandemly repeated sequences [3]. The importance of repetitive DNAs is underlined by the simple fact that repeated sequences account for more than half of the human genome.

The initial idea to this book was born during a visit at the University of Granada (Spain) where Manuel A. Garrido-Ramos of the Department of Genetics convincingly exposed the need of reviewing more recent research on these fascinating classes of DNA. He has done a remarkable job in selecting and coordinating authorities in the field to write ten chapters covering a wide range of subjects. I express my gratitude to him and all the authors for all the time they invested. The constant support of Thomas Karger with this timely book series is again highly appreciated.

Michael Schmid
Würzburg, March 2012

References

1 Kit S: Equilibrium sedimentation in density gradients of DNA preparations from animal tissues. J Mol Biol 1961;3:711–716.

2 Sueoka N: Variation and heterogeneity of base composition of deoxyribonucleic acids: a compilation of old and new data. J Mol Biol 1961;3:31–40.

3 Platzer M: The upcoming genome and its upcoming dynamics; in Volff J-N (ed): Vertebrate Genomes, Genome Dynamics Vol 2. Basel, Switzerland, Karger Publishers, 2006, pp 1–16.

Preface

The seventh volume of *Genome Dynamics* is dedicated to 'Repetitive DNA'. Eukaryotic genomes are composed of a plethora of different types of DNA sequences repeated from a few to hundreds of thousands times, either dispersed or arranged in tandem. The experimental data compiled by the new molecular techniques associated with the completion of genome projects has led to changes in our understanding of the structural features, functional implications and evolutionary dynamics of these repetitive DNA sequences. These recent developments have opened new insights into the knowledge of mechanisms involved in gene expression, organization, and evolution of multigene families, the fraction of the eukaryotic repetitive DNA which has an undisputedly clear function. Also, we have a comprehensive view today on the structure and functionality of telomeres and centromeres, both composed of repetitive DNA sequences. Additionally, these advances have shed light on the most abundant fraction of repetitive DNA, composed of microsatellite DNA, satellite DNA and, above all, transposable elements. Though not long ago these genomic elements were thought to accumulate as junk or, alternatively, as genomic parasites proliferating for their own benefit, today this early view is changing in most cases. Thus, microsatellite DNAs might facilitate an organism's evolvability, satellite DNA transcripts might participate in heterochromatin formation as well as in modulation of gene expression. Also, today there is no doubt about the significant role of mobile elements in shaping the structure and evolution of genes and genomes, generating genetic innovations and regulating gene expression. The present volume offers a timely update of recent developments in the repetitive DNA research, including the study of multigene families, centromeres, telomeres, microsatellite DNA, satellite DNA, and transposable elements.

I would like to thank all authors who have contributed to this volume with their excellent review articles and the referees for their invaluable efforts. I also want to express my gratitude to the Series Editor Dr. Michael Schmid and his team as well as to Karger Publishers for their outstanding assistance during the preparation of this volume.

Manuel A. Garrido-Ramos
Granada, March 2012

Garrido-Ramos MA (ed): Repetitive DNA.
Genome Dyn. Basel, Karger, 2012, vol 7, pp 1–28

The Repetitive DNA Content of Eukaryotic Genomes

I. López-Flores · M.A. Garrido-Ramos

Departamento de Genética, Facultad de Ciencias, Universidad de Granada, Granada, Spain

Abstract

Eukaryotic genomes are composed of both unique and repetitive DNA sequences. These latter form families of different classes that may be organized in tandem or may be dispersed within genomes with a moderate to high degree of repetitiveness. The repetitive DNA fraction may represent a high proportion of a particular genome due to correlation between genome size and abundance of repetitive sequences, which would explain the differences in genomic DNA contents of different species. In this review, we analyze repetitive DNA diversity and abundance as well as its impact on genome structure, function, and evolution.

Copyright © 2012 S. Karger AG, Basel

The Repetitive Fraction of Eukaryotic Genomes

Pioneering work by Britten and Kohne [1] revealed that in addition to unique sequences the eukaryotic genomes contain large quantities of repetitive DNA, classified into moderately or highly repetitive sequences according to their degree of repetitiveness. Later, the repetitive DNA sequences were grouped according to other criteria such as their organization (tandemly arrayed or dispersed) or their functional role. Although repetitive DNA sequences include several types of RNA or protein-coding sequences, most of the repetitive part of the genome was earlier considered 'junk DNA' with no known function. Today, with many genomes completely sequenced and the background research of more than 40 years, we have ample information on the significance of the repetitive DNA within eukaryotic genomes and concepts are changing. Figure 1 shows a classification of the several types of repetitive DNA according to an organizational criterion, which has been followed in this review. Among tandem repetitive DNA, there are moderately repetitive DNAs, such as ribosomal RNA (rRNA) and protein-coding gene families or short tandem telomeric repeats, as well as highly repetitive non-coding microsatellite and satellite

DNAs, including centromeric DNA. Among dispersed repeats, transposable elements (TEs) such as DNA transposons and retrotransposons (mainly long terminal repeat (LTR) retrotransposons and long interspersed elements, LINEs) stand out, constituting a fraction of highly repetitive DNA as a whole. In addition, genomes contain retrotransposed sequences such as short interspersed elements (SINEs; moderately to highly repetitive DNA), retrogenes and retropseudogenes, as well as several gene families composed of dispersed members (moderately repetitive DNA). In addition, many genomes are characterized by segmental duplications (SDs), duplicated DNA fragments greater than 1 kb, with both dispersed and tandem organization.

Gene Families

Gene families are groups of paralogous genes, typically exhibiting related sequences and functions. A gene family is produced when a single gene is copied one or more times by a gene-duplication event, such as whole-genome duplication (ancient polyploidy is common in plant lineages and is considered a key factor in eukaryote evolution) and SD (see below). Over time, duplications may occur several times and produce many copies of a particular gene. Family sizes range from 2 members up to several hundred [2]. Depending on their organization, gene families are classified into dispersed and tandem gene families. Dispersed genes include for example the families of olfactory receptor genes from mammals (forming the largest known multigene family in the human genome: 802 genes, 388 potentially functional and 414 apparent pseudogenes), the MADS box genes, the fatty acid-binding protein genes or the tRNA genes (see [3] for references). Among tandem gene families, some examples are globins, histones, and rRNA genes.

Ribosomal RNA genes (rDNA) are probably the best-known example of a multigene family. rRNA plays a vital role in protein synthesis, as it constitutes the main structural and the catalytic component of the ribosomes. In most eukaryotes, rDNA consists of tandemly arrayed repeat units, containing 3 of the 4 genes encoding nuclear rRNA, located in the nucleolar organizer region (NOR) on 1 or more chromosomes. Each repeat unit contains the 28S large subunit, the 18S small subunit, the 5.8S gene, as well as 2 external transcribed spacers (ETS) and 2 internal transcribed spacers (ITS1 and ITS2) and a large non-transcribed spacer (NTS). Thus, the nuclear rRNA genes are typically arranged as a 5′-ETS-18S-ITS1-5.8S-ITS2-28S-ETS-3′ transcription unit, organized in tandem repeats and separated by the NTS. The ETS plus the NTS constitute the intergenic spacer (IGS). This is known as the major rDNA family. The number of repeat units varies between eukaryotes, from 39 to 19,300 in animals and from 150 to 26,000 in plants [4]. The different components forming rDNA are known to evolve generally at different rates. The 18S rDNA is among the slowest-evolving genes found in living organisms, contrary to the spacers, which are rapidly evolving sequences (they are not the subject to selective constraints) with the NTS evolving faster than the ITSs and ETSs [2]. The 28S rRNA gene also evolves relatively slowly. The evolution of the rRNA gene complex at varying rates has different phylogenetic utilities. The 18S and 28S rRNA genes allow the inference of phylogenetic history across a broad taxonomic range, whereas the spacers can be useful in determining relationships between closely related species, sometimes intraspecific relationships, and at times have been suitable for population studies. Nucleotide sequences of spacers are very similar among repeats of the same species but differ greatly between species. The model of concerted evolution should explain this observation in which the individual repeats do not evolve independently (see below). Instead, the molecular drive force tends to homogenize repeated sequences within genomes and among the genomes of an entire species, leading to divergence between species [5]. However, nucleotide sequences of the rRNA coding regions are almost identical between closely related species, and they are similar even among distantly related species. This similarity should be maintained by strong purifying selection that operates for the coding regions. Thus, we can explain the entire set of observations concerning the rRNA gene family in terms of mutation, homogenization, and purifying selection [3]. The fourth rRNA gene is the gene encoding 5S rRNA, which forms another family known as the minor rDNA family, which comprises tandem repetitions of the gene separated by an NTS. In most eukaryotes, the 5S rRNA genes are found at another location of the nuclear genome, although e.g. in sturgeons, the 2 rDNA families are in the same chromosome pair and in some species of protozoa, fungi, and algae the 5S ribosomal genes are located between the 28S and the 18S genes (within the IGS) [6]. The 5S rRNA genes were also believed to undergo concerted evolution. However, it has been found recently that the 5S genes located at different loci might evolve by the birth-and-death evolution model. This model predicts that new genes in a family are formed by gene duplication (diversification), and some of these duplicate genes specialize (differentiate) and are maintained in the genome for a long period of time, while others are inactivated or deleted in different species (pseudogenization) [3]. In this sense, Freire et al. [7] found that the 5S genes of mussels showed a mixed mechanism, involving the generation of genetic diversity through birth-and-death, followed by a process of local homogenization resulting from concerted evolution in order to maintain the genetic identities of the different 5S genes.

Histone genes provide another widely known example of tandemly arrayed genes. Histones are highly conserved eukaryotic proteins that have a crucial role in the function and formation of the nucleosome. There are 5 major histone genes – H1, H2A, H2B, H3, and H4 – which are separated from each other by non-coding IGSs. Each major histone gene includes some minor variant forms. Some variants originate from changes in only a few amino acids (for example mouse H3.1 and H3.2 differ only in 1 amino acid), while other variants originate from changes affecting larger portions of the protein (e.g. mouse H3.1/H3.2 and H3.3) [8, 9]. The number of histone genes varies between species. For example, the yeast *Saccharomyces cerevisiae* has 2 copies of each major histone gene, whereas some urchin species contain up to 1,000 copies. Although histone genes are generally arranged in tandem arrays, in some species they are clustered but not tandemly organized (e.g. the mouse genome contains 2 clusters located on different chromosomes) or found scattered across different chromosomes (e.g. in *Caenorhabditis elegans* and in *Zea mays*) [8]. In *Drosophila melanogaster*, the 5 major genes are arranged in a repeating unit which is tandemly repeated 110 times on chromosome 2L. In addition, variant histone genes are located in other parts of the fly genome [3].

Among higher eukaryotic species, H4 and H3 proteins are highly conserved and even distantly related species such as animals and plants have very similar protein sequences. For example, only 3 out of 135 residues differentiate animal and plant H3 protein [3]. This high sequence identity might indicate that multigene families encoding histones evolve by concerted evolution. Nevertheless, histone genes as well as other multigene families (such as the major histocompatibility complex or MHC, immunoglobulin, and olfactory receptor genes) evolve primarily by the birth-and-death model of evolution [3, 8, 10]. This model promotes genetic diversity under recurrent gene duplication events and strong purifying selection acting at the protein level, this latter not being systematically required, which would eventually lead to the functional differentiation of the new gene copies through a process of neofunctionalization or subfunctionalization [3].

Microsatellite DNA

Microsatellites are tandem repeats of <9 nucleotides found in arrays of <1 kb, distributed throughout the genome of every organism studied so far. They are also known as simple sequence repeats (SSRs) or short tandem repeats (STRs). The number of repeat units at a given microsatellite locus typically varies between 5 and 40, but longer series of repeats are possible. For example, we have frequently found microsatellites consisting of a dinucleotide repeat up to 80 times in ferns (unpublished results). This variation in repeat number results in alleles of varying lengths. In addition, variation exists between loci. Microsatellites can be found in both protein-coding and noncoding regions, including regulatory sequences. Dinucleotides are the foremost type of microsatellite repeats for many species, although repeats with units containing a multiple of 3 nucleotides (trinucleotide and hexanucleotide repeats) are the most abundant in coding regions, presumably because they do not cause a frameshift mutation. The most common dinucleotide repeat type in the human genome is $(CA)_n/(GT)_n$, while $(GA)_n/(CT)_n$ and $(AT)_n/(TA)_n$ repeats are the most common in plants [11], and $(CT)_n/(GA)_n$ is the most common in some invertebrate species [12, 13]. The proportion of microsatellite sequences within genomes tends generally to increase from invertebrates and fungi to plants and vertebrates. For example, estimates for *C. elegans* and *S. cerevisiae* are 0.21% and 0.30%, the range in plants is between 0.37% for *Z. mays*, and

0.85% for *Arabidopsis thaliana*, while for the fish species *Fugu rubripes* and *Tetraodon nigroviridis* the estimates are 2.12% and 3.21%, respectively [14].

Microsatellites have a characteristic mutational behavior. Their mutation rates are 10 to 100,000 times higher than average mutation rates in other parts of the genome [15]. Moreover, mutations are due not only to point mutations, but rather to variation in the number of repeat units. That is, different individuals in a population differ in repeat number at each microsatellite locus and thus, microsatellite loci are typically highly polymorphic. Molecular mechanisms currently considered responsible for expansion (addition of repeat units) and contraction (deletion of repeat units) of a tandem repeat are strand-slippage during DNA replication and unequal crossing over. Microsatellites are inherently unstable, although mutation rates vary between different microsatellites, depending on the number of repeat units, repeat purity, and the length of the repeat unit. The most important factor is the number of units repeated: the more repeated units, the more unstable the region, presumably because longer loci are more likely to mispair during DNA replication. Further, recombination between longer loci will occur more often than recombination between shorter ones. The purity of the repeat is the second parameter that influences microsatellite stability. Interrupted microsatellite repeats (due to insertion of bases or base substitution) seem to have lower mutation rates than pure repeats, which might be due to a lower rate of mispairing between non-identical repeat units. Finally, microsatellite arrays containing longer repeat units (e.g. tetranucleotide repeats) evolve faster than those containing shorter units (e.g. dinucleotide repeats). This difference could be attributed to relatively inefficient repair of larger mismatched segments by cell-repair processes. Overall, the mutational process of microsatellites appears to be complex and heterogeneous, with differences between loci and alleles [16].

The characteristics of microsatellites described above, as well as their evolution, provide a tool for the estimation of levels of genetic variability within populations and enable analyses to be performed on the genetic relationships between them. These data are very useful for estimating genetic diversity and inbreeding in populations. Thus, microsatellites provide information on the number of alleles per locus and allele frequencies as well as population data. With these data, genetic distances between populations or between individuals can be estimated, and phylogenetic as well as structural analysis on each population can be conducted. Different statistical models, based on estimates of allele frequencies, enable descriptions of the mutational process of microsatellites in a population. The 2 main models are the infinite allele model (IAM) and the stepwise mutation model (SMM). The IAM is the simplest and most general model, which assumes that a mutation event can involve any number of tandem repeats and always create new alleles that did not previously exist in the population. The SMM assumes that alleles add or subtract 1 repeat unit at some constant rate. Traditionally, it was considered that microsatellites were modeled by the SMM, but the experimental analysis showed that microsatellite evolution was not so simple. Apparently, depending on the size of the repeat unit, evolutionary changes of a microsatellite are more suited to one model or another. Therefore, microsatellites with repeat units of 3–5 bp in length seem to follow the SMM, while those of 1–2 bp in length are modeled by IAM, or a proposed intermediate model called the two-phases model (TPM), which is similar to the SMM but includes larger mutations.

Microsatellites also provide a valuable approach to analysis of parentage. Their high mutation rates lead to a large number of alleles existing in a single locus, so unrelated individuals will be unlikely to share alleles, and they are codominant, which allows for exact genotyping and more precise genetic comparisons between individuals, because heterozygotes can be distinguished from homozygotes [17].

In contrast to their historical definition as nonfunctional DNA, microsatellites are currently considered unstable genomic elements that may facilitate an organism's evolvability [15]. On the one hand, an expansion of the number of repeats located in the coding region of specific genes, or in untranslated or regulatory regions of specific genes, can cause several human diseases, such as Huntington disease, spinobulbar muscular atrophy (Kennedy disease), spinocerebellar ataxias, fragile X syndrome, Friedreich ataxia, and myotonic dystrophy type 1 and 2 (for a review, see [15]). Frequently, the expansion of a trinucleotide repeat (CAG, CGG, GAA and CTG in these diseases) over a particular threshold number is responsible for the development of the disease in carriers of expanded alleles. In these diseases, it is frequently assumed that there is a mutation bias and the number of repetitions always increases. Only rare cases of reduction are observed. Effects of the trinucleotide expansion vary in each case. Both loss-of-function and gain-of-function, in the case of microsatellites located in a protein-coding region, are possible mechanisms underlying the disease. If an expanded repeat is located in the coding region of a gene, then its translation will probably affect protein structure and therefore function. If an expanded repeat is located in a non-coding region it will probably trigger indirect effects by altering mRNA (i.e. the repeat is not translated but it is transcribed) (reviewed in [18]). On the other hand, variation in the length of microsatellites can be adaptative, such as the case of the microsatellite in the promoter of the vasopressin 1a receptor *(avpr1a)* that contributes to variation in gene expression and social-behavioral traits [19]. Growing evidence points towards a beneficial role of some microsatellite repeats which confer genotypic variability by facilitating emergence of gene and non-coding DNA variants. This hypothesis is of great interest, given that 10–20% of eukaryotic genes and promoters contain an unstable repeat series. Taken together, these 2 opposite effects of microsatellite series on phenotype would indicate that unstable repeats confer evolutionary benefits when they are present in a 'tolerable' range [15, 18].

Minisatellite DNA

Minisatellites are tandem repeats with units sized ≥9 nucleotides in length. Minisatellite size polymorphisms were used in individual identification and parentage tests, as well as in other forensic studies. Later on, microsatellites began to be described and used in a similar way. However, the abundance of variable microsatellites compared to minisatellites made microsatellite DNA the genetic marker of choice for this kind of studies. Micro- and minisatellite sizes change by similar molecular mechanisms, but they are classified as 2 different types of tandem repeats based not only on their repetitive unit size, but also on their distributions and potential functions in eukaryotic genomes. Minisatellite DNA has been located in subtelomeric regions of chromosomes from human, *C. elegans* and *T. nigroviridis*, whereas in *A. thaliana* these tandem repeats tend to cluster in the pericentromeric region [20, 21], and no obvious centromeric or subtelomeric distribution was observed in the *S. cerevisiae* chromosomes [22]. Recent studies in the yeast *Candida glabrata* and the fungus *Aspergillus fumigatus* revealed minisatellites included in cell wall genes and potential cell surface proteins, making them good candidates to play a role in pathogenicity by generating diversity in these pathogens. In *S. cerevisiae* tri- and hexanucleotide microsatellite repeats are found in ORFs of genes often encoding transcription factors and regulators of gene expression [23–25].

Satellite DNA

Satellite DNAs (satDNAs) are highly repetitive DNA sequences that constitute a considerable part of eukaryotic genomes. Monomers (repeat units) are tandemly repeated sequences, generally over 200 nucleotides in length and typically organized in long arrays of hundreds or thousands of copies which occupy up to several megabases within genomes. SatDNAs are the major component of heterochromatin. Monomers differ in size, and can vary from dozens to several hundred of nucleotides. SatDNAs do not only differ from micro- or minisatellite DNAs by maximum repeat unit size and maximum array length, but also according to the genomic location and specifically to the dominant mechanisms involved in their proliferation. In fact, several satDNAs have shorter, simple monomers, sometimes even as little as micro- or minisatellites, such as in *Drosophila* or hermit crab [26, 27]. And mini- and microsatellite DNAs are not just small satellites (regardless of their name), but rather represent another 2 types of tandem repeats.

SatDNA can represent up to 20% of plant nuclear DNA, around 50% of some insect and rodent genomes [28] and less than 5% in humans [29]. However, the total abundance of satellites in most genomes is probably higher than the estimated quantities. The lower amounts estimated might result both from the indirect procedures used to quantify satDNA and from genomic projects in which satDNAs are underrepresented.

Several different satDNAs can be present in a species, and related species may share a collection or a library of satellite sequences. In terms of their location at specific chromosome positions, satDNAs are located mainly in the centromeric, pericentromeric and subtelomeric regions of chromosomes, but may also be found in heterochromatin covering sex chromosomes and as intercalary DNA in autosomes. Among centromeric satDNAs, the best-known is the human α-satellite, composed of 171-bp repeats and conserved in primates. Several distinct centromeric satDNAs have been described in the genomes of fish [30], cattle and other ruminants [31] as well as in insects [32]. Both in plants and in animals, centromeric satDNAs have been identified as essential components of centromeres such as the human α-satellite or the *CentO* and the *CentC* centromeric satDNAs of rice and maize [33]. Subtelomeric satDNAs have been described for example in fishes of the Sparidae family [34] and in the clam *Donax trunculus* [35]. In plants, subtelomeric/intercalary satellites have been characterized for many species, such as *Silene latifolia*, which has a subtelomeric satDNA of 313 bp [36], *Aegilops tauschii* (the pAs1 family, a 340-bp subtelomeric family occurring in genera of the tribe Triticeae) and the cultivated as well as wild rye species *Secale cereale, S. montanum, S. silvestre*, and *S. africanum* (the pSc119 family, a 118-bp subtelomeric and interstitial family that occurs widely in the tribe Triticeae and outside in the genus *Avena*) [37]. In *S. cereale* other subtelomeric repetitive families include the 350-bp family (a major subtelomeric family accounting for 2.5% of the rye genome) and the pSc250 family (a 550-bp repeat that accounts for 1% of the genome) [37]. Besides satellites located in specific chromosome regions, there are others specific to certain chromosomes, such as the MCSAT satDNA, which is restricted to a single chromosome pair of *Muscari commosum* [38] or those specific to the sex chromosomes or B chromosomes. For example, the larger satellite from *D. melanogaster* (359 bp) is found essentially within heterochromatin, covering about half of the X chromosome, whereas the almost entirely heterochromatic Y chromosome carries 3 smaller satellites, $(AATAC)_n$, $(AATAAAC)_n$, $(AATAGAC)_n$, only present in this chromosome (see [32]). In plants, the RAYSI family has been amplified specifically in the Y chromosomes of the dioecious species *Rumex acetosa* and relatives [39]. A noteworthy example is the RAE180 satDNA present in species of the genus *Rumex*, which has accumulated differentially, showing a distinct distribution pattern in different species, such as the massive amplification in the Y chromosomes of XX/XY_1Y_2 species [40]. The species-specific satellite pSsP216 is predominantly localized in B chromosomes in *Drosophila subsilvestris* [41], as well as the 180-bp pericentromeric satDNA in the grasshopper *Eyprepocnemis plorans* [42].

The preferential monomer length described both in animals and in plants is 150–180 bp and 300–360 bp, the DNA length corresponding to mono- or dinucleosomes [37, 43]. Nevertheless, exceptions do exist, as for example, the human satellite III formed by 5-bp monomers as well as many of the *Drosophila* satellites [44], the 24-bp satDNA from *Musca domestica*, the 35-bp repeat of *Scilla siberica* and, on the other end, the 2.5-kb satDNA from the ant *Monomorium subopacum* [32, 37] or the satDNA from the cultivated rye *S. cereale* with units of repetition of 3.9 kb [45].

SatDNAs are diverse not only in monomer length, but also in nucleotide sequence and copy number or genomic abundance. Within a species, repeat units are not strictly identical but exhibit sequence polymorphisms. However, they are more similar than when compared with repeats of other different species according to a pattern known as concerted evolution [46]. Satellite repeats are among the most dynamic components of eukaryotic genomes. Dover [5] proposed molecular drive as the 2-step process leading to concerted evolution. First, molecular mechanisms of non-reciprocal exchange (unequal crossing over, gene conversion, rolling-circle replication and re-insertion, and transposon-mediated exchange) act to spread new sequence variants appearing in individual repeat units through a family of sequences. Second, the changes are fixed in a population of random mating individuals by sexual reproduction. Thus, molecular drive links phylogeny with satDNA divergence [47]. However, this link is not always established since the molecular-drive process depends on several intrinsic and extrinsic factors [48–50]. Therefore, there are different levels of intraspecific sequence diversity, depending on those factors influencing the process. In addition, rapid changes in their sequences often result in emergence of new satDNA subfamilies independently homogenized [39]. Intraspecific homogenization is accompanied by rapid divergence between repeat sequences of different species and usually leads to species-specific satDNAs with repeats that completely differ in sequence. On the other hand, there are satDNA families extended to a whole family of species, even a whole order, or those that are long-lived, being preserved for more than 90 million years, such as the *Pst*I or *Hin*dIII satDNAs of sturgeons [48]. Different satDNAs can coexist in a genome, with different copy numbers due to differential amplification from a common library shared with related species. Examples include satellite I and satellite II, of which 30% and 4%, respectively, exist in the *Tribolium madens* genome [51]. Different families present in a genome can be independent in origin, as in the subfamilies CLsat-I, CLsat-II, CLsat-III and CLsat-IV of lizards from the genus *Darevskia* [52], or formed at the junction of 2 satellite families such as the TMADhinf repeats (originated from the junction of satellite I and satellite II) characterized in the beetle *T. madens* [51]. Within a library, one or a few families may be present at high copy numbers (then called major satellite/s) and the remaining families are present at lower copy number (called minor satDNAs). Because different satDNAs within a library can be considered as independent evolutionary units, they change in copy number and sequence with particular dynamics. Thus, a major satellite within a species (or genome) may be present as a homologous minor satellite within other related species [53]. For example, within beetles of the genus *Pimelia,* the major satellite PIM357 comprises 25–45% of the whole genome of 26 examined taxa [54].

Monomers of satDNAs showing different complexity have been described. Some satDNAs present different structural elements of functional importance while others are composed of simple repeats. Some examples include elements associated with human α-satellite, murine γ-satellite and avian centromeric satellites. Human α-satellite (171-bp monomers) repeats contain a 17-bp motif known as CENP-B box [44]. Human centromere protein B (CENP-B), which binds to the CENP-B box, is

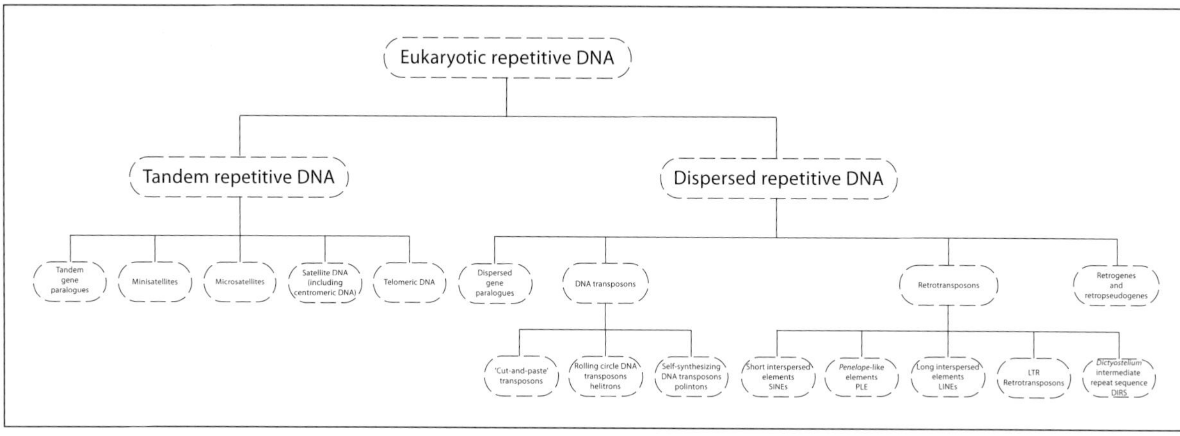

Fig. 1. Classification of the different types of repetitive DNA sequences found within eukaryotic genomes.

important for recruiting the centromere-specific histone H3 variant CenH3 (see below) during de novo centromere assembly and for proper phasing of centromeric nucleosome [55]. Transcription factor YY1 is associated with murine γ-satellite (234-bp monomer), and because YY1 belongs to a group of proteins involved in the repression of homeotic genes (Polycomb proteins) that interacts with heterochromatin, a possible link between the 2 silencing states has been hypothesized [44]. In avian, centromeric satellite motifs with sizes of 3–10 bp are conserved with respect to sequence and location in satellites of 6 species belonging to different families. These motifs have been associated with curvature of the DNA helix, which has been reported for many different satDNAs showing regular phasing of tracts of A+T and dyad structures [44]. Complexity may also come from the origin of monomers. There are monomeric units composed by duplication and divergence events of an initial shorter motif, as found in the satDNAs of animals [30, 54, 56] and plants [28, 49]. In regions of low recombination rates, repeat monomers tend to form higher-order repeats (HORs). HORs are the result of adjacent monomer variants that are homogenized together and form longer repeats while maintaining a high sequence similarity between HORs, but not within them, and preserving a repetitive structure [43]. Examples are the satDNA from the beetle *Pholeuon proserpinae,* which has a 532-bp HOR composed of 2 types of 266-bp monomers, and the human α-satellite of chromosome 7 that presents 2 HORs based on divergent subfamilies of the 171-bp monomer: a 6-monomer HOR and a dimer. In both cases, HORs are highly homogeneous (98.7% in the beetle and 97–100% in human chromosome 7), while subunits within them show lower identity (96.6% in the beetle and 72% on average in human) [43]. In addition, it has been suggested that TEs and some repeated genes may contribute to the formation and spread of satDNAs. For example, a MITE-like TE has been described as a potential generating element of satDNAs in bivalve mollusks [57] as well as in *Drosophila*; also, a mariner-like element has been proposed to take part in the expansion of satDNA between chromosomes in ants from the genus *Messor* [32]. In addition, duplicate copies of 5S rDNA have been proposed as origin of the 5S*Hin*dIII satDNA in the neotropical fish *Hoplias malabaricus* [58], and a tRNA gene ancestor was described as probably being responsible for the formation of small, tandemly repeated DNA sequences of higher plants [59].

In the last few decades, results from different studies point to a functional significance of satDNAs, in contrast to early hypotheses that considered them as junk DNA accumulating in genomes. These functions include a role in the establishment and maintenance of chromatin states by promoting heterochromatin assembly, influencing gene expression, and contributing to epigenetic regulatory processes, as satellite repeats transcribe and are a source of siRNA (see for example [44]). It is also noteworthy that satDNA is a major constituent of the centromeres, and it has been shown that subtelomeric satDNA has a role related to its location (see below).

Centromeres and Telomeres

Eukaryotic chromosomes have 2 main longitudinal differentiations, the centromere and the telomere, which are both responsible for maintaining the integrity of the chromosomes and for conserving and transmitting the genetic material. The centromere is a primary constriction perceived as a gap dividing the chromosome into 2 arms and defining its morphology. The telomeres are located at the tips of chromosomes. Additionally, 1 or several pairs of chromosomes in the karyotype have a secondary constriction at the NOR, which contains the rRNA genes (see above).

The centromere is the locus on each chromosome that maintains sister chromatid cohesion and regulates accurate chromosome segregation during cell division. The centromere nucleates the kinetochore, a proteinaceous structure that regulates chromosome attachment to the spindle microtubules to guide chromosome movement during cell division. These functions are common for all eukaryotic species, but the DNA sequences of the centromere differ greatly between them, even between closely related species (reviewed in [33, 55, 60, 61]). In *S. cerevisiae*, centromeric function is accomplished by a single sequence of 125 bp, which contains 3 conserved functional elements (CDEI, CDEII, and CDEIII) and it is assembled into a single Cse4 (CenH3 histone variant) nucleosome that captures a single microtubule. However, this simple organization is not conserved in the rest of the eukaryotes so far analyzed, from fission yeast *(Schizosaccharomyces pombe)* to humans. The structure of the centromeres in *S. pombe* and *Candida albicans* is similar, being composed in each species of non-homologous central-core sequences flanked by direct or inverted repeat elements. Human centromeres are composed of large arrays of satDNA sequences. In fact, in most animal and plant species, the centromere contains large arrays of tandem repeats which might be interrupted by TEs, such as those in *Drosophila* or in several plant species (reviewed in [33, 55, 60, 61]). However, centromeric DNA sequences are not conserved between species, suggesting that the DNA sequence is not the main determinant of centromere identity and function. Centromere identity and function is regulated epigenetically through the formation of a specialized chromatin structure [33, 55, 60, 61]. In particular, all eukaryotic centromeres, from *S. cerevisiae* to humans, are characterized by the presence of the centromere-specific histone H3 variant CenH3 (or CENP-A). With the exception of those of *S. cerevisiae* and *C. albicans,* that lack heterochromatin, centromeres are organized as euchromatic pocket domains of CenH3 and H3K4me2 (a modification normally associated with transcriptionally active chromatin domains), flanked by heterochromatic domains bearing the heterochromatin marks H3K9me2. Only a portion of centromeric repetitive elements is assembled in CenH3 chromatin, the rest being embedded in heterochromatin. In the case of *S. pombe*, the central domain is assembled in CenH3 chromatin and is flanked by heterochromatin domains assembled on outer repeat sequences. These 2 distinct domains mediate different functions: CenH3 chromatin is responsible mainly for kinetochore assembly while the surrounding heterochromatin domain seems to have a determinant function in sister-chromatid cohesion [33, 55, 60, 61]. Despite the lack of a conserved centromeric sequence, it is possible that all centromeres share common features that make them permissive for CenH3 deposition, such as the sequence composition (AT-rich) and its structure or the length of satellite repeat units, low gene density, transcription of non-coding RNAs, chromatin status, or vicinity to heterochromatin domains [33]. Once assembled, specific chaperones and assembly factors contribute to the maintenance of centromeric domains [33].

The telomeres are ribonucleoprotein complexes characterized by particular protein and DNA sequences. The telomeres protect chromosomes from degradation and repair activities and prevent the chromosome shortening resulting from replication of the end of the linear chromosomes [62]. Different telomere-specific proteins are involved in these functions [62]. Telomeric DNA is composed of short tandem repeats of ~6 bp, a eukaryotic ancestral structure. The way telomeres replicate is also ancestral. The first cloned telomeres were those of *Tetrahymena* and the repeated sequence of these telomeres was 5′-TTGGGG-3′ [63]. Afterwards, the telomeres of several other species of protozoa, fungi and animals were studied and proved to be composed of similar repeats or of slight variants thereof (reviewed in [64]). Soon, the highly similar 5′-TTAGGG-3′ repeat sequence of human telomeres was isolated and found to be common for all vertebrates analyzed. The telomere of *A. thaliana* and most plants is composed of 7-bp repeats with the sequence 5′-TTTAGGG-3′ (reviewed in [64]). The telomeric repeats are added by a telomerase enzyme, rather than by DNA polymerase through semi-conservative replication, solving the replication problem at the ends of a linear DNA double helix. The telomerase is composed of several protein subunits and a short RNA containing the template telomeric repeat sequence, which the RNA-dependent DNA polymerase or reverse transcriptase (RT) domain of the telomerase uses for the addition of telomeric repeats [65]. The telomerase gene is phylogenetically related to the RT genes of the non-LTR and *Penelope* retrotransposons. It appears to constitute an example of a retrotransposon gene 'domesticated' for a cellular role (see below), although an alternative view is that retroelements originated from telomerase [66]. *Drosophila melanogaster* lacks the canonical telomere structure. This species has overcome the lack of telomere repeats by the recruitment of non-LTR retrotransposons (HeT-A, TART and TAHRE) to perform the cell function of capping the ends either as a secondary domestication process or as taking advantage of the retention of an ancestral mechanism [66]. Under telomeres, there are commonly subtelomeric or telomere-associated sequences, which are tandem repeats that sometimes contain intercalated degenerate telomeric motifs and are species-specific, often chromosome-specific, with a variety of lengths and degrees of repetitiveness [64]. Telomere-associated sequences might not have an essential role in telomere function but could facilitate chromosome pairing in meiosis or act to buffer terminal genes against the dynamic processes of loss and addition at the ends [64].

Transposable Elements

Transposable elements are DNA sequences that are able to move from one chromosomal position to another within the same genome. They are highly ubiquitous elements found in all kingdoms of living organisms. Most eukaryote genomes contain TEs. Only small eukaryotic genomes of parasitic apicomplexa and the microsporidia intracellular parasite *Encephalitozoon cuniculi* have been found to be devoid of TEs, while in prokaryotes only about 20% of the genomes sequenced so far lack TEs (reviewed in [67], see also [68]). Ultimately, accumulation of mutations over time may lead to the loss of detectable traces of the extinct TE families, as proposed for *Encephalitozoon, Cryptosporidium* and *Plasmodium* species [68]. The accumulation of TE vestiges in cellular genomes is also a remarkable evolutionary driving force, providing heterogeneous building blocks to create new cellular functions. This is exemplified in *Leishmania* spp. by the acquisition of the ability to use short extinct retrotransposons to post-transcriptionally coordinate gene regulation, whereas the closely related trypanosomes utilize other strategies to fulfill this critical cellular function [69]. In addition, TEs are highly abundant in some genomes, representing 45% of the human genome [29] or 52% of the opossum genome [70] and reaching up to 85% of some large plant genomes, such as that of maize [71]. TE diversity and abundance is highly variable from one species to another, and reflects their specific genome-TE history [67]. For example, TEs are very rare in the *T. nigroviridis* genome (comprising less than 0.5% of the genome), but there are 73 different families [72]. The human genome contains around 170 different families of TEs and there are hundreds of thousands, even millions, of copies of a few of them [29]. Some 3% of the yeast genome is composed of TEs, all of them belonging to 1 of 5 families of the same superfamily [73]. Further, in most species, independently of TE diversity, usually a few types of elements dominate [67]. There is a correlation between genome size and TE abundance that should explain the paradoxical differences between C values (the DNA contents) of different species, which in no way reflects the organism's complexity. Thus, the TE content can differ greatly even between related species as, for example, between the *Takifugu* and *Tetraodon* genomes [72], between *Oryza sativa* and its wild relative *Oryza australiensis* [74] or between *A. thaliana* and *A. lyrata* [75].

TEs are classically divided into 2 classes according to their transposition mechanism (figs. 1 and 2): (1) retrotransposons or Class I elements; and (2) DNA transposons or Class II elements. Retrotransposons are transposed through an RNA intermediate. The RNA is transcribed from the element, then reverse transcribed into a complementary DNA (cDNA), which is integrated into a new location in the genome. These elements are thus replicative in nature and prone to amplify in number within the host genome. However, their expansion, as well as the expansion of DNA transposons, is regulated in 2 ways, by the elements themselves and by the host genome [73]. Reverse transcription is catalyzed before or during cDNA integration into a new position by a RNA-dependent DNA polymerase or reverse transcriptase encoded by autonomous elements. By contrast, DNA transposons are transposed by moving their genomic DNA copies from one chromosomal location to another without any RNA intermediate. The transposition of most, but not all, DNA transposons is conservative, although DNA transposons can increase their number in the host genome under certain circumstances. Most retrotransposons and DNA transposons are flanked by target-site duplications (TSDs) resulting from filling of staggered nicks generated at the DNA target site upon insertion of TEs. With some exceptions, Class I elements are the most abundant TEs in eukaryotic genomes. In addition to full-length autonomous elements, a high proportion of the elements of each TE family in one genome are usually incomplete, deletion derivative, non-autonomous elements. The non-autonomous elements, originating from internal deletions, can mobilize using the transposition machinery of intact elements. For a detailed classification of the different TEs belonging to each of these 2 classes the reader can turn, among others, to previous detailed reviews [67, 76–83]. Below is a brief description of each one.

Class I Elements

All currently known eukaryotic retrotransposons can be divided into 5 types or orders according to a recent classification by Wicker et al. [76] (figs. 1 and 2): (1) LTR retrotransposons; (2) *Dictyostelium* intermediate repeat sequence *(DIRS)* retrotransposons; (3) non-LTR retrotransposons or LINEs; (4) *Penelope*-like retrotransposons *(PLE)*; and (5) SINEs. LTR and non-LTR retrotransposons are the most widespread and abundant retroelements as well as the most abundant TEs in eukaryotes. The LTR retrotransposons are less abundant in animals. The percentage of these elements in mammals ranges between 4 and 10% [29, 70, 84–87]. Mammals, however, are characterized by large genomes (2.3–3.5 Gb) and high amounts of repetitive DNA (36–52% of their genomes), whereas other vertebrate and invertebrate genomes sequenced so far have less LTR content, e.g. 1.3% in the chicken genome (1.1 Gb), with a repetitive DNA content of 9% in the genome [88]. By contrast, the LTR retrotransposons are the predominant type in plants, representing 54.5% of the sorghum genome [89] or 75% of the maize genome [65], and showing a correlation between genome size, amount of repetitive DNA and amount of LTR elements. However, LINEs, which are less common in plants, predominate in most animals, and they are specifically abundant in birds (6.4% of the chicken genome and 66% of TEs [88]) and mammals. In the genomes of mammals sequenced so far, LINEs represent between 18% of the dog genome and 29% of the opossum genome, accounting for around 50% of their TE contents [29, 70, 84–87]. The *Penelope* retrotransposons were first found in *Drosophila virilis*, and subsequently *PLEs* were identified in other animals, in fungi, and in plants, but were not found in several sequenced genomes. These elements are less known and have unusual structures (reviewed in [67, 76, 79, 80]). *DIRS* elements were first found in *Dictyostelium discoideum* and subsequently in diverse species, ranging from green algae to animals and fungi (reviewed in [67, 76, 79, 80]). Finally, SINEs are non-autonomous elements functionally related to LINEs [76]. They are not deletion derivatives of autonomous LINEs, but they originate from retrotransposition of RNA polymerase III (Pol III) transcripts by the reverse transcriptase encoded by LINEs. SINEs are widespread among eukaryotes, but not so much as other TEs [83]. SINEs represent less than 1% of the genomes of plants and animals other than mammals in which bursts of SINE expansions have led to percentages ranging between 7 and 13% of the genome, being the most abundant type of repetitive DNA in the *Platypus* genome [90].

LTR Retrotransposons

These Class I elements have from a few hundred base pairs up to, exceptionally, 25 kb and are flanked by characteristic LTRs with a size ranging from a few hundred base pairs to more than 5 kb [76]. The LTRs contain promoter sequences and features associated with the transcription of the elements. LTR retrotransposons are transcribed from the promoter within the 5′ LTR. It is common to find solo-LTRs within genomes as a by-product of recombination between the terminal repeats from an element. LTR retrotransposons are similar to retroviruses except for the absence of the *env* gene in most, but not all, elements. They contain a *gag* gene, which encodes a nucleic-acid-binding protein which might be involved in the reverse transcription process, and a *pol* gene that encodes various enzymatic domains: proteinase (PR), reverse transcriptase (RT), RNase H (RH), and integrase (INT) (fig. 2). The LTR retrotransposons can be divided into 3 major lineages or superfamilies: (1) the *Ty1/copia* group (the oldest lineage); (2) the *Bel-Pao* group; and (3) the *Ty3/gypsy* group. Wicker et al. [76] also included 2 additional superfamilies within the LTR retrotransposons: (1) the vertebrate retroviruses, as it has been suggested that they evolved from the

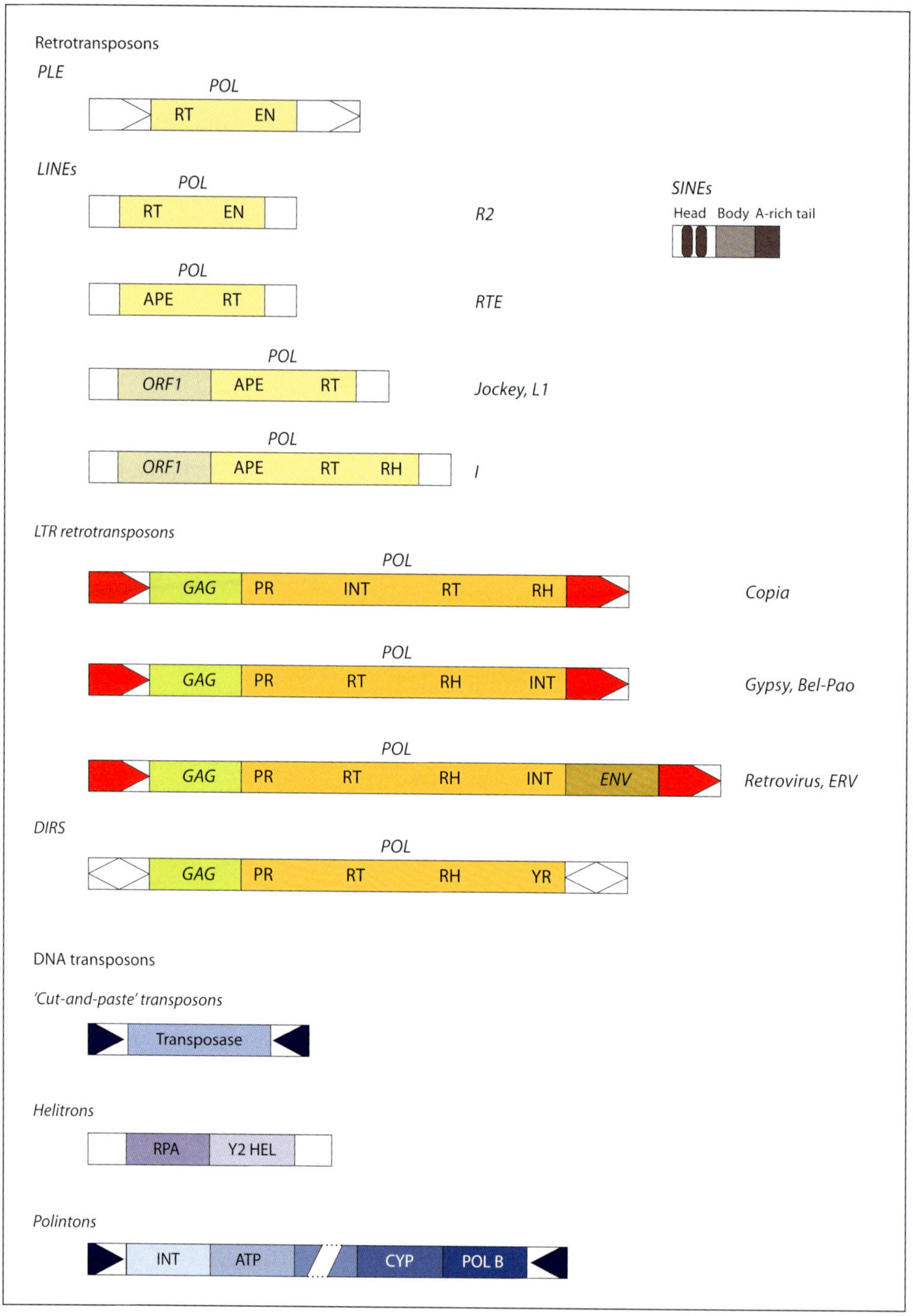

Fig. 2. Schematic representation of the different eukaryotic transposable elements. All retrotransposons have a *pol* gene *(POL)* that encodes various enzymatic domains, which vary depending on the element: RT, reverse transcriptase; EN, endonuclease; APE, apurinic endonuclease; RH, RNase H; PR, proteinase; INT, integrase; YR, tyrosine recombinase. LINEs have an additional ORF *(ORF1)* that encodes a protein which binds to the LINE RNA to form ribonucleoprotein complexes, considered to

Ty3/gypsy LTR retrotransposons by the acquisition of an *env* domain that encodes an envelope protein; and (2) the endogenous retroviruses *(ERVs)* or sequence derivatives of past retroviral infections; today most of them are defective forms (mostly solo-LTR) which lack their replicative capability. The mechanism of retrotransposition for the LTR retrotransposons is similar to that of retroviruses and involves a tRNA molecule which anneals to the primer binding site at the 3′ end of the retrotransposon RNA to prime the reverse transcription in the cytoplasm. Once the 2 DNA strands of the complementary DNA have been synthesized, the cDNA is transferred to the nucleus, in which integration occurs. Although LTR retrotransposons could derive from non-LTR retrotransposons, it has been proposed that LTR retrotransposons might be mosaic in origin, coming from the fusion of bacterial transposons and bacterial retroelements such as Group II introns [73]. Due to their chimerical nature and their modular evolution, it seems today difficult to imagine a general evolutionary scenario for TEs [91, 92].

DIRS Elements

It has been suggested that *DIRS* elements evolved from a *gypsy*-like ancestral LTR retrotransposon (although it cannot be ruled out that the present-day LTR retrotransposons derived from *DIRS* elements) [79, 80]. However, *DIRS* elements have a number of properties that differ from LTR retrotransposons (fig. 2). For example, they contain a tyrosine recombinase domain (YR) instead of integrase and have LTRs, but these are inverted in orientation, although this is not always true (some of them are direct), and a segment of the LTR sequences is repeated within the element giving rise to an internal complementary repeat. These features indicate a mechanism of integration that is different from the rest of retrotransposons. Some *DIRS* elements retain introns in ORFs.

LINEs

The LINEs lack LTRs. At their 3′ end they can display either a poly(A) tail, a tandem repeat or merely an A-rich region. There are 5 different superfamilies *(R2, RTE, Jockey, L1*, and *I)* which were subdivided into 28 different clades [82], but they can be ascribed to 1 of 2 major types of non-LTR retroelements (fig. 2). One type encodes a single ORF coding for a reverse transcriptase (RT) and an endonuclease (EN). This type is believed to be the most ancient type of non-LTR retrotransposons to which the *R2* elements belong. The EN domain in *R2* is similar to different

be transposition intermediates. LTR retrotransposons also contain a *gag* gene *(GAG)*, which encodes a nucleic acid-binding protein which might be involved in the reverse transcription process. 'Cut-and-paste' DNA transposons bear a transposase gene. *Helitrons* encode a Y2-type tyrosine recombinase with an helicase domain (Y2 HEL), and can also encode other proteins, e.g. the replication protein A (RPA) found only in plants [76]. *Polintons* have a coding capacity for multiple proteins, including a B-type DNA polymerase (POL B), and a retroviral-like integrase (INT). They can encode up to 11 proteins, e.g. packaging ATPase (ATP) and cysteine protease (CYP), among others [76].

restriction enzymes and is always preceded by the RT domain. The *RTE* superfamily also possesses a single ORF, but in this case encodes an apurinic-apyrimidinic endonuclease (APE), which precedes the RT domain. The second type of non-LTR retrotransposons usually encodes 2 ORFs. ORF1 may have functional similarity to the *gag* gene. Its protein binds to the LINE RNA to form ribonucleoprotein complexes, considered to be transposition intermediates. ORF2 encodes the RT domain as well as the APE, preceding the RT domain. In addition to RT and EN, all retrotransposons from the *I* superfamily code for RNase H (RH). Analogously, diverse plant *L1* retrotransposons also code for RNAse H [82]. The most representative and most thoroughly studied TE of this type of retrotransposons is the mammalian LINE-1 *(L1)* retrotransposon, a 6–8-kb element, which in humans reaches up to 516,000 copies and represents some 17% of its genome [29, 93]. The non-LTR retrotransposons are transcribed from a promoter that lies within the transcription unit. The transcribed LINE RNA reverse transcribes during its integration into a new site by a process known as target-primed reverse transcription (reviewed in [93]). For this to occur, the endonuclease activity of ORF2 makes a nick in the double-stranded DNA at target AT-rich regions, followed by the annealing of the poly(A) tail of the 3′ LINE RNA to the poly(T) at the 5′ nick and cDNA synthesis by the reverse transcriptase activity of ORF2 using the 3′ OH released by the cleavage to prime the reverse transcription reaction. The process ends with the degradation of the LINE RNA and the second-strand DNA synthesis. This involves a second nick at the other target genomic DNA strand (a few nucleotides away from the first nick) to utilize the 3′ end as a primer for the second-strand DNA synthesis followed by ligation. This mechanism is reminiscent of that of a Group II intron and it is therefore possible that mitochondrial Group II introns are distant ancestors of mammalian LINEs (reviewed in [73]). Premature termination of reverse transcription is common in the target-primed reverse transcription mechanism, and this might explain why most LINEs have truncated 5′ ends.

PLE Retrotransposons

The *Penelope*-like elements encode a single ORF composed of the RT domain preceding the EN domain (fig. 2). It appears that the *Penelope* RT is closer to telomerases and bacterial RTs than RTs encoded by non-LTR retrotransposons [94]. Some members have LTR-like sequences that can be in a direct or an inverse orientation, some contain an additional ORF, and some lack the EN domain. Some elements contain an intron that is retained after their retrotransposition. A subset of *PLEs* found in bdelloid rotifers, basidiomycete fungi, stramenopiles and plants are located at telomeres in an orientation consistent with the utilization of a free chromosomal end to prime reverse transcription. Many of these telomere-associated *PLEs* occupy a basal phylogenetic position close to the point of divergence from the telomerase-*PLE* common ancestor and may descend from the missing link between early eukaryotic retroelements and present-day telomerases [94].

SINEs

SINEs are retrotransposed elements that originated by the reverse transcription of Pol III transcripts. There are 3 main groups of SINEs according to the retrotranscribed RNA [76, 83]: SINEs might come from tRNA, from 5S rRNA, or from 7SL RNA (the RNA involved in protein secretion as a component of the signal recognition particle, SRP). SINEs encode no proteins and use LINE reverse transcriptase for their retrotransposition through a mechanism basically similar to that of LINEs [83]. Genes of all these RNAs, as well as the corresponding SINEs, have an internal Pol III promoter. The presence of the promoter within the transcribed sequence might be critical for SINE amplification, as the promoter is preserved in new SINE copies. Most tRNA-derived SINEs with intact internal promoter can be transcribed. However, some SINEs, such as those derived from 7SL RNA, are rarely transcribed because they lack the necessary flanking regions for autonomous transcription (see below) and only a few elements called 'source' or 'master' genes, seem to be transcribed [93]. SINEs originating from tRNAs are the most common SINEs in invertebrates, vertebrates, and many flowering plants [83, 95]. 7SL RNA-derived SINEs have been identified only in rodents, primates, and scadentians (tree shrews) [83, 96]. However, 5S rRNA-derived SINEs have been found in some fishes and in a few mammals [83].

Most SINEs consist of 3 modules (reviewed in [83]): (1) 5′-terminal 'head'; (2) SINE internal region or 'body'; and (3) 3′-terminal 'tail' (fig. 2). However, this pattern is reduced to the head and the tail in the so-called simple SINEs, the 7SL RNA-derived SINEs being an example. The head bears the Pol III promoter and defines SINE types, revealing their origin (tRNA, 7SL RNA and 5S rRNA), while the body is family-specific, is of variable origin and can comprise a LINE-related segment. The whole sequence ends with an A-rich 3′-terminal variable tail. Additionally, more complex structures can be found. Thus, 2 or more SINEs can combine into homodimeric or homotrimeric structures which are further amplified as such, as well as 7SL RNA/tRNA or 5S rRNA/tRNA complex heteromeric elements [83]. The best studied SINEs are the 7SL RNA-derived *Alu* elements of primates which have reached up to 1.1 million copies within the human genome, comprising about 11% of its total genomic content and 81% of the total SINE content, for which there are 3 families in humans [29]. *Alu* elements are composed of two 130-bp monomers, 7SL RNA-derived, separated by a short A-rich linker region and ending in an A-rich tail of variable length. The total length of each *Alu* sequence is ~300 bp, depending on the length of the 3′ A-rich tail.

The presence/absence of a SINE (as well as of a LINE) at a given locus has been used as a criterion to evaluate phylogenetic relationships among species (see e.g. [97]). It is assumed that all organisms carrying a particular SINE (or LINE) insertion are derived from a unique irreversible event that happened in their common ancestor. In addition, *Alu* element insertion polymorphisms in humans have been used to investigate the human origins, as well as human population structure and demography, and have also been used as forensic tools [93].

SINEs, as well as the rest of TEs, can cause damage to the host genome through insertional mutagenesis or through non-allelic homologous recombination, having important implications for human disease (reviewed in [78, 93, 98]. However, SINEs might have played a key role in enhancing the evolutionary potential of their hosts. Exaptation, the acquisition of a new function from previously useless DNA sequences, has recurrently occurred for SINEs through evolution, generating new *cis*-regulatory elements for processes such as alternative splicing, mRNA polyadenylation, and promoter activity [99]. Some conserved non-coding elements, relevant in gene expression as enhancers, are SINE-derived and have been implicated in the morphological innovations specific to certain taxonomic groups such as mammalian brain development [99]. The involvement of transposed SINEs in generating new exons, termed exonization, was analyzed in primates [100]. Some mammalian precursors of microRNAs appear to be derived from ancient SINEs [79].

Class II Elements

Class II transposable elements or DNA transposons are transposed by moving their genomic DNA copies from one chromosomal location to another without any RNA intermediate. They are found in almost all eukaryotes and several DNA transposon superfamilies are found to be related to DNA transposon superfamilies of prokaryotes, suggesting that the divergence of most superfamilies may even predate the split of eukaryotes and prokaryotes [77]. However, this relatedness might be partly explained by horizontal transfers, which occurred in the distant past [67], as prokaryotes are known to be able to integrate foreign DNA into their genome. Moreover, some eukaryotes are devoid of Class II elements like *S. cerevisiae* [67, 77]. The percentage of Class II elements in animal genomes is low when compared with the rest of transposable elements, even in the case of species with large genomes and high percentage of repetitive DNA. For example, in humans, despite of their diversity, the DNA transposons occupy less than 3% of the genome, with no evidence for transposon activity in the past 50 million years [29]. Among animals, there are some notable exceptions such as those of the nematode *C. elegans*, the cnidarian *Hydra magnipapillata* or the amphibian *Xenopus*, in the genomes of which the most abundant TEs appear to be DNA transposons [29, 101, 102]. In the case of plants, there are higher quantities of DNA transposons, specifically in those species with larger genomes, although the LTR retrotransposons surpass them, e.g. 8.6% in the maize genome [71], 7.5% in sorghum or 14% in rice [88]. DNA transposons are divided into 3 main subclasses [77, 81, 103]: (1) 'cut-and-paste' DNA transposons; (2) rolling-circle DNA transposons *(Helitrons)*; and (3) self-synthesizing DNA transposons *(Polintons)* (figs. 1, 2). Most of the identified eukaryotic DNA transposons are 'cut-and-paste' transposons, which excise (cut) as double-stranded DNA and reinsert (paste) in a new location. Thus, the transposition mechanism of these DNA transposons is conservative, although they can increase their number by transposing during chromosome replication from a position that has already been replicated to another in which the replication fork has not yet passed.

They are characterized by the presence of terminal inverted repeats (TIRs) and, in most of them, there is only 1 ORF encoding a transposase that recognizes the TIRs and cuts both strands at each end. This subclass is currently represented by 17 superfamilies, classified according to the transposase which is superfamily-specific [103], although another classification of 12 superfamilies based on TIR sequences and TSD size have been proposed [76]. Some superfamilies such as *Tc1/Mariner* or *Mutator* are ubiquitous in eukaryotes. Three superfamilies *(En/Spm*, *Harbinger*, and *MuDR)* are characterized by the presence of a second transposon-encoded DNA-binding protein required for transposition in addition to transposase. Non-autonomous elements known as miniature inverted-repeat transposable elements (MITEs) are short transposons (100–600 bp) with TIRs that use a transposase encoded by autonomous elements to transpose. They are probably either deletion derivatives of full-length elements or de novo constructions from the fortuitous emergence of TIR with binding site for the tranposase of an autonomous element. *Helitrons* and *Politons* are replicative, 'copy-and-paste', TEs that transpose by replication involving the cleavage of only 1 strand on each side. *Helitron* elements are present in the genomes of plants (where they have been mainly described), fungi, invertebrates, and vertebrates, transposing via replicative rolling-circle transposition and integrating into the genome without introducing TSDs. *Helitrons* lack TIRs. They encode a Y2-type tyrosine recombinase, such as that found in the bacterial *IS91* rolling-circle transposons, with an helicase domain, and can also encode other proteins. *Helitrons* have hairpin structures at the ends [76, 77]. *Polintons*, also known as *Mavericks*, are very large transposons (15–20 kb long) with long TIRs (100–1000 bp). They have a coding capacity for multiple proteins, most of which are related to double-stranded DNA viruses, including a B-type DNA polymerase, and a retroviral-like integrase. *Polintons* appear to propagate through protein-primed self-synthesis by the B-type DNA polymerase through a replicative 'copy-and-paste' process [76, 77].

Transposable Elements as Drivers of Genome Evolution

The most obvious effect of the mobility of TEs is the induction of insertional mutations, usually being detrimental or neutral. In humans, for example, TEs are responsible for several genetic diseases [78, 93] and have been associated with cancer [98]. In addition, the ectopic recombination between non-allelic homologous elements can generate various types of rearrangements and lead to inversions, deletions, translocations or duplications. By contrast, today there is no doubt of the significant impact of mobile elements shaping the structure, function, and evolution of genes and genomes (reviewed in [67, 77–79, 93, 104]). TEs are considered 'genome architects' with influence in centromere function, in the generation of satellite sequences and heterochromatin or in genome compartmentalization [67]. The A-rich tails of LINEs (and SINEs), are one of the major sources for the generation of microsatellites of varying length and complexity in mammalian genomes by nucleotide substitutions and replication slippage. Retrotransposons and DNA transposons are capable of carrying out the transduction of adjacent host sequences by capturing them as part of mobile elements. The mobilization of host-gene sequences by several types of TEs suggests an involvement in exon shuffling and gene duplication. The transcription machinery can transcribe LINEs far downstream of its 3′ end, including exonic sequences in the transcribed RNA, which, once reverse transcribed and inserted in a new location, can generate duplicated genes or new gene combinations [105]. DNA transposons such as MULEs (*Mutator*-like transposable elements) and *Helitrons* have demonstrated a tremendous potential for gene shuffling and duplication in plants [106, 107]. In these latter 2 cases the transduction mechanism is not completely understood. Gene duplication could also result from TE-mediated recombination. Genetic innovation can also be generated in several other ways. TE genes can evolve as new genes with functions beneficial to the host, an event that many authors term 'molecular domestication'. Many protein-coding genes in the mammalian genome evolved from coding sequences of TEs, most of them from different transposase genes (reviewed in [77, 79, 93, 104]). There are 2 conspicuous examples of 'domesticated' genes. One is the telomerase gene, 'domesticated' from a reverse transcriptase-encoding gene. The second is the gene encoding the recombination-activating protein Rag1, which in jawed vertebrates initiates together with Rag2 the V(D)J recombination of immunoglobulin genes. Evidence has been provided that both Rag1 and recombination signal sequences are derived from Transib DNA transposons [104, 108]. However, there are dozens of protein-coding genes derived from TE genes [79, 104]. For example, the *hAT*-like transposase gene *Daysleeper*, essential for development in *A. thaliana,* or the centromere-associated protein CENP-B of mammals derived from a *pogo* transposase (reviewed in [104]). In addition, many microRNA genes appear to have evolved from TEs. Genetic novelties have also arisen from gene retrotransposition [109]. Many host genes have recruited regulatory and partial coding sequences from TEs during evolution. A transposable element can be recruited as a coding sequence and can be integrated into a gene (exonization) [100]. TEs can provide new splice sites that might promote exonization and alternative splicing or provide polyadenylation signals that induce the termination of gene transcripts. Additionally, TEs can act as controlling elements of gene expression. In mammals, retrotransposons have been proposed to act as general modulators of gene expression [110] and to play a role in X-chromosome inactivation [111]. Many non-coding elements involved in gene regulation seem to be derived from ancient TE sequences such as those mentioned above for SINEs, but also for LINEs. Promoter elements in LTRs can influence the transcription of neighbor genes, resulting in transcriptional activation or gene silencing and in changes in tissue specificity of expression. Further, the epigenetic silencing of TE activity could spread and repress the transcription of nearby genes. The role of TEs in heterochromatin formation and epigenetic regulation of gene activity was investigated in *A. thaliana*. Methylated TEs can be regulated during development. TE methylation can bring genes under control when TEs integrate nearby. Relaxation of this control would occur upon excision or deletion of the TE [112].

Segmental Duplications

Segmental duplications are highly similar duplicated DNA fragments >1 kb. They can contain any constituent of genomic DNA, including typical gene sequences with intron-exon structure and common high-copy repeats such as SINEs and LINEs [113]. Initially, they were considered simply rare particularities of pericentromeric and subtelomeric regions within some genomes. Nevertheless, publication of the human genome sequence revealed the presence of an unexpectedly large number of them. Subsequent identification of SDs in other genomes has been performed computationally, using appropriate algorithms which enabled genome-sequence assemblies (mainly whole-genome assembly comparison or WGAC, and whole-genome shotgun sequence detection or WSSD). From these analyses, SDs were considered sequences >1 kb, aligned with at least 90% identity, which constitutes their formal definition. They are also known as low-copy repeats (LCRs) [113].

In humans and chimpanzees, SDs are mainly dispersed repeats, whereas other mammalian genomes contain lower amounts of SDs, predominantly repeated in tandem. SDs account for approximately 5% of the human and chimpanzee genomes, 2.4% of the macaque, 2% of the marmoset, and 2–4% of the rat, mouse and dog genomes. Information from the new field of whole-genome comparative genomics estimates that, in general, SDs from mammalian genomes are larger in size than those from other eukaryotes such as *C. elegans* or *D. melanogaster* [113, 114]. In human, SDs are separated by more than 1 Mb of unique sequences [114] and show a statistical bias in distribution both in chromosomes and in specific positions within chromosomes. Thus, at the chromosome level, chromosome 3 contains the lowest proportion of SDs (1.7%), while chromosomes 22 and Y have the greatest proportion (11.9% and 50%, respectively) [113, 115]. In relation to their location within chromosomes, in mammals they have been described as forming a peculiar clustering near the subtelomeric and pericentromeric regions, and in the euchromatic portions of specific chromosomes [114], and therefore they may be classified as pericentromeric, subtelomeric, and interstitial regions of duplication. Different types and frequencies of SDs are found in each category. In pericentromeric regions SDs vary in length (50–100 kb) as well as in content (from total absence in chromosome 16, to over 6 Mb in chromosome 9). They are present in 29 out of 43 chromosomes, accounting overall for one-third of all SDs in the human genome, and are mainly interchromosomal duplications (ratio 6:1 interchromosomal:intrachromosomal duplications) [113, 115]. In subtelomeric regions, they have the same variation in length as duplications located in pericentromeric regions (50–100 kb), are present in the subtelomeric regions of more chromosomes compared to their presence in pericentromeric regions (30 out of 42 chromosomes), but the global content of these regions in SDs is lower (2.6 Mb), and apparently comes from exchange between subtelomeric regions [113, 115]. Interstitial regions contain the highest amount of SDs, which predominate in some chromosomes (similarly to pericentromeric SD), and are mainly intrachromosomal duplications [113, 116]. The origin and molecular mechanism responsible for the propagation of SDs is still unclear. Recent data suggest that *Alu* repeat clusters have a role as mediators of recurrent chromosomal rearrangements, with different models of SD formation suggested for pericentromeric, subtelomeric, and interstitial duplications [113].

SDs can involve large duplications of several genes, and also represent predisposition sites (hotspots) for the occurrence of unequal crossing over, leading to genomic mutations such as deletion, duplication, inversion or translocation. These structural alterations are sources of new genes and lead to the evolution of genomes, but can also cause dosage imbalances of genetic material or generate new gene products, resulting in different human diseases. Some examples of genomic disorders originating from chromosomal structural rearrangements include α-thalassemia (caused by α-globin gene deletions, which are the outcome of unequal crossing over between repeated segments within the α-globin locus), Prader-Willi/Angelman syndromes (unequal crossing over appears to be involved in the generation of a common deletion found in the majority of patients), the Charcot-Marie-Tooth disease type 1A or CMT1A (associated with a 1.5-Mb tandem duplication in 17p12, which arises from unequal crossing over and homologous recombination between 24-kb flanking repeats termed CMT1A-REP), and hemophilia A (47% of severely affected individuals are afflicted by an inversion of a portion of the gene-encoding factor VIII) [117, 118].

Acknowledgements

The research in our laboratory is currently financed by the Ministerio de Ciencia e Innovación and FEDER founds, grant CGL2010-14856 (subprograma BOS). We apologize to those authors whose work could not be cited here due to space restriction.

References

1 Britten RJ, Kohne DE: Repeated sequences in DNA. Science 1968;161:529–540.
2 Long EO, Dawid IB: Repeated genes in eukaryotes. Ann Rev Biochem 1980;49:727–764.
3 Nei M, Rooney AP: Concerted and birth-and-death evolution of multigene families. Annu Rev Genet 2005;39:121–152.
4 Prokopowich CD, Gregory TR, Crease TJ: The correlation between rDNA copy number and genome size in eukaryotes. Genome 2003;46:48–50.
5 Dover GA: Molecular drive. Trends Genet 2002;18: 587–589.
6 Robles F, de la Herrán R, Ludwig A, Ruiz Rejón C, Ruiz Rejón M, Garrido-Ramos MA: Genomic organization and evolution of the 5S ribosomal DNA in the ancient fish sturgeon. Genome 2005;48:18–28.
7 Freire R, Arias A, Insua AM, Méndez J, Eirín-López JM: Evolutionary dynamics of the 5S rDNA gene family in the mussel *Mytilus*: mixed effects of birth-and-death and concerted evolution. J Mol Evol 2010;70:413–426.
8 Rooney AP, Piontkivska H, Nei M: Molecular evolution of the nontandemly repeated genes of the histone 3 multigene family. Mol Biol Evol 2002;19: 68–75.
9 Ausio J: Histone variants – the structure behind the function. Brief Funct Genomic Proteomic 2006;5: 228–243.

10 González-Romero R, Rivera-Casas C, Ausió J, Méndez J, Eirín-López JM: Birth-and-death long-term evolution promotes histone H2B variant diversification in the male germinal cell line. Mol Biol Evol 2010;27:1802–1812.
11 Tóth G, Gáspári Z, Jurka J: Microsatellites in different eukaryotic genomes: survey and analysis. Genome Res 2000;10:967–981.
12 Estoup AM, Solignac MH, Cornuet JM: Characterization of (GT)n and (CT)n microsatellites in two insect species: *Apis mellifera* and *Bombus terrestris*. Nucleic Acids Res 1993;21:1427–1431.
13 Naciri Y, Vigouroux Y, Dallas J, Desmarais E, Delsert C, Bonhomme F: Identification and inheritance of (GA/TC)n and (AC/GT)n repeats in the European flat oyster *Ostrea edulis* (L.). Mol Mar Biol Biotechnol 1995;4:83–89.
14 Morgante M, Hanafey M, Powell W: Microsatellites are preferentially associated with nonrepetitive DNA in plant genomes. Nat Genet 2002;30:194–200.
15 Gemayel R, Vinces MD, Legendre M, Verstrepen KJ: Variable tandem repeats accelerate evolution of coding and regulatory sequences. Annu Rev Genet 2010;44:445–477.
16 Ellegren H: Microsatellites: simple sequences with complex evolution. Nat Rev Genet 2004;5:435–445.
17 Webster MS, Reichart L: Use of microsatellites for parentage and kinship analyses in animals. Methods Enzymol 2005;395:222–238.
18 Brower JR, Willemsen R, Oostra BA: Microsatellite repeat instability and neurological disease. Bioessays 2009;31:71–83.
19 Hammock EA, Young LJ: Microsatellite instability generates diversity in brain and sociobehavioral traits. Science 2005;308:1630–1634.
20 Vergnaud G, Denoeud F: Minisatellites: mutability and genome architecture. Genome Res 2000;10:899–907.
21 Roest-Crollius H, Jaillon O, Dasilva C, Ozouf-Costaz C, Fizames C, et al: Characterization and repeat analysis of the compact genome of the freshwater pufferfish *Tetraodon nigroviridis*. Genome Res 2000;10:939–949.
22 Richard GF, Dujon B: Molecular evolution of minisatellites in hemiascomycetous yeasts. Mol Biol Evol 2006;23:189–202.
23 Levdansky E, Romano J, Shadkchan Y, Sharon H, Verstrepen KJ, et al: Coding tandem repeats generate diversity in *Aspergillus fumigatus* genes. Eukaryot Cell 2007;6:1380–1391.
24 Thierry A, Bouchier C, Dujon B, Richard GF: Megasatellites: a new class of large tandem repeats discovered in the pathogenic yeast *Candida glabrata*. Cell Mol Life Sci 2010;67:671–676.
25 Young ET, Sloan JS, van Riper K: Trinucleotide repeats are clustered in regulatory genes in *Saccharomyces cerevisiae*. Genetics 2000;154:1053–1068.
26 Bonaccorsi S, Lohe A: Fine mapping of satellite DNA sequences along the Y chromosome of *Drosophila melanogaster*: Relationships between satellite sequences and fertility factors. Genetics 1991;129:177–189.
27 Chambers CA, Schell MP, Skinner DM: The primary sequence of a crustacean satellite DNA containing a family of repeats. Cell 1978;13:97–110.
28 Macas J, Neumann P, Novák P, Jiang J: Global sequence characterization of rice centromeric satellite based on oligomer frequency analysis in large-scale sequencing data. Bioinformatics 2010;26:2101–2108.
29 Lander ES, Linton LM, Birren B, Nusbaum C, Zody MC, et al: Initial sequencing and analysis of the human genome. Nature 2001;409:860–921.
30 Garrido-Ramos MA, Jamilena M, Lozano R, Ruiz Rejón C, Ruiz Rejón M: The *Eco*RI centromeric satellite DNA of the Sparidae family (Pisces, Perciformes) contains a sequence motive common to other vertebrate centromeric satellite DNAs. Cytogenet Cell Genet 1995;71:345–351.
31 Nijman IJ, Lenstra JA: Mutation and recombination in cattle satellite DNA: a feedback model for the evolution of satellite DNA repeats. J Mol Evol 2001;52:361–371.
32 Palomeque T, Lorite P: Satellite DNA in insects: a review. Heredity 2008;100:564–573.
33 Buscaino A, Allshire R, Pidoux A: Building centromeres: home sweet home or a nomadic existence? Curr Opin Gen Dev 2010;20:118–126.
34 Garrido-Ramos MA, de la Herran R, Ruiz-Rejón M, Ruiz-Rejón C: A satellite DNA of the Sparidae family (Pisces, Perciformes) associated with telomeric sequences. Cytogenet Cell Genet 1998;83:3–9.
35 Petrovic V, Pérez-García C, Pasantes JJ, Šatović E, Prats E, Plohl M: A GC-rich satellite DNA and karyology of the bivalve mollusk *Donax trunculus*: a dominance of GC-rich heterochromatin. Cytogenet Genome Res 2009;124:63–71.
36 Garrido-Ramos MA, de la Herran R, Ruiz-Rejón M, Ruiz-Rejón C: A subtelomeric satellite DNA family isolated from the genome of the dioecious plant *Silene latifolia*. Genome 1999;42:442–446.
37 Sharma S, Raina SN: Organization and evolution of highly repeated DNA sequences in plant chromosomes. Cytogenet Genome Res 2005;109:15–26.
38 De la Herrán R, Robles F, Cuñado N, Santos JL, Ruiz-Rejón M, et al: A heterochromatic satellite DNA is highly amplified in a single chromosome of *Muscari* (Hyacinthaceae). Chromosoma 2001;110:197–202.

39 Navajas-Pérez R, Schwarzacher T, de la Herrán R, Ruiz Rejón C, Ruiz Rejón M, Garrido-Ramos MA: The origin and evolution of the variability in a Y-specific satellite-DNA of *Rumex acetosa* and its relatives. Gene 2006;368:61–71.
40 Navajas-Pérez R, Quesada del Bosque ME, Garrido-Ramos MA: Effect of location, organization, and repeat-copy number in satellite-DNA evolution. Mol Genet Genomics 2009;282:395–406.
41 Gutknecht J, Sperlich D, Bachmann L: A species-specific satellite DNA family of *Drosophila subsilvestris* appearing predominantly in B chromosomes. Chromosoma 1995;103:539–544.
42 Abdelaziz M, Teruel M, Chobanov D, Camacho JP, Cabrero J: Physical mapping of rDNA and satDNA in A and B chromosomes of the grasshopper *Eyprepocnemis plorans* from a Greek population. Cytogenet Genome Res 2007;119:143–146.
43 Plohl M, Luchetti A, Mestrovic N, Mantovani B: Satellite DNAs between selfishness and functionality: Structure, genomics and evolution of tandem repeats in centromeric (hetero)chromatin. Gene 2008;409:72–82.
44 Ugarkovic D: Functional elements residing within satellite DNAs. EMBO Rep 2005;6:1035–1039.
45 Langdon T, Seago C, Jones RN, Ougham H, Thomas H, et al: De novo evolution of satellite DNA on the rye B chromosome. Genetics 2000;154:869–884.
46 Dover GA: Molecular drive: a cohesive mode of species evolution. Nature 1982;299:111–117.
47 Pérez-Gutiérrez MA, Suárez-Santiago VN, López-Flores I, Romero AT, Garrido-Ramos MA: Concerted evolution of satellite DNA in *Sarcocapnos*: a matter of time. Plant Mol Biol 2012;78:19–29.
48 Robles F, de la Herrán R, Ludwig A, Ruiz Rejón C, Ruiz Rejón M, Garrido-Ramos MA: Evolution of ancient satellite DNAs in sturgeon genomes. Gene 2004;338:133–142.
49 Navajas-Pérez R, de la Herrán R, Jamilena M, Lozano R, Ruiz Rejón C, et al: Reduced rates of sequence evolution of Y-linked satellite DNA in *Rumex* (Polygonaceae). J Mol Evol 2005;60:391–399.
50 Suárez-Santiago VN, Blanca G, Ruiz-Rejón M, Garrido-Ramos MA: Satellite-DNA evolutionary patterns under a complex evolutionary scenario: the case of *Acrolophus* subgroup (Centaurea L., Compositae) from the western Mediterranean. Gene 2007;404:80–92.
51 Mravinac B, Plohl M: Satellite DNA junctions identify the potential origin of new repetitive elements in the beetle *Tribolium madens*. Gene 2007;394:45–52.
52 Grechko VV, Ciobanu DG, Darevsky IS, Kosushkin SA, Kramerov DA: Molecular evolution of satellite DNA repeats and speciation of lizards of the genus *Darevskia* (Sauria: Lacertidae). Genome 2006;49:1297–1307.
53 Mestrovic N, Plohl M, Mravinac B, Ugarkovic D: Evolution of satellite DNAs from the genus *Palorus* – experimental evidence for the 'library' hypothesis. Mol Biol Evol 1998;15:1062–1068.
54 Bruvo B, Pons J, Ugarkovic D, Juan C, Petitpierre E, Plohl M: Evolution of low-copy number and major satellite DNA sequences coexisting in two *Pimelia* species-groups (Coleoptera). Gene 2003;312:85–94.
55 Stimpson KM, Sullivan BA: Epigenomics of centromere assembly and function. Curr Opin Cell Biol 2010;22:772–780.
56 De la Herrán R, Fontana F, Lanfredi M, Congiu L, Leis M, et al: Slow rates of evolution and sequence homogenization in an ancient satellite DNA family of sturgeons. Mol Biol Evol 2001;18:432–436.
57 López-Flores I, de la Herrán R, Garrido-Ramos M, Boudry P, Ruiz Rejón C, Ruiz-Rejón M: The molecular phylogeny of oysters based on a satellite DNA related to transposons. Gene 2004;339:181–188.
58 Vicari MR, Nogaroto V, Noleto RB, Cestari MM, Cioffi MB, et al: Satellite DNA and chromosomes in Neotropical fishes: methods, applications and perspectives. J Fish Biol 2010;76:1094–1116.
59 Benslimane AA, Dron M, Hartmann C, Rode A: Small tandemly repeated DNA sequences of higher plants likely originate from a tRNA gene ancestor. Nucleic Acids Res 1986;14:8111–8119.
60 Torras-Llort M, Moreno-Moreno O, Azorín F: Focus on the centre: the role of chromatin on the regulation of centromere identity and function. EMBO J 2009;28:2337–2348.
61 Wang G, Zhang X, Jin W: An overview of plant centromeres. J Genet Genomics 2009;36:529–537.
62 Martínez P, Blasco MA: Telomeric and extra-telomeric roles for telomerase and the telomere-binding proteins. Nat Rev Cancer 2011;11:161–176.
63 Blackburn EH, Gall JG: A tandemly repeated sequence at the termini of the extrachromosomal ribosomal RNA genes in *Tetrahymena*. J Mol Biol 1978;120:33–53.
64 Henderson E: Telomere DNA structure; in Blackburn EH, Greider CW (eds): Telomeres. New York, Cold Spring Harbor Laboratory Press, 1995, pp 11–34.
65 Greider CW, Blackburn EH: Identification of a specific telomere terminal transferase activity in *Tetrahymena* extracts. Cell 1985;43:405–413.
66 Pardue ML, Rashkova S, Casacuberta E, DeBaryshe PG, George JA, Traverse KL: Two retrotransposons maintain telomeres in *Drosophila*. Chromosome Res 2005;13:443–453.

67 Hua-Van A, Le Rouzic A, Boutin TS, Filée J, Capy P: The struggle for life of the genome's selfish architects. Biol Direct 2011;6:19.
68 Bringaud F, Ghedin E, Blandin G, Bartholomeu DC, Caler E, et al: Evolution of non-LTR retrotransposons in the trypanosomatid genomes: *Leishmania major* has lost the active elements. Mol Biochem Parasitol 2006;145:158–170.
69 Bringaud F, Ghedin E, El-Sayed NM, Papadopoulou B: Role of transposable elements in trypanosomatids. Microbes Infect 2008;10:575–581.
70 Mikkelsen TS, Wakefield MJ, Aken B, Amemiya CT, Chang JL, et al: Genome of the marsupial *Monodelphis domestica* reveals innovation in non-coding sequences. Nature 2007;447:167–178.
71 Schnable PS, Ware D, Fulton RS, Stein JC, Wei F, et al: The B73 maize genome: complexity, diversity, and dynamics. Science 2009;326:1112–1115.
72 Jaillon O: Genome duplication in the teleost fish *Tetraodon nigroviridis* reveals the early vertebrate proto-karyotype. Nature 2004;431:946–957.
73 Kidwell MG: Transposable elements; in Gregory TR (ed.): The Evolution of the Genome. San Diego, CA, Elsevier Academic Press, 2005, pp 165–221.
74 Piegu B, Guyot R, Picault N, Roulin A, Saniyal A, et al: Doubling genome size without polyploidization: dynamics of retrotransposition-driven genomic expansions in *Oryza australiensis*, a wild relative of rice. Genome Res 2006;16:1262–1269.
75 Hu TT, Pattyn P, Bakker EG, Cao J, Cheng JF, et al: The *Arabidopsis lyrata* genome sequence and the basis of rapid genome size change. Nat Genet 2011;43:476–481.
76 Wicker T, Sabot F, Hua-Van A, Bennetzen JL, Capy P, et al: A unified classification system for eukaryotic transposable elements. Nat Rev Genet 2007;8: 973–982.
77 Feschotte C, Pritham EJ: DNA transposons and the evolution of eukaryotic genomes. Annu Rev Genet 2007;41:331–368.
78 Belancio VP, Hedges DJ, Deininger P: Mammalian non-LTR retrotransposons: for better or worse, in sickness and in health. Genome Res 2008;18:343–358.
79 Jurka J, Kapitonov VV, Kohany O, Jurka MV: Repetitive sequences in complex genomes: structure and evolution. Annu Rev Genomics Hum Genet 2007;8:241–259.
80 Eickbush TH, Jamburuthugoda VK: The diversity of retrotransposons and the properties of their reverse transcriptases. Virus Res 2008;134:221–234.
81 Kapitonov VV, Jurka J: A universal classification of eukaryotic transposable elements implemented in Repbase. Nat Rev Genet 2008;9:411–412.
82 Kapitonov VV, Tempel S, Jurka J: Simple and fast classification of non-LTR retrotransposons based on phylogeny of their RT domain protein sequences. Gene 2009;448:207–213.
83 Kramerov DA, Vassetzky NS: Origin and evolution of SINEs in eukaryotic genomes. Heredity 2011; 107:487–495.
84 Mouse Genome Sequencing Consortium: Waterston RH, Lindblad-Toh K, Birney E, Rogers J, Abril JF, et al: Initial sequencing and comparative analysis of the mouse genome. Nature 2002;420:520–562.
85 Gibbs RA, Weinstock GM, Metzker ML, Muzny DM, Sodergren EJ, et al: Genome sequence of the brown Norway rat yields insights into mammalian evolution. Nature 2004;428:493–521.
86 Li R, Fan W, Tian G, Zhu H, He L, et al: The sequence and de novo assembly of the giant panda genome. Nature 2010;463:311–317.
87 Lindblad-Toh K, Wade CM, Mikkelsen TS, Karlsson EK, Jaffe DB, et al: Genome sequence, comparative analysis and haplotype structure of the domestic dog. Nature 2005;438:803–819.
88 International Chicken Genome Sequencing Consortium: Sequence and comparative analysis of the chicken genome provide unique perspectives on vertebrate evolution. Nature 2004;432:695–777.
89 Paterson AH, Bowers JE, Bruggmann R, Dubchak I, Grimwood J, et al: The *Sorghum bicolor* genome and the diversification of grasses. Nature 2009;457:551–556.
90 Warren WC: Genome analysis of the platypus reveals unique signatures of evolution. Nature 2008; 453:175–184.
91 Lerat E, Brunet F, Bazin C, Capy P: Is the evolution of transposable elements modular? Genetica 1999; 107:15–25.
92 Capy P: Classification and nomenclature of retrotransposable elements. Cytogenet Genome Res 2005;110:457–461.
93 Cordaux R, Batzer MA: The impact of retrotransposons on human genome evolution. Nat Rev Genet 2009;10:691–703.
94 Gladyshev EA, Arkhipova IR: Telomere-associated endonuclease-deficient *Penelope*-like retroelements in diverse eukaryotes. Proc Natl Acad Sci USA 2007;104:9352–9357.
95 Deragon JM, Zhang X: Short interspersed elements (SINEs) in plants: origin, classification, and use as phylogenetic markers. Syst Biol 2006;55:949–956.
96 Kriegs JO, Churakov G, Jurka J, Brosius J, Schmitz J: Evolutionary history of 7SL RNA-derived SINEs in Supraprimates. Trends Genet 2007;23:158–161.
97 Shimamura M, Yasue H, Ohshima K, Abe H, Kato H, et al: Molecular evidence from retroposons that whales form a clade within even-toed ungulates. Nature 1997;388:666–670.

98 Konkel MK, Batzer MA: A mobile threat to genome stability: The impact of non-LTR retrotransposons upon the human genome. Semin Cancer Biol 2010; 20:211–221.
99 Okada N, Sasaki T, Shimogori T, Nishihara H: Emergence of mammals by emergency: exaptation. Genes Cells 2010;15:801–812.
100 Krull M, Brosius J, Schmitz J: Alu-SINE exonization: en route to protein-coding function. Mol Biol Evol 2005;22:1702–1711.
101 Chapman JA, Kirkness EF, Simakov O, Hampson SE, Mitros T, et al: The dynamic genome of *Hydra*. Nature 2010;464:592–596.
102 Hellsten U, Harland RM, Gilchrist MJ, Hendrix D, Jurka J, et al: The genome of the Western clawed frog *Xenopus tropicalis*. Science 2010;328:633–636.
103 Bao W, Jurka MG, Kapitonov VV, Jurka J: New superfamilies of eukaryotic DNA transposons and their internal divisions. Mol Biol Evol 2009;26:983–993.
104 Volff J-N: Turning junk into gold: domestication of transposable elements and the creation of new genes in eukaryotes. BioEssays 2006;28:913–922.
105 Moran JV, DeBerardinis RJ, Kazazian HH Jr: Exon shuffling by L1 retrotransposition. Science 1999;283: 1530–1534.
106 Jiang N, Bao Z, Zhang X, Eddy SR, Wessler SR: Pack-MULE transposable elements mediate gene evolution in plants. Nature 2004;431:569–573.
107 Morgante M, Brunner S, Pea G, Fengler K, Zuccolo A, Rafalski A: Gene duplication and exon shuffling by helitron-like transposons generate intraspecies diversity in maize. Nat Genet 2005;37:997–1002.
108 Kapitonov VV, Jurka J: RAG1 core and V(D)J recombination signal sequences were derived from Transib transposons. PLoS Biol 2005;3:e181.
109 Pan D, Zhang L: Burst of young retrogenes and independent retrogene formation in mammals. PLoS One 2009;4:e5040.
110 Han JS, Boeke JD: LINE-1 retrotransposons: modulators of quantity and quality of mammalian gene expression? Bioessays 2005;27:775–784.
111 Lyon MF: LINE-1 elements and X chromosome inactivation: a function for 'junk' DNA? Proc Natl Acad Sci USA 2000;97:6248–6249.
112 Vaughn MW, Tanurdžić M, Lippman Z, Jiang H, Carrasquillo R, et al: Epigenetic natural variation in *Arabidopsis thaliana*. PLoS Biol 2007;5:e174.
113 Bailey J, Eichler EE: Primate segmental duplications: crucibles of evolution, diversity and disease. Nat Rev Genet 2006;7:552–564.
114 Marques-Bonet T, Girirajan S, Eichler EE: The origins and impact of primate segmental duplications. Trends Genet 2009;25:443–454.
115 She X, Horvath JE, Jiang Z, Liu G, Furey TS, et al: The structure and evolution of centromeric transition regions within the human genome. Nature 2004;430:857–864.
116 She X, Jiang Z, Clark RA, Liu G, Cheng Z, et al: Shotgun sequence assembly and recent segmental duplications within the human genome. Nature 2004;431:927–930.
117 Lupski JR: Genomic disorders: structural features of the genome can lead to DNA rearrangements and human disease traits. Trends Genet 1998;14:417–422.
118 Bailey J, Yavor AM, Massa HF, Trask BJ, Eichler EE: Segmental duplications: organization and impact within the current human genome project assembly. Genome Res 2001;11:1005–1017.

Dr. Manuel A. Garrido-Ramos
Departamento de Genética, Facultad de Ciencias
Universidad de Granada
Avda. Fuentenueva, s/n, ES–18071 Granada (Spain)
Tel. +34 958 243 260, E-Mail mgarrido@ugr.es

Garrido-Ramos MA (ed): Repetitive DNA.
Genome Dyn. Basel, Karger, 2012, vol 7, pp 29–45

Telomere Dynamics in Mammals

D.C. Silvestre · A. Londoño-Vallejo

Telomeres and Cancer Laboratory, Institut Curie – CNRS UMR3244 – UPMC, Paris, France

Abstract

Telomeres are specialized structures found at the end of linear chromosomes. Telomere structure and functions are conserved throughout evolution and are essential for genome stability, preventing chromosome ends from being recognized as damaged DNA and from being fused or degraded by the DNA repair machinery. The structure of telomeres is intrinsically dynamic and affected by multiple processes that impact their length and nucleoprotein composition, thus leading to functional and structural heterogeneity. We review here the most significant facets of telomere metabolism and its dynamics, with an emphasis on human biology.

Copyright © 2012 S. Karger AG, Basel

During evolution, the organization of genomes as linear chromosomes, probably an adaptation to the increasing genome complexity, was also likely a prerequisite to the emergence of meiosis. Such linearization was achieved through the development of specialized structures, the telomeres, at the ends of the linearized ancestor chromosome. Telomeres were first recognized in *Drosophila* as endowed with distinctive properties because native chromosome ends did not fuse to each other or to artificially induced double strand breaks [1]. More than 4 decades later, the first telomere structure was identified in *Tetrahymena* and was shown to consist of short, repetitive, guanine-rich (G-rich) DNA sequences with a strong strand bias [2]. Remarkably, these structural characteristics prevail at telomeres of most organisms, including the presence of specialized proteins that specifically bind telomeric sequences to form a nucleoprotein complex whose function is to preserve chromosome ends. Interestingly, an important exception to this rule is found precisely in *Drosophila*, where long, repeated retroelement units are placed at the tips of the chromosomes and protection is ensured by a specialized group of proteins that perform this task in a sequence independent manner [3].

In a striking contrast to their amazing evolutionary stability, telomeres are the most dynamic structures of linear genomes, as they undergo shortening/lengthening events as well as changes in nucleoprotein composition depending on the phase of the cell cycle or cell differentiation state. This dynamics has direct and critical consequences, both at the cell and organismal levels. In fact, no other genomic structure has been so directly implicated in such fundamental aspects as longevity and aging, particularly in humans.

Telomere Structure

In vertebrates, telomeres consist of double-stranded tandem repeats of the hexamer 5′-TTAGGG-3′, present in several thousands of copies and ending with a 3′ G-rich protruding overhang of a few tens to few hundreds of bases (at least in humans, figs. 1 and 2). This overhang is essential for telomere function and has been proposed to mediate the formation of a protective higher-order structure, referred to as the T-loop, in which telomere DNA loops back, allowing the G-strand overhang to invade the doubled-stranded portion thus forming a recombination-like D-loop structure [4] (fig. 2). The T-loop hides the chromosome end, thus preventing it from being recognized as a break by the DNA repair machinery leading to its fusion and/or degradation. While G-rich overhangs constitute not only the predominant feature of normal telomeres but are also absolutely required for telomere function, cytosine-rich (C-rich) overhangs may be also present in some organisms as different as worms and mammals. With the exception of *Caenorhabditis elegans*, where C-overhangs are as abundant as G-overhangs, the amount of detectable C-overhangs is low or very low in mouse or human cells, respectively, and they have been proposed to derive from some sort of recombination event [5]. The fact that such structure is more frequently seen in mammalian cells using alternative lengthening of telomeres (ALT) mechanisms (see below) to maintain telomere length supports this contention. Interestingly, C-rich overhangs can also invade double-stranded telomere sequences and therefore could potentially form T-loop-like structures. However, it is not known whether a C-rich-mediated T-loop can function as a protective mechanism for chromosome ends.

Maintaining Telomere Lengths: the End Replication Problem

One critical parameter of telomere function is the number of telomeric repeats as telomeres require a minimum length to exert their protective role. Therefore, mechanisms to maintain telomere length are fundamental to maintain genome stability.

The primary mechanism of telomere length maintenance is telomere replication. It is generally admitted that telomere sequences are devoid of replication origins and that replication forks must proceed from the subtelomeric region into the telomeric repeats and progress uninterrupted all the way to the end of the chromosome. This

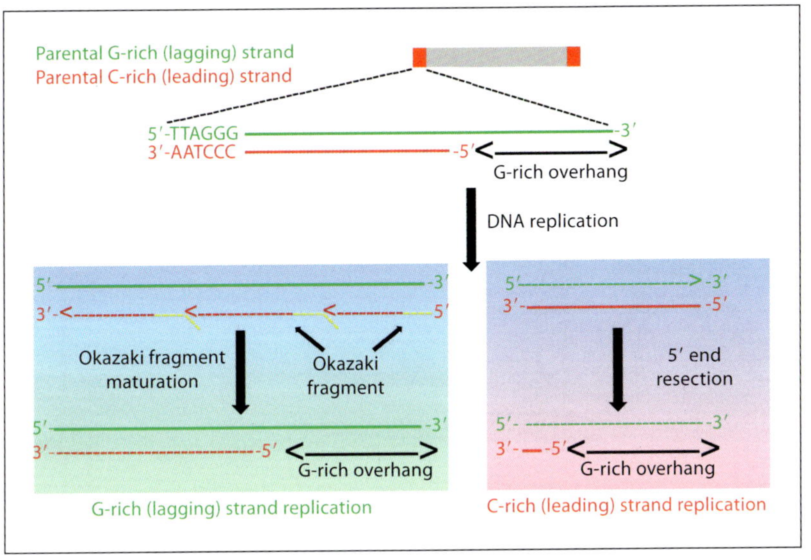

Fig. 1. The end replication problem is a leading strand problem. Telomeric G-rich strands are replicated by lagging mechanisms, while telomeric C-rich strand is replicated by leading mechanisms. The RNA primer (in yellow) giving rise to the last Okazaki fragment will be removed, leaving the end of the G-strand unreplicated, thus reconstituting the normal overhang (eventually elongated by telomerase activity or by further degradation of the C-strand). The leading replication of the C-rich strand, on the other hand, leads to the production of a blunt end, and therefore must undergo resection to produce a functional 3′ overhang, thus suffering a net length loss with regard to the parental telomere.

directionality imposes that the G-rich strand will always be replicated by a lagging mechanism and that the C-rich strand will always be replicated by a leading mechanism. When the fork reaches the end, the last Okazaki fragment on the lagging strand will be positioned more or less close to the 3′ end of the parental strand, inevitably leaving a portion of the G-rich overhang unreplicated (fig. 1). This incomplete replication is actually advantageous since it perfectly duplicates the telomere structure with an identical overall length. However, the telomere on the sister chromatid replicated by leading mechanisms not only ends up being shorter than its counterpart (because it used a receded C-rich template as a template), but the template itself needs to be further shortened in order to create the required G-rich 3′ overhang for the new telomere to become functional (fig. 1). Thus, the end replication problem is a leading strand problem.

Incomplete replication is not the only physiological mechanism leading to telomere shortening. With the discovery of the T-loop, it was proposed that the D-loop may represent an ideal substrate for recombination, leading to its resolution with the excision of a nicked circle and a net loss of telomere length, a phenomenon called telomere rapid deletion (TRD). Such mechanism appears to be responsible for the

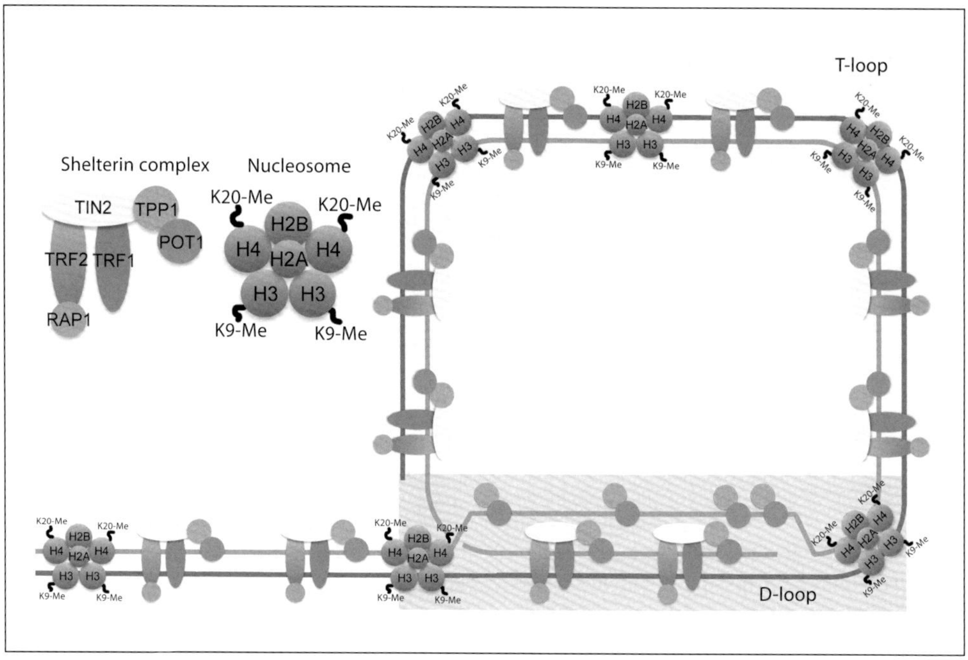

Fig. 2. Telomere nucleoprotein structure: Shelterin and chromatin share the place. A telomeric loop (T-loop) may be formed when the G-rich overhang invades the double-stranded portion of the same telomere. The displacement of the G-rich strand allows the overhang to hybridize to the C-rich strand to form a D-loop. Another possibility (not represented here) is for the G-rich overhang to form a G-quadruplex structure with the displaced G-rich strand. The Shelterin complex promotes the formation of the T-loop, which renders the chromosome extremity transparent to the cell, thus preventing DDR. Nucleosomes with histone modifications typical of heterochromatin are also represented, but the actual arrangement of Shelterin complexes and nucleosomes at telomeric DNA remains unknown.

emergence of extrachromosomal telomeric circles (T-circles) in cells that either use ALT or else carry abnormally long telomeres [6]. However, it has been recently observed that T-circles can also be detected in the human male germ cell line as well as in normal peripheral blood lymphocytes that had undergone telomere elongation as part of the normal response to stimulation. This mechanism of trimming thus contributes to the telomere length homeostasis of the organism [7].

Telomere Length Homeostasis: the Long and the Short of It

The average length of telomeres is a characteristic of every organism and is the result of the equilibrium between shortening due to replication, TRD or accidental telomere loss on one hand, and lengthening events, mediated in most cases by a dedicated enzymatic mechanism, telomerase, on the other. Telomerase is a specific reverse transcriptase that synthesizes de novo telomere repeats using the 3′ end overhang as the substrate and a specific RNA as a template. In some organisms, telomerase is constitutively expressed and telomeres are maintained at a stable length. In humans, telomerase activity is highly regulated and mostly restricted to stem cell compartments, with loss of telomerase expression as cells differentiate [8]. Since differentiation is often associated with proliferation, telomere shortening with tissue generation/regeneration, and therefore with organismal age, is the rule. At the cell level, chromosome ends harbor telomeres of heterogeneous lengths, and this heterogeneity is best explained by allelic length polymorphisms that are inherited and maintained throughout life in spite of an age-related absolute shortening. This observation points to the possibility that polymorphic juxtatelomeric sequences in *cis* actively influence the efficiency of telomere maintenance mechanisms such as telomere replication efficiency, elongation by telomerase or stimulation of TRD [9]. Experimentally, short and long telomeres shorten at similar paces in telomerase-negative cells growing in vitro under physiological conditions, suggesting that replication mechanisms are equally efficient and that spontaneous TRD events are rare [10]. This is probably not the case in telomerase-positive lymphocytic cells that respond to stimuli by elongating telomeres. As noted above, TRD acts in these cells probably as a homeostatic mechanism bringing telomere lengths back to an equilibrium point.

Shelterin and Telomere Dynamics

As noted above, telomeres are bound by telomere-specific proteins. In mammals, there is a 6-protein complex, called Shelterin, composed of TRF1 (telomeric repeat binding factor 1) and TRF2, TIN2 (TRF1 interacting protein 2), POT1 (protection of telomeres 1), TPP1 (TINT1/PIP1/PTOP1) and RAP1 (repressor/activator protein 1) (fig. 2). TRF1, TRF2 and POT1 bind directly to telomeric DNA repeats, with TRF1 and TRF2 binding to telomeric double-stranded DNA and POT1 to the 3′ single-stranded G-overhang. There is direct interaction between TRF1 and TRF2, but TIN2 binds both proteins through independent domains, thus bridging these DNA binding subunits of Shelterin [11]. TIN2 also binds to TPP1 and helps in the recruitment of the TPP1-POT1 complex to telomeres. RAP1 binds to telomeres through TRF2, and this association is, at least in the mouse context, essential for RAP1 stability [12]. Although striking analogies can be drawn between mammalian telomeres and those of lower organisms (such as yeasts and worms), we will focus here mainly on studies made on each of the components of Shelterin in human cells and transgenic mice, with emphasis on their role in telomere dynamics.

TRF1 and TRF2

TRF1 and TRF2 function as negative regulators of telomere length, controlling the access of telomerase to its 3′ G-rich overhang substrate [13, 14]. This may be accomplished through the stabilization of the T-loop structure or by promoting 3′ end occlusion by POT1. The expression of a dominant-negative *TRF1* allele results in the elongation of telomeres [13]. In contrast, the expression of a dominant-negative *TRF2* allele results in an actual loss of G-tail DNA sequences from telomeres, the activation of DNA damage response factors and chromosome end-to-end fusion [15]. TRF2 specifically recognizes telomeric single/double-stranded DNA junctions, thus likely facilitating the formation of the T-loop [4]. Loss of TRF2 from telomeres results in an approximate 50% reduction of the single-stranded telomeric repeat signal, while the duplex part of the telomere remains intact and undergoes fusions [15]. Both TRF1 and TRF2 are required for the normal replication of telomeres. In the absence of TRF1, replication forks stall and telomere fragility is increased [16]. TRF2, on the other hand, is required, together with its interactor Apollo, to alleviate the topological constraints that arise during fork progression [17]. Apollo is a 5′ exonuclease, also implicated in the post-replicative processing of the sister telomere replicated by leading mechanisms [18].

POT1

POT1 binds directly the 3′ overhang, thus controlling telomerase access to its substrate. POT1 interacts with the TRF1 complex via TPP1/TIN2, and this interaction is thought to modulate POT1 loading on the single-stranded telomeric DNA [19]. In human cells, overexpression of a POT1 allele unable to bind DNA leads to telomerase-dependent elongation of telomeres. However, POT1's functions go beyond telomerase control. While the human genome contains only 1 *POT1* gene, mice express 2 *POT1* paralogs, *Pot1a* and *Pot1b*. The duplication of the gene in the mouse naturally resulted in separation of functions, which has allowed a better comprehension of the roles of this protein [20]. In mice, lack of POT1a results in embryonic lethality, whereas POT1b KO mice are viable and fertile [21, 22]. At the cell level, the binding of both POT1 proteins to mouse telomeres depends on TPP1 [23], and both are apparently required for chromosome stability. The specific functions ascribed to mouse POT1a/b may also vary depending on the experimental setting. POT1a has been shown to prevent a DNA damage signal at telomeres, whereas POT1b has been shown to regulate C-rich strand degradation [21]. Lack of POT1b leads to end-to-end fusions and chromosome instability, increased telomere sister chromatid exchanges (T-SCE) and formation of T-circles, suggesting an exacerbation of recombination activity at telomeres [24]. POT1b KO mouse embryonic fibroblasts display a strong DNA damage response at telomeres leading to p53-dependent senescence. In vivo, POT1b KO, combined or not with a telomerase deficiency, profoundly affects bone marrow cell proliferation [25], suggesting an intrinsic requirement of this protein for cell maintenance.

It is likely, but still not proven, that POT1 binds to the G-rich single-stranded DNA that accumulates during normal replication fork progression. POT1 thus may compete with RPA, the normal single-strand DNA binding protein that travels with the fork. However, whether or not POT1 displays higher affinity for the telomeric G-rich strand than RPA is controversial. Independent of their relative affinities measured in vitro, it has been shown in vivo that under conditions where there is excessive accumulation of replicative G-rich (lagging) single-strand DNA (for instance, in the absence of the helicase WRN), POT1 is required to allow full replication of the C-rich (leading) strand [26]. When POT1 is limiting in the cell, the replication of the leading strand is also affected and RPA accumulates at telomeres. These experiments suggest that, at least under particular conditions, POT1 is able to compete out RPA, and that this activity allows the uncoupling of the fork and the progression of the leading replication.

TPP1

TPP1 is required for the protective function of telomeres. Deletion of TPP1 in mice, when homozygous, results in early embryonic lethality [23]. Hypomorphic mutations in the mouse give rise to adrenocortical dysplasia with pleiotropic phenotypes related to telomere dysfunction and genome instability [27]. In a conditional context, TPP1 deletion results in the release of POT1a and POT1b from chromatin and loss of these proteins from telomeres, indicating that TPP1 is required for the telomere association of POT1a and POT1b but not for their stability [28]. The telomere dysfunction phenotypes associated with deletion of TPP1 were identical to those of POT1a/POT1b double KO cells [23]. TPP1 interacts directly with POT1, influencing the functions of the latter with regard to telomerase activity [29]. In particular, and opposite to the activity of POT1 alone, the complex TPP1-POT1 is a processivity factor for telomerase in vitro. In the mouse, TPP1 is required for recruitment of telomerase to telomeres and telomere elongation during nuclear reprogramming [30].

RAP1

The functions of mammalian RAP1 are not fully understood. While in yeast RAP1 is the archetypal telomeric DNA binding protein, in mammals it binds to telomeres exclusively through its interaction with TRF2, where it may function as an adaptor protein [31]. However, RAP1 has been shown to be dispensable for TRF2 function, in particular the repression of ATM signaling and the non-homologous end joining pathway [32], and its absence does not affect the binding of other Shelterin members to telomeres. The requirement of RAP1 itself in the repression of the DNA damage response (DDR) remains controversial. Similarly, different viability outcomes have been observed in mice KO for this telomeric protein further obscuring the role of RAP1 at telomeres. Also in the mouse, a role for RAP1 in the inhibition of homology-directed repair and fragility at telomeres has been demonstrated [32], while in humans RAP1 appears to be implicated in length homeostasis [33]. On the other hand, it has been found that RAP1 participates in intracellular signaling and transcription control [34], similar to the yeast RAP1 activity.

TIN2
This protein constitutes an important connecting block in the Shelterin building since it interacts with TRF1, TRF2, and TPP1. Through this last partner, TIN2 mediates POT1 recruitment to the complex. Consistent with these multiple interactions, depletion of TIN2 or the expression of particular TIN2 mutants has a profoundly destabilizing effect on Shelterin [35]. On the other hand, TIN2 modifies the TRF1 binding to telomeres by controlling the activity of TNK1, a poly-ADP ribose polymerase able to modify TRF1. Poly-APD ribosylated TRF1 has a lower affinity for telomeres and is targeted for degradation. This removal of TRF1 favors telomere elongation by telomerase [36]. TIN2 is also required to establish/maintain sister telomere cohesion after replication. This function requires the recruitment of heterochromatic protein 1γ to telomeres, which binds to a specific domain in the C-terminus of TIN2 [37]. To date, there is just 1 report on the effect of the homozygous deletion of this gene [38]. Inactivation of the mouse *Tinf2* gene results in early embryonic lethality, but whether or not this is solely due to its telomeric role remains to be demonstrated.

Telomeric Chromatin

Double-stranded eukaryotic DNA is assembled into chromatin and therefore nucleosome formation also takes place at telomeres, at least in higher eukaryotes. These nucleosomes appear to be tightly spaced and their mobility is directly affected by the binding of TRF1 and TRF2 in vitro [39]. In vivo, overexpression of TRF2 leads to more spaced nucleosomes and a decrease of heterochromatin marks [40]. On the other hand, lack of TRF2 does not induce obvious changes in nucleosome compaction [41]. Of the 2 types of epigenetic events that govern chromatin compaction, that is DNA methylation and N-terminal modifications of histones, only the latter can occur at telomeric chromatin. However, juxtatelomeric chromatin can bear both heterochromatic marks and actually such marks can influence telomere dynamics. Constitutively, both telomeres and subtelomeres are enriched in heterochromatin marks, including trimethylated H4K20 and H3K9 (fig. 2). In addition, telomeric H3 and H4 histones are underacetylated [42]. Telomeres also contain all the heterochromatin protein 1 isoforms (HP1α, HP1β and HP1γ) [42]. These marks of compacted chromatin state have been demonstrated to impact telomere length homeostasis and stability. It has been shown that the induction of chromatin decondensation leads to an abnormal lengthening of telomeres, mediated by an increase in the rate of telomeric homologous recombination (HR) [43]. A key factor in the regulation of telomeric epigenetics is the retinoblastoma (Rb) family, which is involved in the maintenance of constitutive heterochromatin. Besides their role as transcriptional repressors through interactions with the E2F family of transcription factors, Rb proteins influence global H4K20me3 levels through a direct interaction with trimethylating enzymes. Rb proteins also control the expression of genes coding for DNA methyltransferases, thus influencing the levels of DNA methylation [44]. Furthermore, the aberrant telomere elongation in the context of Dicer1 deficiency is explained by a decrease in the production of a microRNA (miR-290) which targets Rbl2, thus allowing the protein to accumulate and repress the DNA methyltransferase genes with a decrease in DNA methylation and loss of heterochromatin at telomeres [40].

The linker histone H1, which is known to be involved in high-order chromatin compaction [45], plays also an important role at telomeres. Mouse cells defective for H1 showed less compact chromatin [46] and a 4-times higher frequency of HR at telomeres and longer telomeres than wild type cells [47]. On the other hand, there seems to be a counterbalance between telomere elongation and repressive chromatin, since telomeres that have become very long also acquire a more heterochromatic status [48]. On the contrary, progressive telomere shortening in telomerase-deficient mouse embryonic fibroblasts is associated with continuous loss of heterochromatic marks at telomeres and subtelomeres [42].

Transcription at Telomeres: TERRA

Recently, telomere repeats were found to be transcribed by RNA polymerase II, giving rise to UUAGGG-repeat containing non-coding RNAs named TERRA, for telomeric repeat-containing RNA [49]. Approximately 25% of human telomeres contain 3 specific repetitive elements with CpG-rich sequences in their subtelomeric region [50]. DNA fragments comprising these CpG-islands show promoter activity and may therefore drive TERRA transcription at this subset of telomeres. Here also, epigenetics plays an important role: the cytosine methylation state of these DNA repeats negatively correlates with TERRA abundance [42], and telomeric and subtelomeric chromatin hallmarks affect TERRA transcription [51]. In mammals, TERRA molecules range between 100 bp and 9 kb. In vitro, such molecules can form an intermolecular G-quadruplex structure with telomeric DNA repeats, which may negatively affect telomere replication [49]. Several lines of evidence suggest that TERRA may act as a direct regulator of telomerase, either as a competitive inhibitor for telomeric DNA or through direct binding to the telomerase complex without displacing the telomere substrate [52]. Thus, increasing TERRA levels by impairing its degradation is associated with a loss of telomeric DNA repeats [49]. In line with a putative role of TERRA as a negative regulator of telomere length, TERRA levels are particularly high in adult tissues that do not display telomerase activity, low in mouse embryos at E11.5–15.5 where telomerase activity peaks, and low as well in human cancer samples [51]. However, the real relevance of TERRA depletion/overexpression on the overall biology of cells or organisms remains to be established.

An Enlarged View of Telomeres: the Telosome

Recent studies have established an extensive catalog of proteins that interact with the Shelterin complex and that are involved in telomere metabolism at various levels. The reader is invited to visit recent reviews on the subject. Here, it will simply be stressed that many of these proteins play central roles as mediators/effectors of DDR elsewhere in the genome, whereas at telomeres they appear to contribute to the protection of chromosome ends against these activities. Defaults in many of these proteins, such as MRE11 or WRN, have discreet but non-negligible impacts on telomere length maintenance, mainly during replication, but others may have tremendous negative consequences on telomere stability. For instance, elimination of Ku86 in human cells results in overwhelming TRD reactions, leading to rampant telomere deletions and abundant T-circle formation [53]. Defining the precise interactions between components of the telosome, their relationship with Shelterin and their contribution to telomere physiology will be a major task in the field for the next years.

Telomeres and Chromosome Instability and Their Role in Cancer

With shortening, there is an increased risk for telomeres to become unstable. When telomeres become too short, telomeres are impoverished of Shelterin factors, thus loosing their caps and are recognized as DNA damage sites. This response is undistinguishable from the response triggered by a bona fide double strand break elsewhere in the genome, involving the MRN complex, which activates ATM and its signaling cascade including 53BP1 and the accumulation of phosphorylated H2AX at telomeres. The downstream activation of p53/Rb pathways blocks further cell cycle progression. In the classic DDR, this activation is maintained the time for a repair reaction to take place. At uncapped telomeres, however, repair is probably less efficient because most often there is no other uncapped extremity nearby to be fused to. The lesion may then persist, leading to permanent cell senescence or, in some cells, to apoptosis. In fact, it has been suggested that a single dysfunctional telomere is sufficient to trigger senescence in a cell [54]. On the other hand, if uncapping occurs during replication, dysfunctional sister telomeres may be repaired by fusing to each other [55]. However, a conflict will soon appear as fused sister chromatids are pulled apart during mitosis. This conflict may be sufficient to block further progression of cell division (leading to tetraploidization [56]). It may also be resolved either through a DNA double strand break, thus allowing chromatids carrying imbalanced translocations to segregate. Loss of spindle attachment of one of the chromatids may also lead to segregation of a fused, duplicated chromosome into one of the daughter cells. In a context where the p53/Rb pathways are disabled, cells with telomere instability continue to proliferate, which brings further telomere shortening and new chromosome ends become available for breakage-fusion-bridge cycles, leading to rampant genome instability and cell death (crisis) unless a mechanism of telomere maintenance is activated [56]. Interestingly, the type of chromosome instability that is seen in cells that had gone through crisis in vitro resembles the one often seen in cells obtained from human carcinomas, suggesting that telomere uncapping may be implicated in chromosome instability associated with cancer in humans.

Mouse models of short telomeres strongly support the notion that telomere-related chromosome instability directly contributes to the acquisition of tumor phenotypes, perhaps through gains and losses of cancer related genes and regions [57]. In humans, however, the evidence is only correlative, with short telomeres in blood cells being associated with the risk of aggressive cancers in various systems, and extreme telomere shortening being prevalent in tumor cells in vivo [58]. Interestingly, alterations affecting Shelterin components have also been detected in several types of cancers [59], although the relevance of such findings remains to be determined. Once again, mouse models have been quite informative in that respect. Since deletions of either TRF1 or TRF2 in mice are embryonic lethal [12, 60], conditional transgenic mice have been produced in order to study their roles in vivo. Mice in which TRF1 has been conditionally deleted in epithelial cells die perinatally and show reduced skin thickness, reduced skin stratification and predisposition to cancer [61]. On the other hand, mice overexpressing TRF1 and TRF2 in the skin show an accelerated rate of telomere shortening and higher predisposition to cancers. Both mouse models present increased end-to-end chromosomal fusions, multitelomeric signals, and increased telomere recombination [62, 63]. Whether or not alterations in Shelterin components contribute to cancer development in humans remains to be established.

Finally, to acquire fully unlimited proliferation, cancer cells have to reactivate telomere elongation. The mechanism at the base of telomerase reactivation is unclear, but factors most likely contributing to it are the frequent inactivation of p53 (which is able to repress the hTERT promoter) together with the presence of active Myc (which is able to stimulate hTERT expression) in most human tumors [64]. On the other hand, the bases for the reactivation of alternative mechanisms, which are based on some form of homologous recombination and are more frequent in cancers of mesenchymal origin, remain unknown.

Lengthening Telomeres without Telomerase: Alternative Lengthening of Telomeres

Telomerase-independent telomere length maintenance was first described in yeast, where genetic analyses demonstrated its dependence on homologous recombination [65]. The existence of an analogous mechanism in mammalian cells was described soon after in telomerase-negative immortalized human cell lines [66].

This phenomenon has been referred to as alternative lengthening of telomeres, ALT, and its definition encompasses any telomerase-independent telomere maintenance mechanism. ALT cells present unique features such as very long and heterogeneous telomeres and the formation of particular promyelocytic leukemia (PML)-based nuclear structures called APBs (for ALT-associated PML bodies). APBs are formed by the association of PML bodies with telomeric chromatin, recombination factors, as well as proteins participating in DDR [67]. ALT cells also contain abundant extrachromosomal telomeric DNA, which has been found in many different forms: double-stranded telomeric circles (T-circles) [6], single-stranded circles (referred to as C-circles or G-circles, depending on their base composition) [68] and T-complexes, which consist of highly branched T-DNAs with large numbers of internal single-stranded portions [69]. All these telomeric-related extrachromosomal structures are thought to be homologous recombination-related byproducts. ALT telomeres are highly dynamic, undergoing rapid shortening and lengthening events [66] and bearing an elevated level of T-SCEs [70]. This high recombination at telomeres is not associated with a high level of recombination elsewhere in the genome [71]. The factors involved in telomere HR in ALT and the mechanistic behind it are still in debate (for a comprehensive review see [72]). As mentioned above, in telomerase-positive cells global chromatin relaxation induces an increase in telomeric HR, supporting the idea that ALT telomeres have a more open chromatin than telomerase-positive cells. Nevertheless, direct evidence supporting this view is missing. Finally, there seems to be no physiological equivalent of ALT, at least in humans. In the mouse egg, the first post-fertilization divisions occur in the absence of telomerase while telomeres appear to be elongated and T-SCEs become detectable [73]. This observation leaves open the possibility that a recombination-based telomere maintenance mechanism exists under particular physiological settings. On the other hand, ALT is detected in a limited proportion of tumors. It is most frequent in tumors originating from mesenchymal or neuroepithelial tissues (being detected in up to 50% in certain types of sarcomas), although it has been also detected in some types of carcinomas [74]. The reason for such cell-type specificity is not known, but it may be related to the fact that epithelial compartments (giving rise to carcinomas, the most frequent tumors in aging humans) gain more easily telomerase activity (through the re-expression of TERT), whereas the acquisition of ALT phenotypes may follow a more complex pathway.

Telomeres, Aging and Lifespan

Since telomere length predicts the proliferative capacity of a cell, it has been hypothesized that telomere lengths have an impact on life span. However, the analysis of telomere lengths in more than 60 mammalian species and whether or not these organisms show telomere-triggered mitotic senescence has suggested that the telomere/

telomerase system is not always employed to control cell replication. In fact, in many organisms telomere lengths are inversely correlated to life span [75]. At the same time, telomerase expression is correlated to body size [76]. In spite of the poor correlation between telomere length and lifespan in mice, there is evidence from wild type inbred animals that telomeres shorten with age, and perhaps this shortening contributes to aging manifestations [77]. Furthermore, introducing an extra copy of the telomerase gene in mice, in a context where there is increased resistance to cancer, increases life span [78], although it is still possible that this gain is due to extra-telomeric activities of TERT [79].

The discovery that mutations in components of the telomerase holoenzyme (TERT, TERC, DKC1, GAR1, NOP10) or in Shelterin components (such as TIN2) were responsible for the disease dyskeratosis congenita and related syndromes (see [80]) gave strong support to the concept that short telomeres may be responsible for aging manifestations since patients display manifestations of premature aging such as bone marrow failure, pulmonary diseases, skin and mucosa abnormalities, alopecia and higher predisposition to carcinomas.

In a healthy context, though, there is still much debate on whether or not telomere lengths are connected to life span in humans. Longitudinal studies are much needed to know if shortening kinetics throughout life is more important than absolute telomere lengths at birth.

Telomere Dynamics beyond Telomeres

Despite the wealth of basic knowledge that has been accumulated during the last decade concerning the biology of telomeres, the puzzle is still in pieces. There is no doubt that significant gaps remain within every telomere-related research domain (fig. 3), but perhaps more crucial is the fact that there is an enormous deficit in information allowing us to draw relevant connections between them. Clearly, more is needed to be done on telomere dynamics, including the analysis of telomere territories, their place in the nucleus, their relation to other chromosome territories and how telomere homeostasis and localization impacts the dynamics of the rest of the genome. In fact, it is becoming increasingly clear that beyond the direct consequences on chromosome stability and cell proliferation described here, components of the telomere maintenance machinery have direct impact on distant genomic regions and in signaling pathways that take place in the cytoplasm [34, 59, 81–83]. These aspects have just started to be explored at the molecular and cellular level, and a vast flow of information is expected to emerge in the upcoming years. Such information will definitely broaden our possibilities to meet the challenge to comprehensively connect every aspect of telomere biology to whole-cell homeostasis.

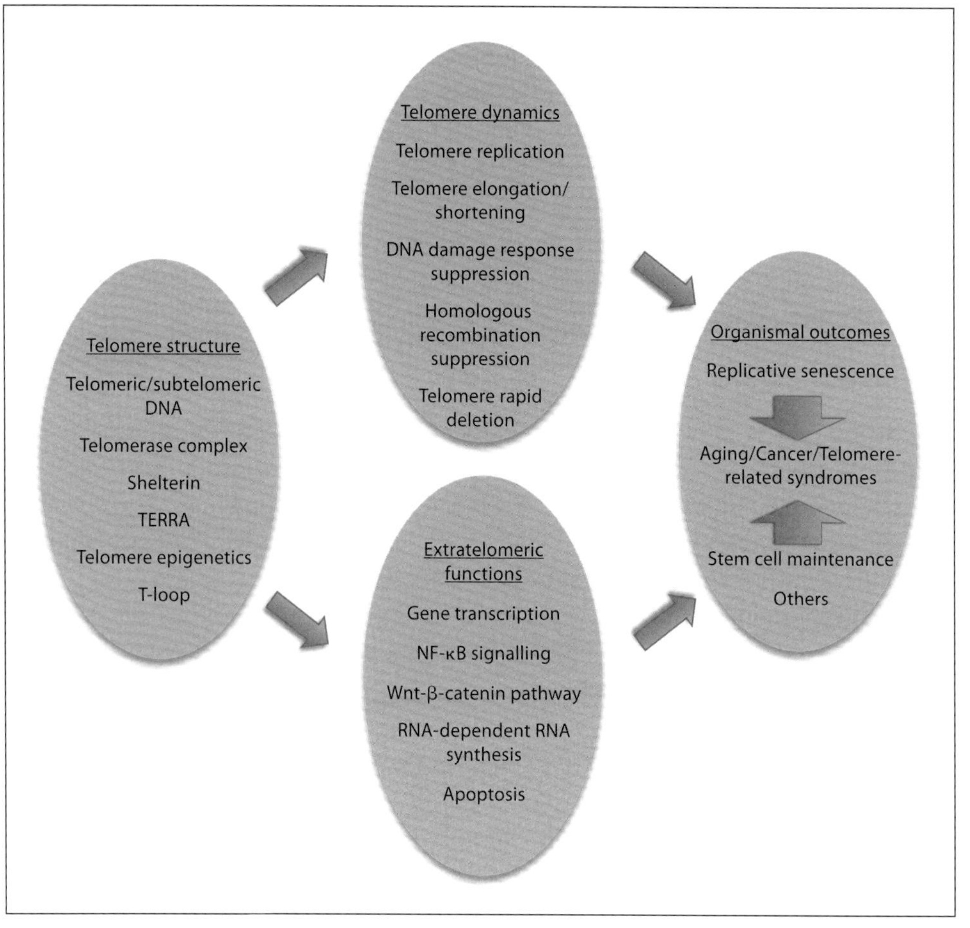

Fig. 3. Connecting telomere dynamics to organismal functions. Different aspects of telomere research are represented in different categories. Connecting arrows represent the targets of future telomere research.

Acknowledgements

The Telomere and Cancer laboratory has been 'Labellisé' by the Ligue contre le Cancer. Work in A.L.'s laboratory has been also supported by the Association pour la Recherche contre le Cancer, the Fondation pour la Recherche Médicale and the Institut Curie PIC program. D.C.S. is a recipient of a post-doctoral fellowship from the Fondation de France.

References

1 Muller H: The remaking of chromosomes. Collecting Net 1938;13:181–198.

2 Blackburn EH: Telomerases. Annu Rev Biochem 1992;61:113–129.

3 Capkova Frydrychova R, Biessmann H, Mason JM: Regulation of telomere length in *Drosophila*. Cytogenet Genome Res 2008;122:356–364.

4 Griffith JD, Comeau L, Rosenfield S, Stansel RM, Bianchi A, et al: Mammalian telomeres end in a large duplex loop. Cell 1999;97:503–514.

5 Oganesian L, Karlseder J: Mammalian 5′ C-rich telomeric overhangs are a mark of recombination-dependent telomere maintenance. Mol Cell 2011; 42:224–236.

6 Cesare AJ, Griffith JD: Telomeric DNA in ALT cells is characterized by free telomeric circles and heterogeneous t-loops. Mol Cell Biol 2004;24:9948–9957.

7 Pickett HA, Henson JD, Au AY, Neumann AA, Reddel RR: Normal mammalian cells negatively regulate telomere length by telomere trimming. Hum Mol Genet 2011;20:4684–4692.

8 Kruk PA, Balajee AS, Rao KS, Bohr VA: Telomere reduction and telomerase inactivation during neuronal cell differentiation. Biochem Biophys Res Commun 1996;224:487–492.

9 Britt-Compton B, Rowson J, Locke M, Mackenzie I, Kipling D, Baird DM: Structural stability and chromosome-specific telomere length is governed by *cis*-acting determinants in humans. Hum Mol Genet 2006;15:725–733.

10 Londono-Vallejo JA, DerSarkissian H, Cazes L, Thomas G: Differences in telomere length between homologous chromosomes in humans. Nucleic Acids Res 2001;29:3164–3171.

11 Chen Y, Yang Y, van Overbeek M, Donigian JR, Baciu P, et al: A shared docking motif in TRF1 and TRF2 used for differential recruitment of telomeric proteins. Science 2008;319:1092–1096.

12 Celli GB, de Lange T: DNA processing is not required for ATM-mediated telomere damage response after TRF2 deletion. Nat Cell Biol 2005; 7:712–718.

13 van Steensel B, de Lange T: Control of telomere length by the human telomeric protein TRF1. Nature 1997;385:740–743.

14 Smogorzewska A, van Steensel B, Bianchi A, Oelmann S, Schaefer MR, et al: Control of human telomere length by TRF1 and TRF2. Mol Cell Biol 2000;20:1659–1668.

15 van Steensel B, Smogorzewska A, de Lange T: TRF2 protects human telomeres from end-to-end fusions. Cell 1998;92:401–413.

16 Sfeir A, Kosiyatrakul ST, Hockemeyer D, MacRae SL, Karlseder J, et al: Mammalian telomeres resemble fragile sites and require TRF1 for efficient replication. Cell 2009;138:90–103.

17 Ye J, Lenain C, Bauwens S, Rizzo A, Saint-Leger A, et al: TRF2 and Apollo cooperate with topoisomerase 2alpha to protect human telomeres from replicative damage. Cell 2010;142:230–242.

18 Lam YC, Akhter S, Gu P, Ye J, Poulet A, et al: SNMIB/Apollo protects leading-strand telomeres against NHEJ-mediated repair. EMBO J 2010;29: 2230–2241.

19 Loayza D, De Lange T: POT1 as a terminal transducer of TRF1 telomere length control. Nature 2003; 423:1013–1018.

20 Palm W, Hockemeyer D, Kibe T, de Lange T: Functional dissection of human and mouse POT1 proteins. Mol Cell Biol 2009;29:471–482.

21 Hockemeyer D, Daniels JP, Takai H, de Lange T: Recent expansion of the telomeric complex in rodents: Two distinct POT1 proteins protect mouse telomeres. Cell 2006;126:63–77.

22 Wu L, Multani AS, He H, Cosme-Blanco W, Deng Y, et al: *Pot1* deficiency initiates DNA damage checkpoint activation and aberrant homologous recombination at telomeres. Cell 2006;126:49–62.

23 Kibe T, Osawa GA, Keegan CE, de Lange T: Telomere protection by TPP1 is mediated by POT1a and POT1b. Mol Cell Biol 2010;30:1059–1066.

24 He H, Multani AS, Cosme-Blanco W, Tahara H, Ma J, et al: POT1b protects telomeres from end-to-end chromosomal fusions and aberrant homologous recombination. EMBO J 2006;25:5180–5190.

25 Wang Y, Shen MF, Chang S: Essential roles for Pot1b in hematopoietic stem cell self-renewal and survival. Blood 2011;118:6068–6077.

26 Arnoult N, Saintome C, Ourliac-Garnier I, Riou JF, Londono-Vallejo A: Human POT1 is required for efficient telomere C-rich strand replication in the absence of WRN. Genes Dev 2009;23:2915–2924.

27 Vlangos CN, O'Connor BC, Morley MJ, Krause AS, Osawa GA, Keegan CE: Caudal regression in adrenocortical dysplasia (*acd*) mice is caused by telomere dysfunction with subsequent p53-dependent apoptosis. Dev Biol 2009;334:418–428.

28 Hockemeyer D, Palm W, Else T, Daniels JP, Takai KK, et al: Telomere protection by mammalian Pot1 requires interaction with Tpp1. Nat Struct Mol Biol 2007;14:754–761.

29 Wang F, Podell ER, Zaug AJ, Yang Y, Baciu P, et al: The POT1-TPP1 telomere complex is a telomerase processivity factor. Nature 2007;445:506–510.

30 Tejera AM, Stagno d'Alcontres M, Thanasoula M, Marion RM, Martinez P, et al: TPP1 is required for TERT recruitment, telomere elongation during nuclear reprogramming, and normal skin development in mice. Dev Cell 2010;18:775–789.

31 Kabir S, Sfeir A, de Lange T: Taking apart Rap1: an adaptor protein with telomeric and non-telomeric functions. Cell Cycle 2010;9:4061–4067.

32 Sfeir A, Kabir S, van Overbeek M, Celli GB, de Lange T: Loss of Rap1 induces telomere recombination in the absence of NHEJ or a DNA damage signal. Science 2010;327:1657–1661.

33 O'Connor MS, Safari A, Liu D, Qin J, Songyang Z: The human Rap1 protein complex and modulation of telomere length. J Biol Chem 2004;279:28585–28591.
34 Yang D, Xiong Y, Kim H, He Q, Li Y, et al: Human telomeric proteins occupy selective interstitial sites. Cell Res 2011;21:1013–1027.
35 Kim SH, Beausejour C, Davalos AR, Kaminker P, Heo SJ, Campisi J: TIN2 mediates functions of TRF2 at human telomeres. J Biol Chem 2004;279:43799–43804.
36 Smith S, Giriat I, Schmitt A, de Lange T: Tankyrase, a poly(ADP-ribose) polymerase at human telomeres. Science 1998;282:1484–1487.
37 Canudas S, Houghtaling BR, Bhanot M, Sasa G, Savage SA, et al: A role for heterochromatin protein 1gamma at human telomeres. Genes Dev 2011;25:1807–1819.
38 Chiang YJ, Kim SH, Tessarollo L, Campisi J, Hodes RJ: Telomere-associated protein TIN2 is essential for early embryonic development through a telomerase-independent pathway. Mol Cell Biol 2004;24:6631–6634.
39 Galati A, Rossetti L, Pisano S, Chapman L, Rhodes D, et al: The human telomeric protein TRF1 specifically recognizes nucleosomal binding sites and alters nucleosome structure. J Mol Biol 2006;360: 377–385.
40 Benetti R, Gonzalo S, Jaco I, Munoz P, Gonzalez S, et al: A mammalian microRNA cluster controls DNA methylation and telomere recombination via Rbl2-dependent regulation of DNA methyltransferases. Nat Struct Mol Biol 2008;15:268–279.
41 Wu P, de Lange T: No overt nucleosome eviction at deprotected telomeres. Mol Cell Biol 2008;28:5724–5735.
42 Benetti R, Garcia-Cao M, Blasco MA: Telomere length regulates the epigenetic status of mammalian telomeres and subtelomeres. Nat Genet 2007;39:243–250.
43 Schoeftner S, Blasco MA: Chromatin regulation and non-coding RNAs at mammalian telomeres. Semin Cell Dev Biol 2010;21:186–193.
44 Gonzalo S, Blasco MA: Role of Rb family in the epigenetic definition of chromatin. Cell Cycle 2005;4: 752–755.
45 Woodcock CL, Skoultchi AI, Fan Y: Role of linker histone in chromatin structure and function: H1 stoichiometry and nucleosome repeat length. Chromosome Res 2006;14:17–25.
46 Fan Y, Nikitina T, Zhao J, Fleury TJ, Bhattacharyya R, et al: Histone H1 depletion in mammals alters global chromatin structure but causes specific changes in gene regulation. Cell 2005;123:1199–1212.
47 Murga M, Jaco I, Fan Y, Soria R, Martinez-Pastor B, et al: Global chromatin compaction limits the strength of the DNA damage response. J Cell Biol 2007;178:1101–1108.
48 Tham WH, Zakian VA: Transcriptional silencing at *Saccharomyces* telomeres: implications for other organisms. Oncogene 2002;21:512–521.
49 Azzalin CM, Reichenbach P, Khoriauli L, Giulotto E, Lingner J: Telomeric repeat containing RNA and RNA surveillance factors at mammalian chromosome ends. Science 2007;318:798–801.
50 Nergadze SG, Farnung BO, Wischnewski H, Khoriauli L, Vitelli V, et al: CpG-Island promoters drive transcription of human telomeres. RNA 2009;15:2186–2194.
51 Schoeftner S, Blasco MA: Developmentally regulated transcription of mammalian telomeres by DNA-dependent RNA polymerase II. Nat Cell Biol 2008;10:228–236.
52 Redon S, Reichenbach P, Lingner J: The non-coding RNA TERRA is a natural ligand and direct inhibitor of human telomerase. Nucleic Acids Res 2010;38: 5797–5806.
53 Wang Y, Ghosh G, Hendrickson EA: Ku86 represses lethal telomere deletion events in human somatic cells. Proc Natl Acad Sci USA 2009;106:12430–12435.
54 Zou Y, Sfeir A, Gryaznov SM, Shay JW, Wright WE: Does a sentinel or a subset of short telomeres determine replicative senescence? Mol Biol Cell 2004; 15:3709–3718.
55 Soler D, Pampalona J, Tusell L, Genesca A: Radiation sensitivity increases with proliferation-associated telomere dysfunction in nontransformed human epithelial cells. Aging Cell 2009;8:414–425.
56 Der-Sarkissian H, Bacchetti S, Cazes L, Londono-Vallejo JA: The shortest telomeres drive karyotype evolution in transformed cells. Oncogene 2004;23: 1221–1228.
57 Artandi SE, DePinho RA: A critical role for telomeres in suppressing and facilitating carcinogenesis. Curr Opin Genet Dev 2000;10:39–46.
58 Willeit P, Willeit J, Mayr A, Weger S, Oberhollenzer F, et al: Telomere length and risk of incident cancer and cancer mortality. JAMA 2010;304:69–75.
59 Martinez P, Blasco MA: Telomeric and extra-telomeric roles for telomerase and the telomere-binding proteins. Nat Rev Cancer 2011;11:161–176.
60 Karlseder J: Telomere repeat binding factors: keeping the ends in check. Cancer Lett 2003;194:189–197.
61 Martinez P, Thanasoula M, Munoz P, Liao C, Tejera A, et al: Increased telomere fragility and fusions resulting from TRF1 deficiency lead to degenerative pathologies and increased cancer in mice. Genes Dev 2009;23:2060–2075.

62 Munoz P, Blanco R, de Carcer G, Schoeftner S, Benetti R, et al: TRF1 controls telomere length and mitotic fidelity in epithelial homeostasis. Mol Cell Biol 2009;29:1608–1625.
63 Munoz P, Blanco R, Flores JM, Blasco MA: XPF nuclease-dependent telomere loss and increased DNA damage in mice overexpressing TRF2 result in premature aging and cancer. Nat Genet 2005;37:1063–1071.
64 Shin JS, Hong A, Solomon MJ, Lee CS: The role of telomeres and telomerase in the pathology of human cancer and aging. Pathology 2006;38:103–113.
65 Lundblad V, Blackburn EH: An alternative pathway for yeast telomere maintenance rescues *est1*⁻ senescence. Cell 1993;73:347–360.
66 Murnane JP, Sabatier L, Marder BA, Morgan WF: Telomere dynamics in an immortal human cell line. EMBO J 1994;13:4953–4962.
67 Yeager TR, Neumann AA, Englezou A, Huschtscha LI, Noble JR, Reddel RR: Telomerase-negative immortalized human cells contain a novel type of promyelocytic leukemia (PML) body. Cancer Res 1999;59:4175–4179.
68 Henson JD, Cao Y, Huschtscha LI, Chang AC, Au AY, et al: DNA C-circles are specific and quantifiable markers of alternative-lengthening-of-telomeres activity. Nat Biotechnol 2009;27:1181–1185.
69 Nabetani A, Ishikawa F: Unusual telomeric DNAs in human telomerase-negative immortalized cells. Mol Cell Biol 2009;29:703–713.
70 Londoño-Vallejo JA, Der-Sarkissian H, Cazes L, Bacchetti S, Reddel R: Alternative lengthening of telomeres is characterized by high rates of inter-telomeric exchange. Cancer Res 2004;64:2324–2327.
71 Bechter OE, Shay JW, Wright WE: The frequency of homologous recombination in human ALT cells. Cell Cycle 2004;3:547–549.
72 Cesare AJ, Reddel RR: Alternative lengthening of telomeres: models, mechanisms and implications. Nat Rev Genet 2010;11:319–330.
73 Liu L, Bailey SM, Okuka M, Munoz P, Li C, et al: Telomere lengthening early in development. Nat Cell Biol 2007;9:1436–1441.
74 Heaphy CM, Subhawong AP, Hong SM, Goggins MG, Montgomery EA, et al: Prevalence of the alternative lengthening of telomeres telomere maintenance mechanism in human cancer subtypes. Am J Pathol 2011;179:1608–1615.
75 Gomes NM, Ryder OA, Houck ML, Charter SJ, Walker W, et al: Comparative biology of mammalian telomeres: hypotheses on ancestral states and the roles of telomeres in longevity determination. Aging Cell 2011;10:761–768.
76 Gorbunova V, Seluanov A: Coevolution of telomerase activity and body mass in mammals: from mice to beavers. Mech Ageing Dev 2009;130:3–9.
77 Flores I, Canela A, Vera E, Tejera A, Cotsarelis G, Blasco MA: The longest telomeres: a general signature of adult stem cell compartments. Genes Dev 2008;22:654–667.
78 Tomas-Loba A, Flores I, Fernandez-Marcos PJ, Cayuela ML, Maraver A, et al: Telomerase reverse transcriptase delays aging in cancer-resistant mice. Cell 2008;135:609–622.
79 Chung HK, Cheong C, Song J, Lee HW: Extratelomeric functions of telomerase. Curr Mol Med 2005;5:233–241.
80 Armanios M: Syndromes of telomere shortening. Annu Rev Genomics Hum Genet 2009;10:45–61.
81 Simonet T, Zaragosi LE, Philippe C, Lebrigand K, Schouteden C, et al: The human TTAGGG repeat factors 1 and 2 bind to a subset of interstitial telomeric sequences and satellite repeats. Cell Res 2011;21:1028–1038.
82 Park JI, Venteicher AS, Hong JY, Choi J, Jun S, et al: Telomerase modulates Wnt signalling by association with target gene chromatin. Nature 2009;460:66–72.
83 Kabir S, Sfeir A, de Lange T: Taking apart Rap1: an adaptor protein with telomeric and non-telomeric functions. Cell Cycle 2011;9:4061–4067.

Arturo Londoño-Vallejo
Telomeres and Cancer Laboratory
Institut Curie – CNRS UMR3244 – UPMC
26 rue d'Ulm, FR–75005 Paris (France)
Tel. +33 156 246 611, E-Mail Arturo.Londono@curie.fr

Garrido-Ramos MA (ed): Repetitive DNA.
Genome Dyn. Basel, Karger, 2012, vol 7, pp 46–67

Drosophila Telomeres: an Example of Co-Evolution with Transposable Elements

R. Silva-Sousa · E. López-Panadès · E. Casacuberta

Institute of Evolutionary Biology, IBE (CSIC-UPF), Barcelona, Spain

Abstract

Telomeres have a DNA component composed of repetitive sequences. In most eukaryotes these repeats are very similar in length and sequence and are maintained by a highly conserved specialized cellular enzyme, telomerase. Some exceptions of the telomerase mechanism exist in eukaryotes of which the most studied are concentrated in insects, and from these, *Drosophila* species stand out in particular. The alternative mechanism of telomere maintenance in *Drosophila* is based on targeted transposition of 3 very special non-LTR retrotransposons, *HeT-A, TART* and *TAHRE*. The fingerprint of the co-evolution between the *Drosophila* genome and the telomeric retrotransposons is visible in special features of both. In this chapter, we will review the main aspects of *Drosophila* telomeres and the telomere retrotransposons that explain how this alternative mechanism works, is regulated, and evolves. By going through the different aspects of this symbiotic relationship, we will try to unravel which have been the necessary changes at *Drosophila* telomeres in order to exert their telomeric function analogously to telomerase telomeres, and also which particularities have been maintained in order to preserve the retrotransposon personality of *HeT-A, TART* and *TAHRE*. *Drosophila* telomeres constitute a remarkable variant that reminds us how exceptions should be treasured in order to widen our knowledge in any particular biological mechanism.

Copyright © 2012 S. Karger AG, Basel

Telomeres are specialized structures composed of DNA and proteins that protect the end of the chromosome and whose function is essential for several cellular processes like aging, senescence or tumorigenesis. Telomeres become shorter with each cell division because of the end replication problem, which refers to the inability of all DNA polymerases to move in the 3′ to 5′ direction. Thereby, every living organism with linear chromosomes requires a specialized mechanism that replenishes the telomere when it becomes critically short [1]. Early work by H.J. Muller [2] showed that the end of the chromosome cannot be simply a blunt end and must involve a specialized structure. Later, Barbara McClintock [3] demonstrated that, if left unprotected, the ends of chromosomes could fuse to each other and enter in a breakage-fusion cycle with deleterious consequences for the cell. In most eukaryotes, telomere replication is achieved by a specialized polymerase, telomerase, that carries its own RNA template. The RNA template of telomerase is highly conserved in most eukaryotes, resulting in telomeres composed of similar short repeats (6–10 bp) that are in general G/T-rich [1]. Unexpectedly, *Drosophila*, the organism where H.J. Muller first described telomeres, lacks the telomerase holoenzyme. *Drosophila* telomeres are also composed of repeated sequences, although in this species the repeats are at least 3 orders of magnitude longer than telomerase repeats. The telomeric repeats of *Drosophila* are also reverse transcribed at the end of the chromosome, but in this case the enzyme that likely performs this reaction is encoded by some of the telomere repeats. The telomere repeats in *Drosophila* have their own personality since they correspond to multiple copies of 3 different non-LTR retrotransposons, *HeT-A, TART* and *TAHRE* [4]. Retrotransposons belong to Class I transposable elements (TEs), and their mechanism of transposition involves an RNA intermediate implying that each new successful transposition will result in an increase in the copy number of the element. Although TEs have been historically referred to as 'junk' DNA, pioneering work by Barbara McClintock in the 1950s pointed out that genomes containing TEs had a good reservoir of genetic material that could be used in stressful situations [5]. Because recombination based methods have been found to maintain the telomeres in some organisms, insects among them [6], it is possible that *Drosophila* telomeres depended on recombination for certain time. With time, retrotransposon telomeres in *Drosophila* could have arisen as an alternative solution, constituting a nice example that corroborates McClintock's hypothesis.

While retrotransposons might seem very different from telomerase repeats, in the next sections we will show how these 2 types of telomeres are equivalent when it comes to functionality, have similar chromatin characteristics and share some protein complexes for protection. Moreover, in both cases, the telomeres are elongated by reverse transcription of an RNA template by enzymes that may have evolved from the same ancestor. We will now review some of the main aspects of these telomeric features for *Drosophila* and compare them, when possible, with telomerase telomeres in general.

The Telomeric Retrotransposons: *HeT-A, TART* and *TAHRE*

The study of *Drosophila* telomeres shows how genome and TEs adapt to each other for the benefit of both. On one hand, the genome of *Drosophila* recognizes the telomeric retrotransposons as a mechanism that performs an essential function and, although tightly regulated, allows their transcription and the entrance of some of their proteins into the nucleus [4]. On the other hand, the telomeric retrotransposons, while maintaining the main hallmarks of non-LTR retrotransposons, have adapted to their telomeric role by developing certain unusual features that are conserved across *Drosophila* species [7].

Shared Features of HeT-A, TART and TAHRE with Their non-LTR Counterparts

HeT-A, TART and *TAHRE* are composed of unusually long 5′ and 3′ untranslated regions (UTRs) flanking the coding regions, responsible for the proteins that the elements need in order to transpose, although *HeT-A* lacks the second protein encoded by the other 2 elements (see below). All 3 elements end with an oligo(A) sequence at one junction of the insertion site, as expected by the reverse transcription of a poly(A)+ RNA [4]. Although not directly demonstrated for the telomeric transposons, it is assumed that *HeT-A, TART* and *TAHRE* transpose by the mechanism known as target-primed reverse-transcription [8]. In target-primed reverse-transcription, the transposition RNA intermediate is directly reverse transcribed onto an internal nick in the chromosome performed by the retrotransposon endonuclease. The reverse transcriptase of the element uses the 3′ OH of the nicked DNA and the poly(A) to prime synthesis of cDNA beginning in the 3′ poly(A) of the RNA. The second strand of DNA could be synthesized by the same reverse transcriptase or by regular DNA synthesis. This mechanism of transposition is relevant for the telomere function of the telomeric retrotransposons because it is actually analogous to the mechanism that telomerase uses to elongate the end of the chromosome copying the telomerase RNA template. By successive transpositions to the end of the chromosome, *HeT-A, TART* and *TAHRE* form long arrays, always oriented in a head-to-tail direction, with the poly(A) oriented towards the centromere [4, 9]. Phylogenetically, by comparing the DNA and amino acid sequences of the encoded proteins, the 3 telomeric retrotransposons belong to the *Jockey* clade of non-LTR elements in *Drosophila* [10]. Further comparisons across *Drosophila* species show that this classification is still valid in species at least as far as 120 million years (Myr) of genetic distance [7].

Special Features Shared by the Three Telomeric Retrotransposons

The first of the unusual features of the telomeric retrotransposons is their genomic distribution. Retrotransposons in general integrate in different places in the genome according to the specificity of their endonuclease or integrase, which recognizes a specific DNA sequence or a particular chromatin structure [11]. The case of the telomeric retrotransposons is unique because they only successfully transpose to a specific genomic compartment, the telomere domain at the telomeres (see section 'Telomere Domains in *Drosophila*'). Although some fragments of *HeT-A* and *TART* have been found in heterochromatic regions at the centromere and pericentromere of the Y chromosome [12], it seems most likely that these fragments reached these non-telomeric regions by recombination, chromosomal reorganization events or some other method other than their usual transposition mechanism. It is intriguing that the telomere retrotransposons are never found in euchromatic regions, although the promoter of the *HeT-A* element is able to function in euchromatin [13].

Secondly, all 3 telomere retrotransposons have exceptionally very long 3′ UTRs, which can account for more than half of the length of the element. The fact that most orthologues of the telomere retrotransposons conserve this unusual feature (see fig. 1) demonstrates evolutionary pressure and suggests functionality [7, 14]. The possible function of the long 3′ UTRs may be related to the establishment of telomere chromatin or specific interactions with telomere or chromosomal proteins. Interestingly, the DNA sequence of the telomere retrotransposons has a strong sequence bias, as the strand that runs 5′ to 3′ towards the centromere is extremely G-poor, resembling the same strand bias shown by telomerase repeats [4]. Maybe because this composition bias is important, we should mention that comparisons at DNA and amino acid level among the orthologues of the telomere retrotransposons showed a higher conservation at the DNA than at the amino acid level for most of the length of the telomeric retrotransposon, suggesting a strong evolutionary pressure to maintain certain characteristics at the nucleotide level [7].

Thirdly, expression from the antisense strand has been demonstrated for both *HeT-A* and *TART* retrotransposons [4, 15]. No report on *TAHRE* antisense transcription is known, but because *TAHRE* and *HeT-A* actually share the 3′ UTR region, where the antisense promoter and the start site are located, it is possible that *TAHRE* is also transcribed from the antisense strand [16, 17]. The antisense transcripts of both *HeT-A* and *TART* are processed, and the splicing sites are strongly conserved between members of the different subfamilies of the elements [18, 66]. The functionality of these antisense transcripts is unknown to date, but the conservation of the splicing process points towards its discovery in the near future.

Particularities of Each of the Three Telomeric Retrotransposons

The special characteristics of the 3 telomeric retroelements, as well as of their main orthologues in other species, are graphically shown in figure 1. We will briefly explain the particular traits that distinguish each of the elements from their telomeric partners.

HeT-A

HeT-A, the main component of *Drosophila* telomeres, is actually a non-autonomous transposon. *HeT-A* only encodes the gene responsible for structural properties, the *gag* gene, but no gene with polymerase (*pol*) activity is found in its genome. Evolutionary studies have demonstrated that *HeT-A* elements lack a *pol* gene probably since before the separation of *Drosophila* species. Nevertheless, the lack of a *pol* gene has not been a burden for *HeT-A* which is the most effective of the 3 retrotransposons transposing to the telomeres and outnumbers its telomeric partners across distant *Drosophila* species [7, 19]. A second distinguishable feature of the *HeT-A* retrotransposon is the location of its promoter. Usually, the promoter in non-LTR elements is located at the 5′ UTR, but in the case of *HeT-A* the promoter is found at the 3′ UTR and drives transcription of the element immediately downstream in

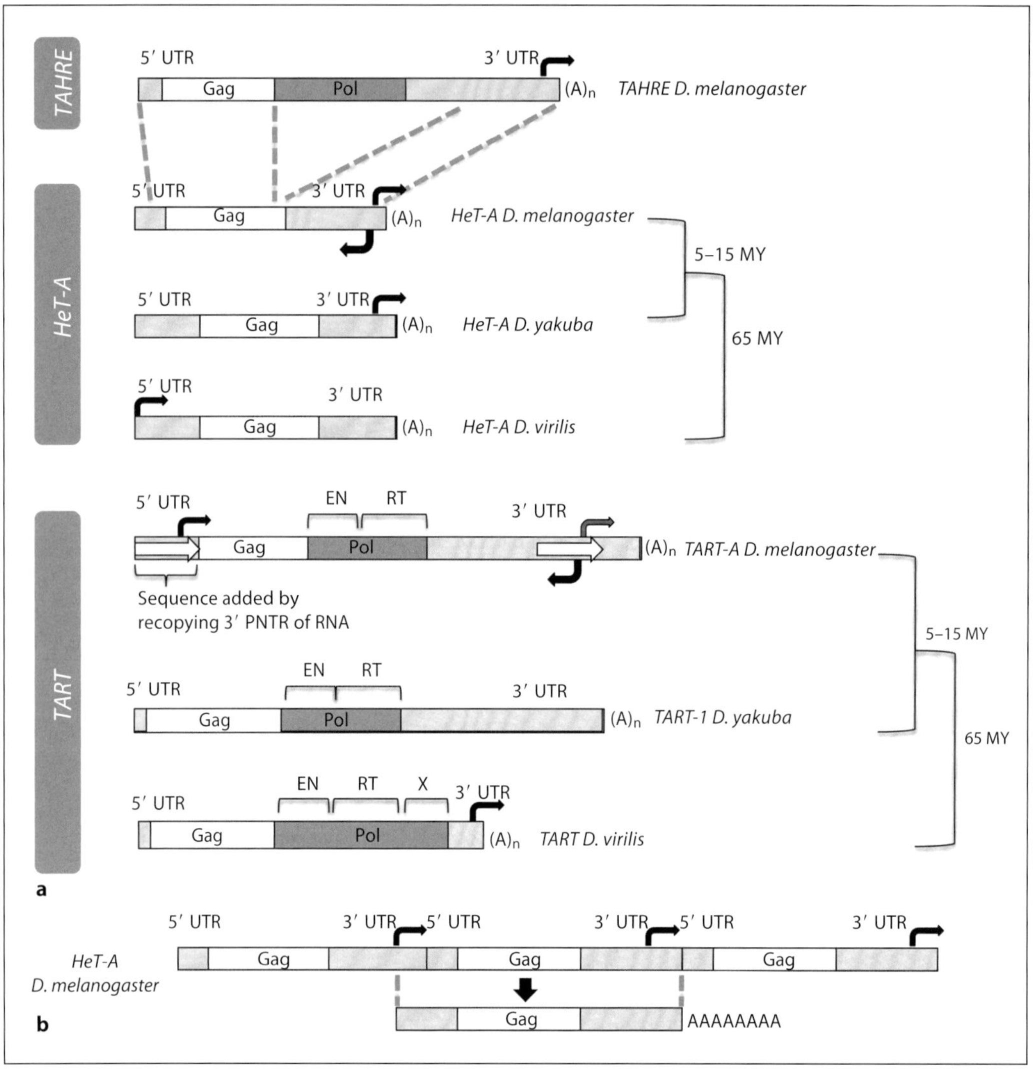

Fig. 1. The telomeric retrotransposons. **a** Telomeric retrotransposons *TAHRE, HeT-A* and *TART* from *D. melanogaster*, *D. yakuba* and *D. virilis* (drawn approximately to scale). Solid bars on the right indicate the phylogenetic relationships. MY, million years. Dotted grey lines show conserved regions of *TAHRE* and *HeT-A* DNA sequences. Bright grey boxes, non-coding 5′ and 3′ UTR sequences; white boxes, Gag ORF; dark grey boxes, Pol ORF; EN, endonuclease domain; RT, reverse transcriptase domain; X, extra domain of Pol coding region. White arrows, PNTRs; $(A)_n$, 3′ oligo(A); black arrows, transcription start sites for full-length sense and antisense strand RNA; grey arrow, start site for short sense strand RNA. **b** Representation of a telomeric fragment of assembled *D. melanogaster HeT-A,* showing the analogy of an element plus its upstream sense strand promoter to an LTR retrotransposon when containing the sense promoter. AAAAA indicates 3′ poly(A) on RNA.

the array [4]. Therefore, each copy of *HeT-A* depends on a successive transposition upstream, probably selecting for multiple transpositions at one time. Interestingly, if a complete *HeT-A* copy with its own promoter (the 3′ end of the upstream element) is extracted from the telomeric array, the resulting sequence matches the structure of an LTR retrotransposon (see fig. 1b). This special feature of *HeT-A* opens questions about its origin and suggests an intermediate step in evolution between non-LTR and LTR retrotransposons.

TART

Besides the unusual features described above, *TART* is the telomeric retrotransposon that more closely resembles a canonical non-LTR retrotransposon. Nevertheless, *TART* elements from *D. melanogaster* also show a feature that is reminiscent of evolutionary intermediates between LTR and non-LTR retrotransposons. The 3 different *TART* subfamilies in *D. melanogaster*, *TART-A*, *TART-B* and *TART-C*, each have perfect non-terminal repeats (PNTRs). The PNTRs are located inside the 5′ and 3′ UTRs and are 100% identical within each individual element, and around 70% among subfamilies (fig. 1) [4]. The perfect conservation of the sequence of both PNTRs in a particular copy suggests that the PNTRs are evolving together much as the 2 LTRs on each LTR retrotransposon do. It is proposed that this concerted evolution results from extending the 5′ end of the element by a second copy of the 3′ UTR when the element is reverse transcribed onto the chromosome during transposition.

PNTRs are not exclusive to *TART* elements, they have also been found in unrelated non-LTR retrotransposons *TRE5-A* in *Dictyostelium* and *TOC1* in *Chlamydomonas* [11]. Interestingly, these 2 retrotransposons also produce substantial antisense transcripts as is the case for *TART*. Further studies in the transposition mechanism of these 3 transposons are necessary to elucidate the importance and functionality of the PNTRs.

TAHRE

The third telomeric element recently discovered, *TAHRE* (telomere associated element HeT-A related), received its name because it shares features with the main component of *Drosophila* telomeres, *HeT-A*. *TAHRE* shares the 5′ UTR, the *gag* coding region and the end of the 3′ UTR with *HeT-A* [16] (fig. 1). In addition, *TAHRE* encodes a Pol protein like *TART* but, although phylogenetically related and therefore with a common ancestor, the 2 *pol* genes are not identical. It is puzzling why a telomeric retrotransposon that seems to combine the best of the other telomeric partners has not been more successful. Only a few copies of *TAHRE* are available and only 1 corresponds to a potentially active element. Although *TAHRE* was first found in *D. melanogaster*, *TAHRE* orthologues have been cytologically detected in other species of the *melanogaster* species group [17], and the draft of the 12 sequenced *Drosophila* species revealed the presence of putative *TAHRE* orthologues in distant *Drosophila* [14]. These studies propose that *HeT-A* might have derived from *TAHRE* in different lineages. The specific characteristics of *TAHRE* in *D. melanogaster* as, for example, the inability of its Gag protein to localize to telomeres without the help of *HeT-A* Gag, may offer a clue of why this third telomere retrotransposon has not been more successful in transposing onto the ends of the *Drosophila* chromosomes [20].

Why Three? What Is the Nature of Their Relationship?

An intriguing question about *Drosophila* telomeres is why there are 3 retroelements devoted to this function and/or why any of the 3 has not been able to outcompete the other two. Maybe, the secret resides in the particularities of each member, resulting in a collaborative threesome to ensure their successful transposition at telomeres. *HeT-A* would be the one that most benefits from this arrangement, since it is by far the most abundant of the 3 elements at the telomeres. This is particularly interesting because, as mentioned above, *HeT-A* is by itself a non-autonomous element and must rely on a source of this activity in *trans*. On the other hand, the Gag protein of *HeT-A* is the only of the 3 telomeric Gags with the ability to localize at the telomeres [4]. The telomere targeting of the *HeT-A* Gag protein has been conserved in different species and also across species [21]. *TART* and *TAHRE* depend on *HeT-A* for telomere targeting and *HeT-A* likely relies on the 2 autonomous telomeric elements for polymerase activities [4, 20]. *TAHRE* seems the perfect partner for *HeT-A* because it shares part of its genome, the pattern of transcription in germline cells and is controlled similarly by the *rasi* pathway [17]. Nevertheless, *TAHRE* is present in only a few copies in most analyzed stocks, while *TART*, although not as abundant as *HeT-A*, is present in several functional copies in all *Drosophila* stocks that have been analyzed [19]. In summary, a collaborative scenario would explain a relationship where *HeT-A* would choose *TART* or *TAHRE* as a source of polymerase activities and in exchange *HeT-A* would provide telomere targeting to *TART* and/or *TAHRE*. Depending on the cell type or developmental stage where the elements are transposing, *HeT-A* would choose *TART* or *TAHRE* as partner [6]. Because *HeT-A* is by far more abundant than its partners, it must be more successful in transposing to telomeres or, alternatively, telomeres with more *HeT-A* elements might provide a better telomere function and may be positively selected.

Telomere Elongation in *Drosophila*

Mechanisms

The telomeres in *Drosophila* are mainly maintained by specific transposition onto the ends, but recombination by terminal gene conversion (non-reciprocal recombination) can act as a backup mechanism as it does in telomerase organisms [1]. Recombination is often used as alternative lengthening of telomeres (ALT) when immortal human cancer cells fail to reactivate telomerase. Terminal gene conversion maintains telomere length by replicating the end sequence when a template from the same, or homologous, chromosome is available. In *Drosophila*, this mechanism was observed when a still uncharacterized mutation, *E*(*tc*) [22], showed a telomere length double that of a wild type strain without affecting the expression level of the telomere retrotransposons or its transposition rate.

Regulators

Regulation of telomere length in *Drosophila* means regulation of the expression of the telomeric retrotransposons *HeT-A*, *TART* and *TAHRE*. Interestingly, no positive regulators of telomere length have yet been found, although a mechanism to promote expression and transposition should be in place in order to ensure telomere maintenance [Sousa R., López-Panadès E., Piñeyro D. and Casacuberta E., unpublished]. In the following, we briefly describe some regulatory factors that have been found to affect telomere length (we do not include mutations *Tel* and *E*(*tc*) that, although affecting telomere length in *Drosophila*, are still uncharacterized) [22, 23].

Ku70/80

As in other organisms, the heterodimer Ku70/80 binds to telomeres in a sequence-independent manner and is involved in telomere protection in *Drosophila* where it acts as a negative regulator of telomere length [1, 24]. Mutations in either gene *ku70* or *ku80* increase the rate of telomere transposition without changing the expression level of the telomeric retrotransposons. Therefore, the mechanism by which the Ku heterodimer regulates telomere transposition is likely by controlling the accessibility to the end of the chromosome for the telomere transposons. Depending on the organism, mutations in *ku* have opposite effects on telomere length, likely reflecting different structures at the end of such chromosomes [1].

HP1

Heterochromatin protein 1 (HP1) has been recently renamed to HP1a, HP1b and HP1c due to the existence of 3 paralogs in *Drosophila* [25]. All 3 proteins have a chromo domain and a chromo shadow domain linked by a hinge domain [25]. In *Drosophila*, HP1a is present at telomeres, chromocenter and in many interbands in the polytene chromosomes [26]. HP1a has a dual role at *Drosophila* telomeres [27]. On one hand, HP1a is one of the basic components of the capping complex that protects the ends of the chromosomes, and therefore, its presence might be related with end accessibility. On the other hand, HP1a is an important silencer of the telomeric retrotransposons. Through the chromo domain HP1a binds to the modified histone H3, H3K9me3, along *Drosophila* telomeres. The presence of this modification at the *HeT-A* promoter is directly linked to a low level of *HeT-A* expression [28]. The presence of HP1 and H3K9me3 at the *HeT-A* promoter changes with the presence of *HeT-A* piRNAs as well as with mutations in the DNA methylase (*dnmt2*) gene, suggesting an important role of HP1 in both transcriptional and posttranscriptional regulation of the telomeric retrotransposons [29].

PROD

PROD is a protein that has been localized at the promoter of the *HeT-A* element and is necessary to negatively regulate the expression of the telomeric retrotransposons. Mutations in *prod* exhibit an increase in *HeT-A* transcription but not in telomere length, suggesting that PROD does not control end accessibility [30].

PIWI and *rasi* Pathways

Due to their potential deleterious effects, eukaryotes have evolved a combination of transcriptional and posttranscriptional methods to silence TEs. The posttranscriptional silencing relies mainly on the RNA interference (RNAi) machinery, where the dicer enzyme cleaves a double-stranded RNA into small RNAs (21–26 nt). These small RNAs will guide the Argonaute proteins and degrade protein complexes through complementarity to an mRNA from the TE, avoiding the production of the TE proteins and the synthesis of a transposition intermediate [31]. Often, the small RNAs from the TE can also target different silencing complexes to their DNA copies and silence them transcriptionally by epigenetic changes. The telomeric retrotransposons are not an exception and have been found to be regulated both transcriptionally and posttranscriptionally by the PIWI and the *rasi* pathways in *Drosophila* germline tissues [28, 32]. Moreover, the production of piRNAs from the *HeT-A* transposon has recently been linked to the proper assembly of the capping complex that protects the telomeres, relating in that way 2 different and apparently separate telomere functions in *Drosophila*; regulation of the telomere retrotransposons and protection of the chromosome ends [33]. Interestingly, our laboratory has found a 28-nt sequence at the 3′ UTR of *HeT-A*, which is at the same time a piRNA target and one of the *HeT-A* sequences with higher similarity inside the *HeT-A* orthologues in the *melanogaster* species group, the HeT-A_pi1. Because such remarkable conservation is not expected for a piRNA target, we suggest a possible functional role for HeT-A_pi1 still to discover [Petit N., Piñeyro D., López-Panadès E., Casacuberta E. and Navarro A., unpublished].

Telomere Protection in *Drosophila*

If unprotected, the ends of the telomeres are recognized as double strand breaks by the DNA damage machinery which will repair the telomere by a telomere-telomere fusion, as a consequence opening a cascade of events that may result in genomic instability [34]. All eukaryotes have solved this problem by organizing a nucleoprotein complex that masks the end of the telomere, exerting a protective function named capping. In mammals, the shelterin complex, which contains several proteins that recognize the telomerase repeats, is responsible for the capping function [1]. The telomeric DNA binding proteins serve as a platform to assemble a complex network of interactions of telomere-specific proteins and other proteins that have also additional functions elsewhere in the genome [1]. Among these are the DNA repair proteins which are necessary for proper telomere function but, paradoxically, are also a potential danger if the end stands naked [34]. The shelterin complex should be loaded for protection and unloaded for telomere replication whenever needed. In some eukaryotes, the last few kilobases of the telomeres are folded in a specialized structure known as the T-loop because of its fold-back structure [1]. One component of the shelterin complex is specialized in binding only at certain positions in the loop by being a single-strand DNA-binding protein; others bind the T-loop where this is double-stranded, making a very organized structure. Telomeres recede with each cell duplication and division, the T-loop disappears, the shelterin disassembles, and, as a consequence, exposes the telomere sequence and chromatin marks that signal for telomere elongation or for DNA damage repair and/or cell cycle check point [35].

In *Drosophila*, *HeT-A*, *TART* and *TAHRE* with their well-differentiated sequences are randomly mixed in the telomeres. With this scenario, it is not surprising that the capping function in *Drosophila* turned out to be DNA sequence independent. This unique characteristic of *Drosophila* has been demonstrated by different examples which show how telomeres with non-telomeric sequence at the very end were able to remain stable for several generations [36–38] and recruit capping proteins [39, 40]. With time, these telomeres would acquire telomere-specific sequences (*HeT-A, TART* or *TAHRE*), demonstrating that, in *Drosophila*, telomere capping and telomere elongation are separate functions [11, 36]. The ability to assemble the capping complex independent of a particular sequence suggests that structural or chromatin determinants define the end of the chromosome and points toward an epigenetic mechanism for telomere protection in *Drosophila*. Below we will only briefly describe the main proteins that have been found to be important for telomere protection in *Drosophila*.

HP1

HP1a has been shown to be responsible for a wide repertoire of functions besides the ones concerning the telomeres [25]. In-depth characterization of different mutant alleles of HP1a revealed a dual role for this protein in *Drosophila* telomeres. The chromo domain is responsible for binding HP1a to H3K9me3, a histone modification that has been found on the telomeres. Mutation in the chromo domain resulted in silencing the release of telomeric retrotransposons (see above) but did not affect the capping function, while mutations outside the chromo domain resulted in telomere fusions but no change in the expression of the telomeric retrotransposons [27]. These experiments suggested dual and independent roles for HP1a at *Drosophila* telomeres. In addition to the cap domain, HP1a has been found to extend into the telomeric domains towards the centromere [41] (see section 'Telomere Domains in *Drosophila*').

HOAP

HP1-ORC associated protein (HOAP) is a protein that, as its name indicates, binds HP1 as well as the Origin of Recognition Complex (ORC) [39, 42] and is found almost exclusively at telomeres, where it is significantly abundant. Mutations in the gene that encodes HOAP, *Caravaggio* (*cav*), together with mutations of *hiphop* (see below) result in the strongest phenotype of unprotected telomeres. Although HP1 and HOAP physically interact, mutants of both proteins are still able to partially recruit the other partner to the telomeres, indicating that both proteins should have more than 1 mechanism of telomere binding. Interestingly, for both HP1 and HOAP, DNA binding properties that would explain this alternative binding to the telomeres have also been suggested [27, 42].

HipHop and K81

HipHop is, with HOAP and HP1, an essential protein for the capping function in *Drosophila* [43]. HipHop is present specifically at mitotic telomeres through the cell cycle in significant abundance. The gene encoding HipHop has been the subject of a duplication event in recent evolutionary history (inside the *melanogaster* group, 5–20 Myr) [44]. The 2 genes resulting from this duplication, *hiphop* and *k81*, have undergone specific changes that allowed specification of function, *hiphop* being necessary for telomere protection in somatic tissues and *k81* specifically needed for protecting the telomeres in male germ cells [44, 45]. Genetic assays showed how K81 could replace HipHop in somatic cells, but HipHop could not carry out the K81 function in testes. The determinants for HipHop or K81 loading seem to be epigenetic and cell-specific. The genes encoding HipHop, K81 and HOAP are rapidly evolving genes, which may have facilitated the exploration of possible new functions after the duplication events [44, 45].

Modigliani

Modigliani (Moi) has also been subject to a recent genomic reorganization in *Drosophila*. While in *D. melanogaster* and a few more species, Moi is produced from a bicistronic mRNA encoding 2 different proteins, in other species it is produced from an independent gene [46]. Modigliani physically interacts with HOAP and HP1 and, as its partners, *moi* mutants fail to protect the telomeres. Moi specifically localizes at telomeres in mitotic and polytene chromosomes, as is the case for HOAP and HipHop [39, 43]. Maurizzio Gatti and collaborators [47] have recently proposed that those *Drosophila* proteins that (1) are specifically enriched at the telomeres, (2) bind to the telomeres throughout the cell-cycle, (3) cause telomere fusions if lost, and (4) do not have homologues in telomerase telomeres constitute the terminin complex. The terminin complex would be analogous to the shelterin complex in humans [1]. HipHop and K81, together with Verocchio (see below), should be now considered part of the terminin complex [44, 45]. Interestingly, *moi* (as *hiphop, k81* and *cav*) is also a rapidly evolving gene [43–46].

Verrocchio

Verrocchio (Ver) is another protein that is specifically enriched at telomeres in *Drosophila* [47]. Ver binds Moi and HOAP and is necessary to prevent telomere fusions. Ver contains an oligonucleotide/oligosaccharide OB-fold domain that structurally resembles the OB-fold domain from the human Rpa2/Stn1 proteins. Rpa2/Stn1 proteins together with Cdc13 form the CST complex that protects human telomeres in addition to the shelterin complex. All the proteins of the CST complex contain OB-fold domains [48]. Interestingly, a search in the *Drosophila* genome identified Ver as the only protein with an OB-fold domain in this organism [47]. Ver would be the only member of the terminin complex with certain resemblance to proteins involved in telomere protection in humans.

UbcD1

The first mutation discovered to result in telomere fusions in *Drosophila* was in the *ubcd1/eff* gene [49]. The *eff* gene encodes a highly conserved protein of the class I ubiquitin-conjugating enzymes (E2), UbcD1. The need of UbcD1 for telomere protection in *Drosophila* suggests that regulation by ubiquitination is important for telomere capping. Nevertheless, the possible substrates for UbcD1 at *Drosophila* telomeres are still unknown.

Woc

Without children (Woc) is an 8 zinc finger protein with a role in gene regulation [50]. Woc is not a telomere-specific protein since it localizes to many internal sites in the chromosomes. Woc mutants produce telomeric fusions in mitotic chromosomes in *Drosophila*, demonstrating its role in telomere protection [51]. Mutants for HP1 and HOAP show normal Woc accumulation at telomeres, and vice versa [51]. Therefore, the capping mechanism by Woc and the one governed by HP1-HOAP should be considered as independent. Interestingly, a genetic study of the different mutant alleles of *woc* uncovered a point mutation that shows a decrease in telomere binding, suggesting a possible telomere-specific targeting mechanism by protein-protein or, more likely, protein-DNA interaction.

ATM, ATR and the MRN Complex

Several of the proteins involved in DNA damage repair are also involved in telomere protection in *Drosophila*. The ATR and ATM kinases seem to have overlapping functions in telomere protection since mutations in *mei-41* (ATR) do not seem to result in telomere fusions, but mutations in both *mei-41* and *tefu* (ATM) enhance the mutant phenotype of the single *tefu* mutations. Mutations in the *nbs* or the *mre11* genes from the MRN complex also result in telomere protection defects, although in this case these genes seem to belong to the same pathway as the ATM kinase. Moreover, ATM and the MRN complex are necessary for the loading and maintenance of HOAP at telomeres (reviewed in [34]).

Telomere Domains in *Drosophila*

Chromatin Characteristics

The presence of a highly compacted chromatin structure at the telomeres of several organisms was suspected years ago because the transgenes inserted close to telomeres were subjected to position effect variegation. Position effect variegation occurs when transgenes are silenced because they have been inserted in a highly packed chromatin region [52]. In the case of telomeres, this is referred to as telomere position effect variegation. Recently, studies at the molecular level demonstrated that the telomeres in most eukaryotes are composed of 2 domains differing in their chromatin characteristics; the distal domain, composed of the telomerase repeats or the retrotransposon array *HeT-A, TART, TAHRE* (HTT) in the case of *Drosophila*, and the proximal domain, composed of highly repetitive sequences usually longer and more complex, referred to as the telomere associated sequences (TAS). Generally, TAS nucleate a compacted chromatin structure not permissive of gene expression [35]. The strong silencing potential and the low resolution of telomeres in most species resulted in the notion that telomeres are in general heterochromatic.

In *Drosophila*, it was suspected that the retrotransposon array HTT should have different chromatin characteristics than the flanking TAS domain for several reasons: (1) while in TAS insertions of different TEs are frequent, they are significantly less abundant in the HTT array [4]; (2) transgenes inserted into TAS are strongly silenced, while the few transgenes that have been inserted into the HTT arrays, show an intermediate level of silencing, depending on their position in the array [9]; (3) the promoter of the *HeT-A* element (located at the HTT array) is capable of driving transcription when inserted in euchromatic regions [13]; (4) mutations in genes from the polycomb repressive complex suppress the silencing of reporter genes inserted into the TAS domain but do not affect the expression of genes inserted into the HTT array [26, 53]; (5) mutations in HP1 strongly suppress the silencing of telomeric retrotransposons but do not affect transgenes inside TAS [26, 27]. Several investigations have contributed to a clearer picture of the chromatin characteristics of the different telomeric domains in *Drosophila* [41, 54]. Below we explain the main characteristics of the HTT, the TAS and the cap domain at the very end of the telomere. See also figure 2 for a complementary explanation.

Andreyeva et al. [54] took advantage of the *Tel* mutant strain of *D. melanogaster*, which has telomeres 10 times the wild type length [23], to compare the resolution of a wild type telomere with an extended one by the influence of the *Tel* mutation in the same fly. The immunolocalization experiments for several candidate proteins into the 3 telomere domains of the *Tel* strain gave the first differential picture for each of the telomeric domains in *Drosophila* (see fig. 2). Although, the *Tel* stock has telomeres 10 times longer than a wild type strain, and this feature by itself could influence the presence or absence of some of the identified proteins, these experiments demonstrated that the 3 telomeric domains in *Drosophila* have a particular

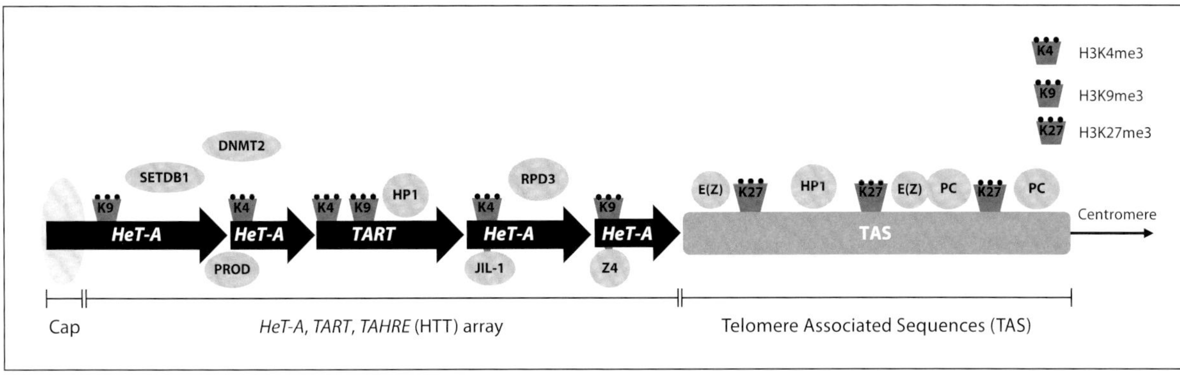

Fig. 2. Telomeric domains in *Drosophila*. In *Drosophila*, the telomeres are composed of 3 different domains: the cap, the HTT (array of *HeT-A, TART* and *TAHRE*) and the subtelomeric TAS domain. Schematic representation of specific proteins and chromatin marks on the HTT and TAS domains. See text for further explanation and characteristics of the cap domain.

set of chromatin components that in most cases do not overlap. From this work, the capping domain recruits the specific chromosomal proteins HP1, HP2, SUUR and Su(var)3-7; the HTT array shows mixed characteristics of euchromatin, JIL-1, Z4 and H3K4Me3, as well as heterochromatin, H3K9Me3; and the TAS domain recruits Polycomb repressive chromatin, E(Z), PC and H3K27Me3. Studies using CHIP assays have refined these first immunolocalization experiments, demonstrating that actually HP1 is found not just at the capping domain but also inside the HTT and even in the TAS domain [41]. The presence of HP1 inside the HTT array was expected because, as mentioned above, HP1 mutant alleles show a strong derepression of *HeT-A* and *TART* transcription. The binding of HP1 in the HTT array could be regulated by the presence of the H3K9Me3, which would imply the previous action of a histone methyltransferase. Recent studies have found that the SetDB1 (*eggless*) methyltransferase is the one responsible for repressive marks at the promoter of *HeT-A* (see below) [29].

From this picture, the telomeric array would maintain a certain level of mixed chromatin with a major tendency to euchromatin. The compacted chromatin from the TAS domain would spread into the vicinity of the flanking region, into the HTT array [9]. And finally, the protective structure of the capping domain would also influence the chromatin behavior of the HTT array as suggested by the incapacity of communicating between enhancer and promoter sequences of the *yellow* gene when those have been located in a distance shorter than 5 kb from the very end of the telomere [55]. More studies are needed to understand how these well-defined chromatin domains are established and to understand, for example, why the HTT array with a more open chromatin structure is much less prone to acquire TE insertions from non-telomeric elements than its neighbor, the TAS domain.

Epigenetic Regulation

Although we have already devoted one section to telomere regulation in *Drosophila*, we think that because of the special nature of the *Drosophila* telomere repeats, the telomeric retrotransposons, it is important to briefly highlight the main characteristics of the epigenetic regulation of telomeres in *Drosophila*. The epigenetic control of telomeres and TEs is a multilayer process that needs to integrate information at the DNA, RNA and protein levels. In *Drosophila*, telomere regulation reaches one more step in complexity because the genes for telomere length maintenance are embedded inside the telomeric chromatin. Gene expression from the telomeres involves chromatin remodeling to release the repressive marks established constitutively in this domain. Moreover, this release needs to be tightly controlled because the telomeric retrotransposons, although they are fulfilling an essential cellular function, maintain their personality as retrotransposons and their uncontrolled transposition could bring both abnormal telomere elongation and genomic instability.

In most eukaryotes the posttranscriptional silencing of TEs is often linked to modifications at the DNA level which will further silence transcriptionally the target sequences [31]. The telomeric retrotransposons *HeT-A, TART* and *TAHRE* have been shown to be regulated by the PIWI pathway in the germline and the RNAi machinery in somatic cells [15, 32]. The loss of silencing in the HTT array by mutations of components of the PIWI or the RNAi pathway resulted in enrichment in activation marks (H3K4me3) and a decrease of repressive marks (H3K9me3) [15, 32]. Therefore, posttranscriptional silencing and chromatin modification are also linked at *Drosophila* telomeres. The methyltransferase SetDB1 is the enzyme in charge of the repressive marks H3K9me3 and H3K9me2 at the nucleosomes at the *HeT-A* promoter, and, as a consequence, the binding of HP1 further represses these nucleosomes at the HTT array [29]. Finally, the deacetylase Rpd3 has recently been shown to deacetylate the *HeT-A* promoter and bring stability to the telomeres [56]. It is not known which enzymes are responsible for the release of gene silencing or the establishment of activation marks in the HTT array, but surely the regulation of *Drosophila* telomeres should contemplate activation of the telomeric transposons since the expression of these elements is vital in order to maintain telomere length through end transposition. In agreement with this hypothesis, Andreyeva et al. [54] as well as our laboratory [Sousa R., López-Panadès E., Piñeyro D. and Casacuberta E., unpublished] have found the kinase JIL-1, a protein related with activation of gene expression, in the HTT array.

Telomeres are methylated at the subtelomeric repeats in vertebrates and yeast and at the telomeric repeats in *Arabidopsis* and *Drosophila* [29, 57]. Hypomethylation of subtelomeric repeats in both yeast and vertebrates results in increase of recombination rate with fatal consequences for genomic stability. In *Drosophila*, mutation in the gene that encodes the DNA methylase 2 (*dnmt2*) causes a de-repression of the *HeT-A* retrotransposon [29]. *HeT-A* de-repression under a *dnmt2* background does not result in a telomere phenotype (longer, shorter or unstable telomeres), but since Dnmt2 also methylates other TEs in *Drosophila*, the general de-repression of mobile elements causes genetic instability [58]. It is unknown if the TAS domain in *Drosophila* is also methylated and if this methylation contributes to telomere function.

In different organisms, such as yeast or *Arabidopsis*, telomere transcription and epigenetic regulation of telomeres have been related [57, 59]. In both cases, telomere transcription results in negative regulation of telomere length. In *Drosophila*, telomere transcription has been known long ago, and it was not a surprise since telomere elongation in *Drosophila* depends on telomere transcription [60]. The telomeric retrotransposons are transcribed from both strands sense and antisense, antisense transcription being a highly conserved feature in all *Drosophila* species (see above). The role of these long non-coding RNAs at *Drosophila* telomeres is still unknown, but their conservation in several *Drosophila* species suggests they have a function [7].

Evolution of Telomeres

The study of the telomeric retrotransposons in several *Drosophila* species reveals more variability in the sequence of retrotransposon telomeres than in the telomerase telomeres [6], [Piñeyro D., López-Panadès E., Lucena M. and Casacuberta E., unpublished]. Although unexpected, this feature could be extremely useful to understand the minimal requirements for telomere function. Below, we review some of the evolutionary features of retrotransposon telomeres and speculate on their possible origin on the basis of what is known about the evolutionary relationship between telomerase and reverse transcriptases.

Drosophila Telomeres Are Far from Being Static

In this section only *HeT-A* and *TART* will be considered since not enough orthologous sequences from *TAHRE* elements are available to allow conclusions. Because the targeted transposition of the telomeric retrotransposons fulfills an essential function for the cell, one would expect, a priori, that their sequences would change slowly as a result of a strong selective pressure. Studies comparing the *HeT-A* and *TART* orthologues among several *Drosophila* species already showed that their sequence changes faster than cellular genes or even other retroelements which have no apparent role in the same genetic distance [7]. This pattern of rapid sequence change has given rise to multiple subfamilies of *HeT-A* [14, 60]. Recent work from our laboratory has shown that the variability and the resulting number of *HeT-A* subfamilies in different strains are actually higher than previously reported [Piñeyro D., López-Panadès E., Lucena M. and Casacuberta E., unpublished]. In this scenario, at least 2 hypotheses could explain the dynamics of sequence change of the telomeric retrotransposons. The first would consider that little sequence in these retroelements is actually under selection for function, resulting in little restriction for change in most of the sequence. In this case the high variability shown by the *HeT-A* sequences could be the result of the low replication fidelity shown by reverse transcriptases. A second hypothesis would consider this fast pattern of sequence change as a strategy for escaping from genomic control. If the telomere retrotransposons could succeed in escaping genome control, they may transpose more often and to more genomic locations than just the telomeres whenever needed. In this scenario a genetic conflict between the *Drosophila* genome and the telomeric retrotransposons would be in place.

Whichever, if any, of these scenarios is true, the telomeric retrotransposons have been evolving different strategies in order to preserve their transposing capacity in spite of being at the end of the chromosome. Two recent reports from Pardue and collaborators [6, 61] give proof of such strategies. The first study investigated how the sequences of the telomeric retrotransposons present at the pericentromeric region of the Y chromosome evolve under different constraints than the *HeT-A* and *TART* sequences at the telomeres [61]. The second study analyzed how *HeT-A* in *D. melanogaster* and *TART* in *D. virilis* have converged to similar strategies to protect the 5′ ends of their copies with non-essential sequence when these are at the very end of the chromosome [6]. *HeT-A*^{mel} and *TART*^{vir} also share the unusual 3′ promoter that, when driving transcription from the upstream element, adds at the point of transcription initiation non-essential sequence to the sense RNA copy of the downstream element. This non-essential sequence will buffer the retrotransposon copy that sits at the very end of the chromosome from the terminal erosion until a new copy will transpose upstream, preserving their capacity for further transpositions.

Possible Origins of the Retrotransposon Telomeres

The mechanism of telomere maintenance by telomerase and the target-primed reverse transcription by which non-LTR retrotransposons integrate in a new genome location is mechanistically similar. In both cases the catalytic unit of telomerase (TERT) or the reverse transcriptase (RT) reverse transcribes an RNA template directly at the site of integration after priming the reaction with a free 3′ OH generated at the end of the chromosome (in the first case) or in an internal nick at the DNA by the action of the endonuclease also encoded by the transposon (in the second case). Further connections between these mechanisms were demonstrated by the discovery that several retrotransposons in different distant organisms are able to transpose directly onto the telomeres when they are endonuclease-deficient. Analyses of endonuclease-defective *Penelope*-like elements (*PLE*) found at the end of chromosomes in protists were the first of such examples, followed by work with *L1* elements in which the endonuclease has been inactivated. These mutant elements were shown to be able to transpose onto the telomeres of mouse cells if those cells were defective in telomere capping and in non-homologous DNA repair [62]. Actually, connections towards this relationship also exist coming from the other direction, because telomerase has been shown to be able to occasionally reverse transcribe telomere repeats at internal genomic locations resembling a 'transposition' event of telomerase repeats [63]. Together all these functional points indicate a common origin in evolution for TERT and retrotransposon RTs.

Phylogenetic studies showed that the RT from *PLE* elements is similar to the telomerase RT, and they likely descend from the same ancestor [62]. Because of their single copy condition, *PLE* elements likely appeared early in evolution, and the subsequent acquisition of endonuclease by non-LTR retrotransposons freed them from having to rely on the repair of double strand breaks or on the replication forks in order to transpose. Once non-LTR retrotransposons increased their efficiency and their chances of transposition, they became real selfish elements spreading their copies throughout the genome [62]. Paradoxically, the transposons associated with telomere maintenance in *Drosophila* that contain an ORF2 protein, *TART* and *TAHRE*, have both RT and endonuclease activity. The conservation of the amino acid residues important for endonuclease activity across the orthologues of *TART* elements from *D. melanogaster* to *D. virilis* suggests that the endonuclease activity of these elements is necessary for telomere transposition. On the other hand, the presence of non-LTR retroelements in several eukaryote telomeres does not stop with endonuclease-deficient elements. Several examples exist in which non-LTR retrotransposons have acquired specificity for telomere repeats and transpose into the telomeres of several organisms [62, 64, 65]. Although the elements do not directly maintain the telomeres in these organisms, they indirectly contribute to the maintenance of the whole telomere length by buffering the telomere shortening with their transposition into internal telomeric transpositions. Interestingly, in 2 of these organisms, *Bombyx* and *Tribolium*, the TERT subunit of telomerase lacks a functional domain important for processivity (the N-terminal domain) [65].

In summary, many functional and evolutionary connections seem to relate retrotransposons and telomerase RTs, but further studies are needed to understand the chain of events that brought the telomeric retrotransposons to efficiently substitute with time the ancestral telomerase mechanism in an ancestral insect. We should take into account that this transition not only involves the actual mechanism of telomere elongation but many of the different features that constitute the telomere function in *Drosophila*, such as the proteins that exert the capping function as well as the chromatin structure and the consequent epigenetic regulation at those telomeres. A smooth and progressive change from the loss of RT activity from an ancient telomerase, opportunistic insertion of telomeric retrotransposons already in place, and progressive loss of sequence specificity for the shelterin complex, could be one of such chains of events among many variations around these necessary steps (see fig. 3).

Further studies on the involvement of the proteins encoded by the telomeric retrotransposons as well as the study of evolutionary intermediates that provide different points in evolution will be crucial to better understand this spectacular transition that has allowed TEs to adapt and perform an essential cellular role. The cases of *Bombyx mori* and *Tribolium castaneum*, which combine telomerase repeats and retrotransposons with insertion specificity for telomerase repeats, give the opportunity to further

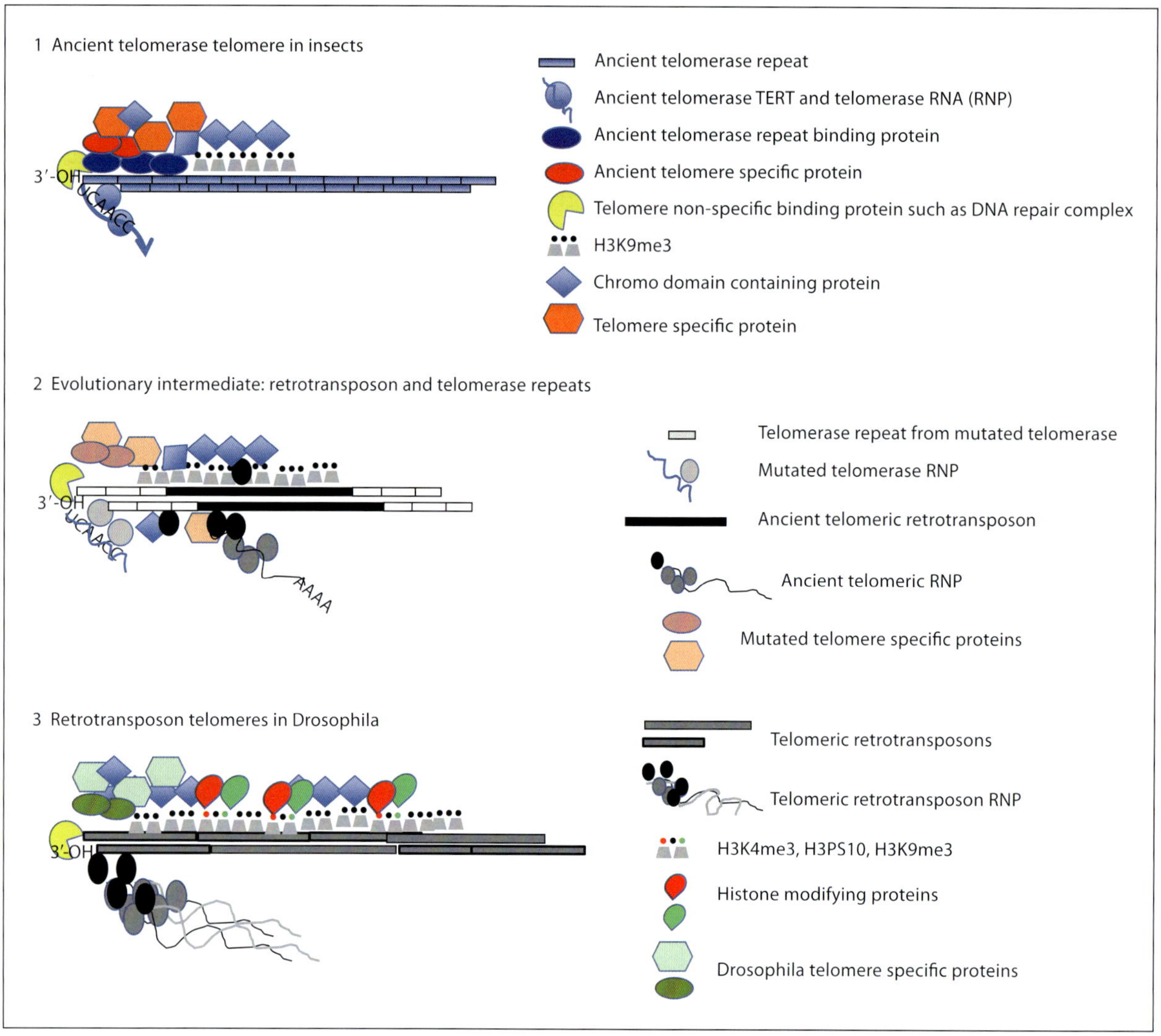

Fig. 3. Origin and evolution of *Drosophila* telomeres. Schematic representation of a possible chain of events from an ancient telomerase telomere towards a retrotransposon telomere. See text for further explanation. The initial step of mutation in the telomerase ribonucleoprotein (RNP) would have resulted in a change in the telomerase repeats at the telomeres. As a consequence, proteins that bind telomere repeats would have evolved or disappeared and their partners, (telomere-specific proteins with no DNA binding) would also have evolved, changing the shelterin complex in place to a sequence-independent capping complex. The transposition of pre-existing telomere-specific retrotransposons into the telomeres accelerated the process. The selection of the telomere-specific retrotransposons for telomere elongation introduced epigenetic changes along the telomere sequence in order to control telomere length.

investigate the balance between the 2 mechanisms and the components that are in charge of the capping function in those organisms.

Conclusions

The particular composition of telomeres in *Drosophila* opens the door to perform studies that will undoubtedly help to understand the complicated and fascinating relationship between genomes and TEs. The combination of subjects such as telomere protection, retrotransposon evolution and epigenetic regulation of both telomeres and retrotransposons in these studies make *Drosophila* a powerful model for telomere and retrotransposon biology. Finally, as more reports on *Drosophila* telomeres reveal more features in common with telomerase telomeres, reports on their composition and evolution are highlighting the extent of their variation. The combination of these remarkable characteristics make *Drosophila* telomeres a particularly interesting model organism to study on one hand the minimal requirements for telomere function and, on the other, the level of diversity that sequences with telomeric function can bear.

References

1 De Lange T, Blackburn EH, Lundblad V (eds): Telomeres. Cold Spring Harbor, Cold Spring Harbor Press, 2006.

2 Muller HJ: The remaking of chromosomes. Collect Net 1938;8:182–198.

3 McClintock B: The behavior in successive nuclear divisions of a chromosome broken at meiosis. Proc Natl Acad Sci USA 1939;25:405–416.

4 Pardue ML, DeBaryshe PG: *Drosophila* telomeres: a variation on the telomerase theme. Fly (Austin) 2008;2:101–110.

5 McClintock B: The significance of responses of the genome to challenge. Science 1984;226:792–801.

6 Pardue ML, DeBaryshe PG: Retrotransposons that maintain chromosome ends. Proc Natl Acad Sci USA 2011;108:20317–20324.

7 Casacuberta E, Pardue ML: HeT-A and TART, two *Drosophila* retrotransposons with a bona fide role in chromosome structure for more than 60 million years. Cytogenet Genome Res 2005;110:152–159.

8 Luan DD, Korman MH, Jakubczak JL, Eickbush TH: Reverse transcription of R2Bm RNA is primed by a nick at the chromosomal target site: a mechanism for non-LTR retrotransposition. Cell 1993;72: 595–605.

9 Mason JM, Frydrychova RC, Biessmann H: *Drosophila* telomeres: an exception providing new insights. Bioessays 2008;30:25–37.

10 Malik HS, Burke WD, Eickbush TH: The age and evolution of non-LTR retrotransposable elements. Mol Biol Evol 1999;16:793–805.

11 Craig NL, Craigie R, Gellert M, Lambowitz AM: Mobile DNA II. Washington, ASM Press, 2002.

12 Abad JP, de Pablos B, Agudo M, Molina I, Giovinazzo G, et al: Genomic and cytological analysis of the Y chromosome of *Drosophila melanogaster*: telomere-derived sequences at internal regions. Chromosoma 2004;113:295–304.

13 George JA, Pardue ML: The promoter of the heterochromatic *Drosophila* telomeric retrotransposon, *HeT-A*, is active when moved into euchromatic locations. Genetics 2003;163:625–635.

14 Villasante A, Abad JP, Planelló R, Méndez-Lago M, Celniker SE, de Pablos B: *Drosophila* telomeric retrotransposons derived from an ancestral element that was recruited to replace telomerase. Genome Res 2007;17:1909–1918.

15 Shpiz S, Kwon D, Rozovsky Y, Kalmykova A: rasiRNA pathway controls antisense expression of *Drosophila* telomeric retrotransposons in the nucleus. Nucleic Acids Res 2009;37:268–278.

16 Abad JP, De Pablos B, Osoegawa K, De Jong PJ, Martín-Gallardo A, Villasante A: *TAHRE*, a novel telomeric retrotransposon from *Drosophila melanogaster*, reveals the origin of *Drosophila* telomeres. Mol Biol Evol 2004;21:1620–1624.

17 Shpiz S, Kwon D, Uneva A, Kim M, Klenov M, et al: Characterization of *Drosophila* telomeric retroelement *TAHRE*: transcription, transpositions, and RNAi-based regulation of expression. Mol Biol Evol 2007;24:2535–2545.

18 Maxwell PH, Belote JM, Levis RW: Identification of multiple transcription initiation, polyadenylation, and splice sites in the *Drosophila melanogaster* TART family of telomeric retrotransposons. Nucleic Acids Res 2006;34:5498–5507.

19 George JA, DeBaryshe PG, Traverse KL, Celniker SE, Pardue ML: Genomic organization of the *Drosophila* telomere retrotransposable elements. Genome Res 2006;16:1231–1240.

20 Fuller AM, Cook EG, Kelley KJ, Pardue ML: Gag proteins of *Drosophila* telomeric retrotransposons: collaborative targeting to chromosome ends. Genetics 2010;184:629–636.

21 Casacuberta E, Marín FA, Pardue ML: Intracellular targeting of telomeric retrotransposon Gag proteins of distantly related *Drosophila* species. Proc Natl Acad Sci USA 2007;104:8391–8396.

22 Melnikova L, Georgiev P: Enhancer of terminal gene conversion, a new mutation in *Drosophila melanogaster* that induces telomere elongation by gene conversion. Genetics 2002;162:1301–1312.

23 Siriaco GM, Cenci G, Haoudi A, Champion LE, Zhou C, et al: Telomere elongation *(Tel)*, a new mutation in *Drosophila melanogaster* that produces long telomeres. Genetics 2002;160:235–245.

24 Melnikova L, Biessmann H, Georgiev P: The Ku protein complex is involved in length regulation of *Drosophila* telomeres. Genetics 2005;170:221–235.

25 Lomberk G, Wallrath L, Urrutia R: The heterochromatin protein 1 family. Genome Biol 2006;7:228.

26 Cryderman DE, Morris EJ, Biessmann H, Elgin SC, Wallrath LL: Silencing at *Drosophila* telomeres: nuclear organization and chromatin structure play critical roles. EMBO J 1999;18:3724–3735.

27 Perrini B, Piacentini L, Fanti L, Altieri F, Chichiarelli S, et al: HP1 controls telomere capping, telomere elongation, and telomere silencing by two different mechanisms in *Drosophila*. Mol Cell 2004;15:467–476.

28 Klenov MS, Lavrov SA, Stolyarenko AD, Ryazansky SS, Aravin AA, et al: Repeat-associated siRNAs cause chromatin silencing of retrotransposons in the *Drosophila melanogaster* germline. Nucleic Acids Res 2007;35:5430–5438.

29 Gou D, Rubalcava M, Sauer S, Mora-Bermúdez F, Erdjument-Bromage H, et al: SETDB1 is involved in postembryonic DNA methylation and gene silencing in *Drosophila*. PLoS One 2010;5:e10581.

30 Török T, Benitez C, Takács S, Biessmann H: The protein encoded by the gene proliferation disrupter *(prod)* is associated with the telomeric retrotransposon array in *Drosophila melanogaster*. Chromosoma 2007;116:185–195.

31 Slotkin RK, Martienssen R: Transposable elements and the epigenetic regulation of the genome. Nat Rev Genet 2007;8:272–285.

32 Savitsky M, Kwon D, Georgiev P, Kalmykova A, Gvozdev V: Telomere elongation is under the control of the RNAi-based mechanism in the *Drosophila* germline. Genes Dev 2006;20:345–354.

33 Khurana JS, Xu J, Weng Z, Theurkauf WE: Distinct functions for the *Drosophila* piRNA pathway in genome maintenance and telomere protection. PLoS Genet 2010;6:e1001246.

34 Rong YS: Telomere capping in *Drosophila*: dealing with chromosome ends that most resemble DNA breaks. Chromosoma 2008;117:235–242.

35 Blasco MA: The epigenetic regulation of mammalian telomeres. Nat Rev Genet 2007;8:299–309.

36 Biessmann H, Champion LE, O'Hair M, Ikenaga K, Kasravi B, Mason JM: Frequent transpositions of *Drosophila melanogaster* HeT-A transposable elements to receding chromosome ends. EMBO J 1992; 11:4459–4469.

37 Ahmad K, Golic KG: The transmission of fragmented chromosomes in *Drosophila melanogaster*. Genetics 1998;148:775–792.

38 Levis RW: Viable deletions of a telomere from a *Drosophila* chromosome. Cell 1989;58:791–801.

39 Cenci G, Siriaco G, Raffa GD, Kellum R, Gatti M: The *Drosophila* HOAP protein is required for telomere capping. Nat Cell Biol 2003;5:82–84.

40 Fanti L, Giovinazzo G, Berloco M, Pimpinelli S: The heterochromatin protein 1 prevents telomere fusions in *Drosophila*. Mol Cell 1998;2:527–538.

41 Frydrychova RC, Mason JM, Archer TK: HP1 is distributed within distinct chromatin domains at *Drosophila* telomeres. Genetics 2008;180:121–131.

42 Shareef MM, King C, Damaj M, Badagu R, Huang DW, Kellum R: *Drosophila* heterochromatin protein 1 (HP1)/origin recognition complex (ORC) protein is associated with HP1 and ORC and functions in heterochromatin-induced silencing. Mol Biol Cell 2001;12:1671–1685.

43 Gao G, Walser JC, Beaucher ML, Morciano P, Wesolowska N, et al: HipHop interacts with HOAP and HP1 to protect *Drosophila* telomeres in a sequence-independent manner. EMBO J 2010;29: 819–829.

44 Dubruille R, Orsi GA, Delabaere L, Cortier E, Couble P, et al: Specialization of a *Drosophila* capping protein essential for the protection of sperm telomeres. Curr Biol 2010;20:2090–2099.

45 Gao G, Cheng Y, Wesolowska N, Rong YS: Paternal imprint essential for the inheritance of telomere identity in *Drosophila*. Proc Natl Acad Sci USA 2011;108:4932–4937.
46 Raffa GD, Siriaco G, Cugusi S, Ciapponi L, Cenci G, et al: The *Drosophila modigliani (moi)* gene encodes a HOAP-interacting protein required for telomere protection. Proc Natl Acad Sci USA 2009;106:2271–2276.
47 Raffa GD, Raimondo D, Sorino C, Cugusi S, Cenci G, et al: Verrocchio, a *Drosophila* OB fold-containing protein, is a component of the terminin telomere-capping complex. Genes Dev 2010;24:1596–1601.
48 Sun J, Yu EY, Yang Y, Confer LA, Sun SH, et al: Stn1-Ten1 is an Rpa2-Rpa3-like complex at telomeres. Genes Dev 2009;23:2900–2914.
49 Cenci G, Rawson RB, Belloni G, Castrillon DH, Tudor M, et al: UbcD1, a *Drosophila* ubiquitin-conjugating enzyme required for proper telomere behavior. Genes Dev 1997;11:863–875.
50 Wismar J, Habtemichael N, Warren JT, Dai JD, Gilbert LI, Gateff E: The mutation *without children(rgl)* causes ecdysteroid deficiency in third-instar larvae of *Drosophila melanogaster*. Dev Biol 2000;226:1–17.
51 Raffa GD, Cenci G, Siriaco G, Goldberg ML, Gatti M: The putative *Drosophila* transcription factor Woc is required to prevent telomeric fusions. Mol Cell 2005;20:821–831.
52 Weiler KS, Wakimoto BT: Heterochromatin and gene expression in *Drosophila*. Annu Rev Genet 1995;29:577–605.
53 Boivin A, Gally C, Netter S, Anxolabéhère D, Ronsseray S: Telomeric associated sequences of *Drosophila* recruit polycomb-group proteins in vivo and can induce pairing-sensitive repression. Genetics 2003;164:195–208.
54 Andreyeva EN, Belyaeva ES, Semeshin VF, Pokholkova GV, Zhimulev IF: Three distinct chromatin domains in telomere ends of polytene chromosomes in *Drosophila melanogaster* Tel mutants. J Cell Sci 2005;118:5465–5477.
55 Melnikova L, Georgiev P: *Drosophila* telomeres: the non-telomerase alternative. Chromosome Res 2005; 13:431–441.
56 Burgio G, Cipressa F, Ingrassia AM, Cenci G, Corona DF: The histone deacetylase Rpd3 regulates the heterochromatin structure of *Drosophila* telomeres. J Cell Sci 2011;124:2041–2048.
57 Vrbsky J, Akimcheva S, Watson JM, Turner TL, Daxinger L, et al: siRNA-mediated methylation of *Arabidopsis* telomeres. PLoS Genet 2010;6:e1000986.
58 Phalke S, Nickel O, Walluscheck D, Hortig F, Onorati MC, Reuter G: Retrotransposon silencing and telomere integrity in somatic cells of *Drosophila* depends on the cytosine-5 methyltransferase DNMT2. Nat Genet 2009;41:696–702.
59 Schoeftner S, Blasco MA: A 'higher order' of telomere regulation: telomere heterochromatin and telomeric RNAs. EMBO J 2009;28:2323–2336.
60 Pardue ML, DeBaryshe PG: Retrotransposons provide an evolutionarily robust non-telomerase mechanism to maintain telomeres. Annu Rev Genet 2003; 37:485–511.
61 DeBaryshe PG, Pardue ML: Differential maintenance of DNA sequences in telomeric and centromeric heterochromatin. Genetics 2011;187:51–60.
62 Curcio MJ, Belfort M: The beginning of the end: links between ancient retroelements and modern telomerases. Proc Natl Acad Sci USA 2007;104:9107–9108.
63 Wells RA, Germino GG, Krishna S, Buckle VJ, Reeders ST: Telomere-related sequences at interstitial sites in the human genome. Genomics 1990;8:699–704.
64 Osanai-Futahashi M, Fujiwara H: Coevolution of telomeric repeats and telomeric repeat-specific non-LTR retrotransposons in insects. Mol Biol Evol 2011;28:2983–2986.
65 Osanai M, Kojima KK, Futahashi R, Yaguchi S, Fujiwara H: Identification and characterization of the telomerase reverse transcriptase of *Bombyx mori* (silkworm) and *Tribolium castaneum* (flour beetle). Gene 2006;376:281–289.
66 Piñeyro D, López-Panadès E, Pérez ML, Casacuberta E: Transcriptional analysis of the *HeT-A* retrotransposon in mutant and wild type stocks, reveals extreme sequence variability at *Drosophila* telomeres and other unusual features. BMC Genomics 2011;12:573.

Elena Casacuberta
Institute of Evolutionary Biology (CSIC-UPF)
Passeig Marítim de la Barceloneta 37–49
ES–08003 Barcelona (Spain)
Tel. +34 93 230 9637, E-Mail elena.casacuberta@ibe.upf-csic.es

Garrido-Ramos MA (ed): Repetitive DNA.
Genome Dyn. Basel, Karger, 2012, vol 7, pp 68–91

The Evolutionary Dynamics of Transposable Elements in Eukaryote Genomes

M. Tollis · S. Boissinot

Department of Biology, Queens College, The City University of New York, Flushing, N.Y., and
The Graduate Center, The City University of New York, New York, N.Y., USA

Abstract

Transposable elements (TEs) are ubiquitous components of eukaryotic genomes. They have considerably affected their size, structure and function. The sequencing of a multitude of eukaryote genomes has revealed some striking differences in the abundance and diversity of TEs among eukaryotes. Protists, plants, insects and vertebrates contain species with large numbers of TEs and species with small numbers, as well as species with diverse repertoires of TEs and species with a limited diversity of TEs. There is no apparent relationship between the complexity of organisms and their TE profile. The profile of TE diversity and abundance results from the interaction between the rate of transposition, the intensity of selection against new inserts, the demographic history of populations and the rate of DNA loss. Recent population genetics studies suggest that selection against new insertions, mostly caused by the ability of TEs to mediate ectopic recombination events, is limiting the fixation of TEs, but that reduction in effective population size, caused by population bottlenecks or inbreeding, significantly reduces the efficacy of selection. These results emphasize the importance of drift in shaping genomic architecture.

Copyright © 2012 S. Karger AG, Basel

The complete or ongoing sequencing of more than 1,000 eukaryotic genomes (www.genomesonline.org) has been an extraordinary source of information for scientists, thereby revolutionizing the field of genetics, development and evolutionary biology. Eukaryote genomes vary considerably in size and structure, and understanding the cause(s) of these differences is fundamental for interpreting meaningfully genomic annotations. Among the genomic features that show the most variation among organisms is the abundance and diversity of transposable elements (TEs). TEs are DNA sequences that can move from one location in the genome to another location. They have considerably affected the size and structure of eukaryotic genomes. In fact, with the exception of polyploidy, the abundance of TEs is the major determinant of genome size differences among eukaryotes. The abundance and diversity of TEs in a genome has important evolutionary implications as TEs constitute an important source of evolutionary novelties by providing a tool-box of sequence motifs on which natural selection can act.

The number and diversity of TEs in a genome result from the interactions between the rate of transposition, the intensity of selection against new inserts and the demographic history of populations. How these different factors interact remains controversial, but the complete sequencing of a multitude of eukaryotic genomes as well as recent population studies have provided new insights on the evolutionary dynamics of TEs in eukaryotes. Understanding the dynamics of TEs is important for 2 main reasons. First, as TEs occupy a significant fraction of genomes, knowing the mechanisms that control their copy number will help understand why eukaryotic genomes differ so much in size, structure and function. Second, the evolutionary dynamics of TEs can help decipher the interplay between selective and neutral factors in the evolution of genomic features, a highly contentious issue in the field of comparative genomics [1].

Classification and Mechanisms of Transposition

TE is a generic term that covers an extraordinary diversity of mobile elements. TEs are usually classified into 2 groups, often referred to as Class I and Class II elements, based on their mode of transposition. Class I elements, also called retrotransposons, mobilize using an RNA intermediate during transposition and encode the enzyme reverse transcriptase. Class II elements, also called DNA transposons, do not have an RNA intermediate during transposition but use a DNA intermediate. Class I elements are further divided into 2 categories based on the presence or absence of long terminal repeats (LTRs).

Retrotransposons Lacking Long Terminal Repeats

This group of TEs includes 2 categories, the non-LTR retrotransposons sensu stricto and the *Penelope* elements. *Penelope* elements constitute the most basal group in the evolution of retrotransposons. They are structurally very diverse, they sometimes retain introns and their reverse transcriptase shows some similarity to telomerases [2]. Although they are widely distributed among eukaryotes, *Penelope* elements remain one of the least studied groups of retrotransposons. Non-LTR retrotransposons sensu stricto constitute an extremely old and diverse component of eukaryotic genomes. A number of very ancient monophyletic lineages of elements, called clades, have been described, from 11 to 25 depending on the authors [3, 4], and it is likely that additional clades will be discovered when more genome sequences become available. These clades can be sorted into 6 groups based on structural differences (fig. 1a): the *R2, RandI, L1, RTE, I* and *Jockey* groups [3].

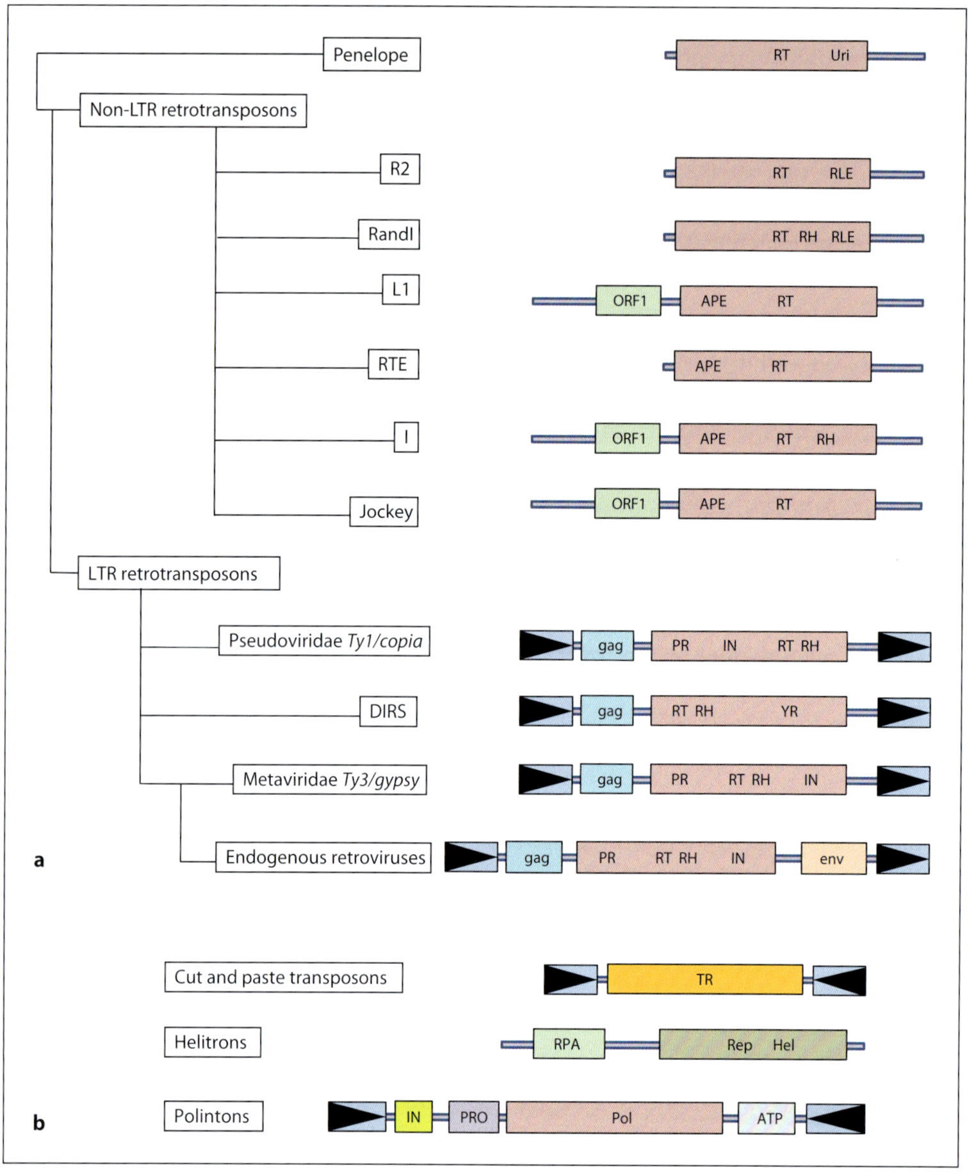

Fig. 1. **a** Schematic classification (left) and structure (right) of autonomous retrotransposons. The elements are not drawn to scale. The following abbreviations are used: APE, apurinic endonuclease; env, envelope gene; gag, gag gene; IN, integrase; ORF1, open-reading frame 1; PR, proteinase; RH, RNase H domain; RLE, restriction-like endonuclease; RT, reverse transcriptase; Uri, endonuclease domain with similarity to group I introns; YR, tyrosine recombinase. The purple lines indicate the non-protein coding regions of the retrotransposons. The boxes represent the open-reading frames and the boxed triangles represent the LTRs. **b** Schematic structure of autonomous class II transposons. The following abbreviations are used: ATP, ATPase; Hel, helicase; IN, integrase; Pol, polymerase; PRO, cysteine protease; Rep, replication initiation domain; RPA, replication protein A; TR, transposase. The structure of the *polintons* can vary considerably from the type represented here. Boxed triangles represent the TIRs.

All these elements encode at least 1, but more often 2 open-reading frames (ORF1 and ORF2). ORF2 codes for reverse transcriptase activity and, logically, it is the ORF all clades have in common. The most basal groups in the evolution of non-LTR retrotransposons are the *R2* and *RandI* groups, which have a single ORF that contains a restriction-like endonuclease domain near the C-terminus, in addition to the reverse transcriptase. These elements tend to insert in a sequence-specific manner, as exemplified by the *R2* element which inserts specifically in 28S rRNA genes [5]. All other clades encode an apurinic-apyrimidinic endonuclease located near the N-terminus of ORF2 and some of them have an RNase H motif downstream of the reverse transcriptase domain. Most clades belonging to the *L1, I* and *Jockey* groups have another ORF, ORF1. ORF1 is poorly conserved among clades and, depending on the clade, contains esterase, CCHC zinc knuckles or RNA recognition motifs. The mammalian ORF1 protein encoded by *L1* has been the most studied, yet its function remains unclear. It contains a conserved RNA recognition motif [6] and a rapidly evolving coiled coil domain [7] which mediates the formation of trimers [8]. The ORF1 protein participates in the formation of ribonucleoprotein particles [9] and encodes nucleic acid chaperone activity [10]. A 5′ untranslated region (UTR), that has been shown to act as an internal promoter in *L1* [11], can be found upstream of the ORFs. A second UTR of unknown function flanks ORF2 in 3′.

The transposition mechanism of non-LTR retrotransposons was first deciphered for the *R2* element in *Bombyx mori* [5] and subsequently the same mechanism was demonstrated for the human *L1* [12]. Following transcription, the retrotransposon mRNA is exported to the cytoplasm where it is translated. The translated proteins remain bound to the RNA and the resulting complex is then re-imported in the nucleus where insertion takes place. A nick is made on the bottom strand of the insertion site by the endonuclease encoded by the element. The 3′ OH released by this cleavage is then used to prime reverse transcription of the mRNA into a cDNA. Because the reverse transcription occurs at the sites of insertion, this reaction has been named target-primed reverse transcription (TPRT). The TPRT reaction lacks processivity, particularly in the *L1* clade, and up to two-thirds of the new insertions are truncated in 5′ [13]. Although the mechanism of transposition of other clades has not been studied in great details, the similarity in structure of insertions belonging to the *L1, L2, RTE* and *CR1* clades suggests that all these elements are mobilized by a mechanism similar or identical to the TPRT reaction [14].

The evolution of non-LTR retrotransposons is quite complex and seems affected by the nature of the interactions with the host. This is particularly true of *L1* (fig. 2). In fish and squamate reptiles the *L1* clade is represented by a multitude of lineages which diverged before the diversification of vertebrates [15–17]. These lineages are represented by very small copy number, but within each family elements are very similar, suggesting they inserted recently. It seems that in fish and reptiles *L1* elements do not accumulate to large numbers and are possibly eliminated by purifying selection. This mode of evolution contrasts drastically with the situation in mammals where *L1* has

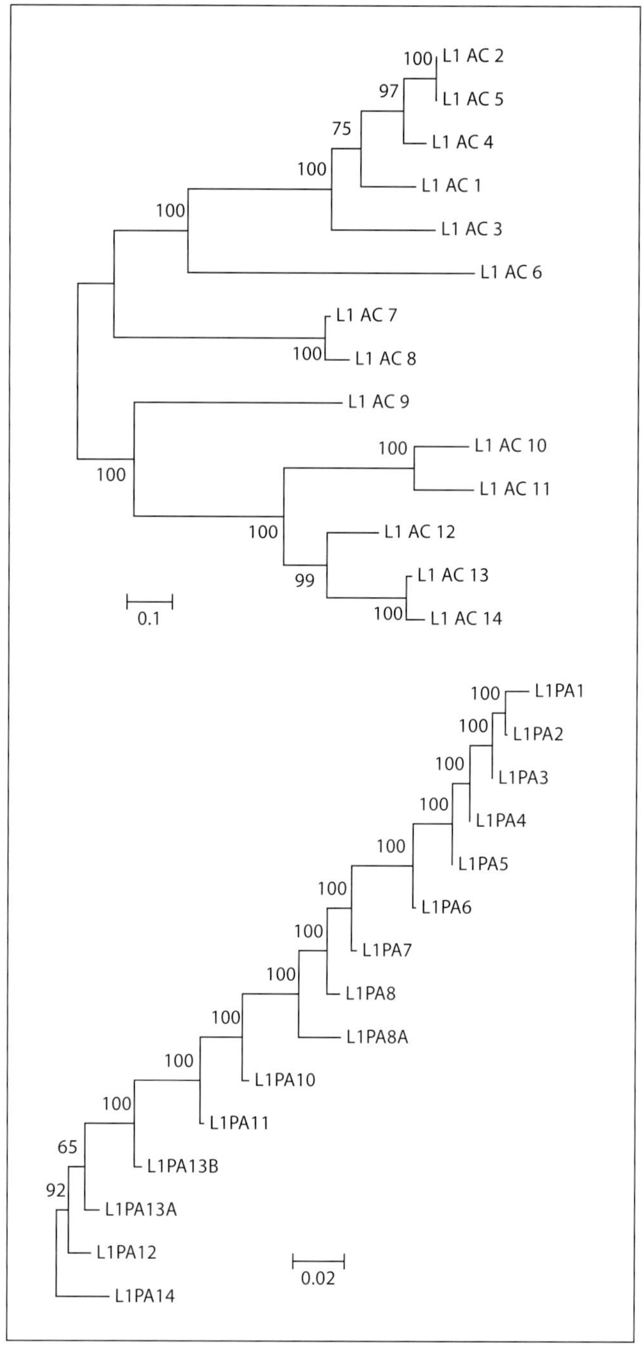

Fig. 2. Phylogeny of L1 families in the lizard *Anolis carolinensis* (top) and in human (bottom). The tree is based on consensus sequences derived by Novick et al. [17] and Khan et al. [22] for *Anolis* and human, respectively. The trees were built using the maximum likelihood method using the TN93+G+I model.

accumulated to extremely large numbers. For instance, the human genome contains 800,000 *L1* copies that account for 21% of its size [18]. As *L1s* are never excised, a host's genome contains a complete repertoire of the families that have been active in the past [19]. Phylogenetic analyses in humans and other mammals revealed that *L1* retrotransposons evolved as a single lineage, meaning that only 1 family of element is active at a time until it is replaced by a most recent family [20–22]. This mode of evolution is extremely unusual and is reminiscent of the evolution of the influenza virus, suggesting it might be driven by repression by the host, a hypothesis supported by the observation that a region of *L1* is evolving adaptively [7, 22]. Interestingly, this single lineage mode of evolution is also observed in *Platypus anatinus* whose genome is not dominated by *L1* but by *L2* [23]. However, up until 40 million years (Myr) ago, multiple lineages of *L1* were concurrently active in primates [22]. It was found that coexisting families always had non-homologous promoter sequences, raising the intriguing possibility that a competition for transcription factor encoded by the host might be limiting the diversification of *L1*. This hypothesis is supported by the fact that coexisting families in mouse [20] and lizard [17] also have non-homologous 5′ UTRs.

Although retrotransposition acts preferentially in *cis* [24], the replicative machinery encoded by non-LTR retrotransposons can also act on other transcripts and is responsible for the amplification of a number of non-autonomous TEs, called SINEs for short interspersed elements, and processed pseudogenes [25]. For instance, the human *Alu* element, which is derived from the 7SL RNA, uses the *L1* replicative machinery for its own benefit and amplified to considerable numbers in primate genomes (~1,000,000 copies in human). Other SINEs are derived from tRNAs, and some of them show similarity at their 3′ end with the autonomous elements that mobilize them as a way to recruit the biochemical machinery necessary for transposition [26]. *L1* is not the only clade to generate SINEs, as elements mobilized by *RTE* and *L2* were recently discovered [27–29].

Retrotransposons with Long Terminal Repeats

This group includes 3 subgroups, the LTR retrotransposons sensu stricto and the endogenous retroviruses (ERV) which have very similar structure and mode of transposition, and the DIRS which differ considerably in structure (fig. 1a) [30, 31]. LTR retrotransposons are evolutionarily more recent than non-LTR retrotransposons and it is believed that they originated by recruitment of the reverse transcriptase domain of a non-LTR retrotransposon by a DNA transposon [32]. They are classified into 2 main families, the Metaviridae, which includes 2 subgroups *Ty3/gypsy* and *Bel*, and the Pseudoviridae, including the *Ty1* and *copia* elements. LTR retrotransposons are widely distributed in eukaryotes, in particular fungi, insects and plants, where they constitute the dominant category of TEs.

LTR retrotransposons have a protein-coding region which is flanked by direct LTRs that regulate transcription and play a critical role during reverse transcription. The protein-coding region contains 2 genes: *gag*, which encodes structural and nucleic acid domains required for reverse transcription, and *pol*, which encodes the enzymatic activities protease, RNase H, reverse transcriptase and integrase. The mechanism of transposition begins with transcription of the element and export of the resulting mRNA to the cytoplasm. The mRNA is then translated and the resulting poly-protein is cleaved by the protease. The gag proteins form a virus-like particle which contains typically 2 RNA molecules as well as the integrase, reverse transcriptase and RNase. The reverse transcriptase and RNase catalyze the reverse transcription of the RNA into a linear double strand cDNA, which is then re-imported inside the nucleus and inserted back into the genome by the integrase. Many LTR retrotransposons have independently acquired an additional gene, *env* (envelope), which, in the case of the *Drosophila gypsy* element, confers the ability to infect oocytes [33, 34]. There are strong reasons to believe that infectious vertebrate retroviruses evolved from Metaviridae after recruitment of the *env* gene.

Eventually, some infectious retroviruses infected the germline of vertebrates and became stable residents of these genomes [35]. Although they lost their infectivity, they have retained their mobility and have multiplied in their host. There are 3 groups of ERV, called ERV Class I, II and III which are derived from different families of retroviruses [36], yet a large number of ERVs are still unclassified because they do not show similarity with any of the currently recognized groups of exogenous retroviruses. Vertebrate genomes often contain more than 1 type of ERVs. For instance, the human genome contains at least 26 distinct ERV families, representing the 3 known classes of ERVs, and the number of independent acquisition of novel ERVs is probably close to 50 [36, 37].

The third group of LTR containing retrotransposons, DIRS, is the least studied [38], although it is quite widespread in nematodes, fish, amphibians, sea urchins, slime mold and fungi [39]. DIRS elements differ from other LTR retrotransposons in structure, as they lack a protease or an integrase. They encode a tyrosine recombinase, suggesting that insertion into the host genome occurs by a recombination reaction catalyzed by the tyrosine recombinase.

It should be noted that the majority of LTR containing retrotransposons in some genomes are not complete and are represented only by LTRs. The loss of the coding region results from homologous recombination between the 2 LTRs, so that a single LTR remains, usually called solo-LTR. Some LTR retrotransposons have successfully multiplied in the absence of protein-coding capacity, such as the *Dasheng* element of rice [40]. These non-autonomous elements have LTRs but no protein coding capacity. It is believed that the LTRs of these elements can still be recognized by the retrotransposition machinery encoded by complete copies and are thus mobilized by their autonomous counterparts.

DNA Transposons

DNA transposons (or Class II) are a general group which includes 3 subclasses, cut-and-paste transposons, *helitrons* and *polintons*, that don't have much in common, except that they do not go through an RNA intermediate during transposition (fig. 1b). The cut-and-paste transposons constitute a very diverse group found in all eukaryotic phyla [41]. Cut-and-paste transposons have a very simple structure, containing a single ORF encoding a transposase flanked by terminal inverted repeats (TIRs). The transposase recognizes the TIRs of the element, excises the transposon and inserts it elsewhere in the host genome. At the time of insertion, duplications of the target site are generated. The length and sequence of the target site duplication, terminal motifs in the TIR and similarity in the transposase domain are used to classify cut-and-paste transposons into 15 superfamilies [3], including the widespread *Tc1/mariner, MuDR/Foldback*, *hAT* and *piggyBac* superfamilies. Most of these superfamilies are widely distributed across eukaryotes, suggesting they were already diversified in the ancestor of all eukaryotes. Although cut-and-paste transposons move through a non-replicative mechanism, they can still amplify in the genome of their host by 2 means: (1) if the transposition occurs during replication and if the transposon moved from an already replicated to a non-replicated chromatid; (2) when the element is excised, the repair machinery might use homologous recombination with a chromosome still containing the insertion to repair the gap [41].

The second subclass, called *helitrons*, transpose by rolling-circle transposition, a mechanism of transposition found in some bacterial transposons [42, 43]. They encode a DNA helicase and a nuclease/ligase, they do not have TIRs and do not generate duplication of the target site. They have now been found in most eukaryotic lineages including plants, invertebrates, vertebrates and fungi. The third subclass, *polintons* (also called *Mavericks*), include some of the longest TEs [44, 45]. It was recently suggested that they evolved from a Mavirus virophage [46]. They encode 5 to 9 genes, including a protein-primed polymerase B, an integrase, a cysteine protease and an ATPase. The *polinton* transposition mechanism is called self-synthesizing: the excised copy serves as a template for synthesis of a double strand DNA copy by the DNA polymerase which is then inserted in the genome by the integrase. *Polintons* are also widespread in fungi, vertebrates, invertebrates and protists. The 3 subclasses exist as autonomous families, i.e. families with protein-coding capacities, but also as non-autonomous families [41, 43, 44]. The non-autonomous families are often derived from autonomous elements that have suffered from internal deletions. In cut-and-paste transposons, these shorter copies still possess TIRs that are recognized by the transposase encoded by complete elements and therefore retained their mobility [47]. Non-autonomous families compete with their progenitors for the transposase and often outnumber greatly their autonomous relatives [41, 48, 49].

Mode of Transmission of Transposable Elements

TEs are components of the genome, and as such they are transmitted vertically from parents to offspring. However, since the invasion of the *P* element into populations of *Drosophila melanogaster* was discovered, it has been known that TEs can also be transmitted horizontally among organisms. Vertical and horizontal transmission leaves drastically different signatures. First, the phylogeny of vertically transmitted TEs is identical to the phylogeny of their hosts whereas horizontal transfer will produce conflicting phylogenies. Second, the sequence divergence between vertically transmitted elements in different species should be similar to the background neutral divergence between the host species; horizontally transferred TEs will be less divergent (as they were inserted after the species split from a common ancestor). Third, the presence of horizontally transferred TEs will be patchy within a group, whereas vertically transmitted TEs should be present in all of the descendants of a common ancestor (minus the possibility of stochastic loss of family).

Numerous evidences indicate that non-LTR retrotransposons are transmitted mostly vertically and that horizontal transfer rarely occurs. Malik et al. [4] showed that the level of divergence between non-LTR retrotransposons in distantly related organism was consistent with a strict vertical model of transmission. Several studies based on a large number of *L1* elements have demonstrated that the phylogeny of *L1* in mammals and other deuterostomes recapitulates perfectly the phylogeny of the host, again supporting the vertical transmission of these TEs [50, 51]. However, there are a few cases of horizontal transfer of non-LTR retrotransposons. One of the best documented cases is found in vertebrates where an element belonging to the *RTE* clade has been horizontally transferred from a squamate genome to bovine genomes [52]. More recently, several instances of horizontal transfer of *RTE* in the opossum genome were demonstrated [29]. In their recent review of the topic, Schaack et al. [53] cited 14 cases of horizontal transfer involving non-LTR retrotransposons; that is about 6% of all known instances of horizontal transfer. Interestingly, this is exactly the proportion of non-LTR retrotransposons believed to have been laterally transferred among *Drosophila* genomes [54]. Thus, non-LTR retrotransposons are the least likely TE to be horizontally transferred. There are several reasons why this might be the case which may be related to the mechanism of transposition of non-LTR retrotransposons and to the instability of the mRNA. Another possibility results from the fact that non-LTR retrotransposons are perfectly adapted to their host and might be unable to successfully replicate in another host. It is interesting to note that more than half of the cases cited by Schaak et al. [53] are involving elements lacking ORF1 which is a region suspected to play a role in host-*L1* interactions [7]. The main mode of transmission of LTR retrotransposons is vertical, although horizontal transfer has been documented in plants and *Drosophila* where it is particularly frequent, accounting for more than half of the known cases of lateral transfer [53, 54].

Most cases of horizontal transfer have been documented in DNA transposons, particularly in cut-and-paste transposons but also in *helitrons* [55]. Horizontal transfer has been documented in 8 out of the 15 cut-and-paste superfamilies and seems particularly common in the *hAT* and *mariner* superfamily [53]. Most cases of horizontal transfer have been detected in animals, including insects, but more surprisingly in reptiles and mammals [56, 57]. It was believed for a long time that the sequestration of the germline in tetrapods presented an insurmountable barrier to horizontal transfer. By now, multiple and independent instances of horizontal transfer have been documented. It seems that some species, such as the little brown bat, the tenrec and several squamate reptiles, may be more prone to horizontal transfer than others [56, 58].

Although the exact mechanism of germline colonization is not yet known, the transmission of transposons seems to be mediated by parasites [59] or by viruses [55, 60]. Horizontal transmission seems to be an important feature of DNA transposon evolution and propagation. Although DNA transposons have become stable residents that are transmitted strictly vertically in some taxa, their amplification is sporadic. Their persistence in genomes over long periods of evolutionary time is not the rule, probably because of vertical inactivation. Thus without horizontal transfer, the diversity of DNA transposons in most genomes would be considerably reduced. Another consideration is that cut-and-paste transposons require only the transposase to be mobile and they have been shown to transpose in heterologous species.

Abundance and Diversity of Transposable Elements in Eukaryotes

The TE profile of a particular organism can be described in terms of abundance, defined as the number of copies, and diversity, defined as the number of different types of TEs. The examination of complete genome sequences has revealed huge differences in the abundance and diversity of TEs among groups of eukaryotes but also within these groups.

Among the unicellular eukaryotes that have been sequenced so far, pathogenic and parasitic forms are over-represented, yet the abundance and diversity of TEs in these organisms is extremely variable. The genome of the parasite *Trichomonas vaginalis* is extremely repetitive (75%) and about 39 of its 160 Mb (~24%) are occupied by mobile elements [61]. *T. vaginalis* harbors a wide diversity of repeated elements, including 7 elements of viral origin, 2 retrotransposons and 19 DNA transposons. All but 3 families are represented by relatively small copy numbers (<1,000 copies). Elements within each family are very similar to each other, as all but 3 families have an average pairwise divergence <5%, suggesting that most elements inserted very recently. In contrast to the TE-rich genome of *T. vaginalis*, TEs seem completely absent from the genome of the malaria-causing parasite *Plasmodium falciparum* [62] and represent only 2–5% of the genome of parasitic trypanosomids *(Trypanosoma* and *Leishmania)* [63, 64]. All TEs in trypanosomids are retrotransposons and no DNA transposons have been detected so far in this group. In *Leishmania*, retrotransposons are highly degenerate and no intact copies are found [65]. Thus, it is likely that this genome does not experience any TE activity. In contrast, *Trypanosoma* genomes contain several conserved and potentially active full-length non-LTR retrotransposons, suggesting these elements are still active in this genus.

Large differences in TE profiles can also be observed among closely related protists, even within the same genus. For instance, 19.7% of the genome of *Entamoeba histolytica* is TE-derived, whereas this fraction is 9.7% in *E. dispar* and 9.9% in *E. invadens* [66]. The genome of the 2 human parasites *E. histolytica* and *E. dispar* is dominated by non-LTR retrotransposons, but is almost completely devoid of DNA transposons [66, 67]. In contrast, the genomes of the free-living *E. moshkovskii* and of the reptile parasite *E. invadens* are almost completely devoid of retrotransposons but harbor a large diversity of DNA transposons from the *mutator, Tc1/mariner, piggyBac* and *hAT* clades [67].

The most extreme amplification of TEs has been observed in plants. About 85% of the maize genome *(Zea mays)* is composed of TE-derived sequences. Close to 1,300 TE families have been discovered, including 406 LTR retrotransposons, 31 non-LTR retrotransposons and 855 DNA transposons [68, 69]. The LTR retrotransposons are by far the most abundant types of TEs as *copia* and *gypsy* account for 23.7 and 46.4% of genome size, respectively. In the genus *Oryza* (the domestic rice and its relatives), which is known for large variations in genome size, the amplification of LTR retrotransposons has also occurred in extreme proportion. For instance, the genome of *Oryza australiensis* has doubled during the last 3 Myr due to the amplification of 3 families of LTR retrotransposons [70]. In fact, polyploidization aside, the amplification of LTR retrotransposons accounts for most of the variation in genome size in *Oryza* [71]. Although LTR retrotransposons are the most dominant TEs in *Oryza*, DNA transposons are quite abundant as they account for 4.3–11.9% of genome size, depending on the species [71]. Compared with other plants, the model *Arabidopsis thaliana* contains a relatively small number of TEs, accounting for only 10% of its 125-Mb genome [72]. Yet a diversity of elements is represented in *A. thaliana*, including Class I (2,109 elements) and Class II elements (1,209 elements). Although LTR retrotransposons are the most abundant TEs (~37% of the total), like in other plants, they have not amplified to large numbers with only ~1,600 copies.

In insects, the best-characterized TE profile is the one of *D. melanogaster*, which has been the premier animal model in TE research. TEs represent as much as ~15% of the *D. melanogaster* genome, although they constitute at most 4% of the euchromatic fraction [73]. Interestingly, the abundance of TEs differs among species of the *D. melanogaster* group (e.g. *D. simulans* has only 2% of its euchromatin derived from TEs) and accounts for significant differences in genome size within this group [74]. TEs in *D. melanogaster* are extremely diverse as 93 families of elements, ranging from 1 to 146 copies, have been discovered [73]. All major groups are represented, including the LTR retrotransposons (49 families), non-LTR retrotransposons (27 families) and cut-and-paste DNA transposons (19 families). Elements within families are very similar to each other, suggesting they inserted recently in this genome. A similar level of diversity is found in the genome of the red flour-beetle *Tribolium castaneum* which comprises 6% TE-derived sequences and hosts numerous families of LTR retrotransposons (49), non-LTR retrotransposons (69) and DNA transposons (48) [75]. In contrast, the genome of the honey bee *Apis mellifera* contains a surprisingly small number of TEs (~1% of genome size) [76]. Only *mariner* elements are found in significant number (70 to 390 copies, depending on the family), but no intact copies have been found. At the other end of the insect spectrum, the genome of *Bombyx mori* contains at least 35% of TE-derived sequences [77, 78]. The *B. mori* genome is dominated by Class I elements which represent close to 90% of the elements. More than 50% of the TEs in *B. mori* are the product of a rapid amplification of the *gypsy* clade of LTR retrotransposons beginning 4.9 Myr ago.

TE profiles are also extremely diverse among vertebrates. Fish genomes tend to be more compact than mammalian genomes and this is due, for the most part, to the smaller fraction of their genome occupied by TEs [79]. For instance, the genomes of the 2 pufferfish species *Takifugu rubripes* and *Tetraodon nigroviridis* are, at 400 Mb, among the smallest vertebrate genomes and not more than 1% consists of TEs [80]. Yet, considering their size, these 2 genomes contain an unexpected diversity of TEs as all major categories of TEs are represented. In particular non-LTR retrotransposons are represented by multiple clades, each of them with multiple families [81]. This level of non-LTR retrotransposon diversity is found in all teleostean fish analyzed so far [82]. In the zebrafish *Danio rerio*, the *L1* clade of non-LTR retrotransposons is represented by more than 30 distinct lineages. Each of these lineages contains a small number (<100 copies) of very similar elements, reminiscent of the situation in *Drosophila* [15, 16].

The genome of the frog *Xenopus tropicalis* is unique among vertebrates as it is dominated by Class II transposons (25% of genome size) and not by Class I transposons (9% of the genome) [83]. DNA transposons are represented by 5 prolific clades *(Kolobok, hAT, Harbinger, mariner* and *piggyBac)* as well as *helitrons* and *polintons*. LTR retrotransposons are more diverse in *Xenopus* than in any other extant vertebrates, but non-LTR retrotransposons are surprisingly rare as they account for less than 5% of the genome. However, non-LTR retrotransposons are represented in the frog by multiple divergent families of similar elements, suggestive of their recent activity [50, 83].

The genome of the lizard *Anolis carolinensis* is similar to fish genomes as it contains a very large diversity of non-LTR retrotransposon clades *(L1, L2, CR1, RTE* and *R4)* and families, each represented by a small number of very similar elements [17, 84]. For instance, 20 *L1* and 17 *L2* families are found in this genome and all but 2 have a level of divergence <2%, indicating that most elements inserted very recently. The *Anolis* genome also contains a wide diversity of recently active DNA transposons from the *mariner, hAT* and *helitron* clades [48] as well as a number of LTR retrotransposons from the Metaviridae and Pseudoviridae families [85]. Although *Anolis* is the only reptile genome sequenced so far, comparison of the repetitive fraction in several reptilian genomes suggests that the diversity and dynamics of TEs might differ drastically among squamate reptiles [85, 86]. The genome of the chicken is unusual as it appears to lack any TE activity [87]. Multiple lineages of the *CR1* non-LTR retrotransposons were once active in chicken, but there is no evidence that this clade, or any other TE, might still be active.

Mammalian genomes tend to be large and dominated by non-LTR retrotransposons and their SINE relatives. This is true of monotremes, metatherians (marsupials) and eutherians (placentals). The only difference between these 3 groups is that placental and marsupial genomes are dominated by the *L1* clade [18, 19, 29, 88] whereas monotremes are dominated by the *L2* clade [23], otherwise they have very similar TE profiles. Among eutherians, the most studied genome is arguably the human genome. The human genome is dominated by the *L1* clade, which accounts for at least 21% of the genome mass, and its non-autonomous counterpart, *Alu*, which has amplified to more than 1 million copies. This extraordinarily large number of *L1* elements results from the activity of a single lineage of family that has been active since the origin of eutherians. Although the TE diversity of the human genome has been limited to *L1* and *Alu* for some time, this was not always the case. DNA transposons were once very active and diverse in ancestral primates and became extinct ~37 Myr ago [89]. LTR retrotransposons have long been extinct in human, but endogenous retroviruses have been quite active in mammalian and primate evolution as they account for 8% of genome size, but apparently they recently became extinct [90, 91].

Other eutherian genomes are also dominated by *L1* and generally resemble the human genome, but some groups deviate significantly from the human model. First, the germline of some species has been colonized recently by retroviruses that became endogenized and subsequently produced a large number of copies, adding to the TE diversity in these taxa. Some of these elements have been extremely successful, such as the *MysTR* element which amplified to 10,000 copies in the rice rat *Oryzomys palustris* [92] and the *IAP* element which is one of the most active TEs in mouse [35]. Second, some mammalian genomes have been colonized by laterally-transferred TEs which subsequently amplified to large numbers (see above). This is the case of bats that have been invaded by several types of DNA transposons [56, 93] and of cows that host an *RTE* family related to a reptile element [52]. Third, some genomes have lost *L1* activity. *L1* activity is known to be sporadic, waves of amplification alternating with periods of low activity [19], but in a very small number of mammalian lineages *L1* elements seem to be completely extinct [94, 95].

What have we learned from this comparison of eukaryote genomes? The first observation is that there is no clear trend: each of the major groups examined (protists, plants, insects and vertebrates) contains species with large numbers of TEs and species with few or no TEs, and species with a large diversity of TEs as well as species with a limited diversity of TEs. There is no evolutionary trend toward larger genomes or more TE-rich genomes. Second, there is no relation between the complexity of organisms and their TE repertoire, suggesting that TE abundance and TE diversity are unlikely to be adaptive traits, although specific insertions might very well be adaptive (see below). Third, rapid increases in genome size usually result from the amplification of a single type of TE (LTR retrotransposons in plants or *L1* in eutherians) and not by the amplification of multiple families.

The Impact of Natural Selection on Transposable Elements

The number of TE copies in a genome is dependent on the rate at which new insertions are generated, i.e. the rate of transposition, and the rate at which insertions accumulate in the genome, i.e. the rate of fixation. The fixation (or loss) of a TE insertion is dependent on its effect on the fitness of the host. If the insertion is deleterious to the host, the insertion will have a lower chance of fixation and will most likely be eliminated from the gene pool by purifying selection. The deleterious effect of TEs has been recognized since the 1970s when a phenomenon called hybrid dysgenesis was discovered in *D. melanogaster*. When females from strains of *D. melanogaster* lacking a DNA transposon called the *P* element are crossed with males that carry *P* elements, the resulting progeny is sterile and suffers from an increase in germline transposition and an elevated mutation and recombination rate [96, 97]. Since the discovery of hybrid dysgenesis, numerous evidences suggesting a negative impact of TEs on host fitness have been described. For instance, the deleteriousness of TEs is apparent when comparing genomes that differ in TE copy number. Pasyukova et al. [98] compared the fitness and egg hatchability among strains of *D. melanogaster* that differ by the number of TE copies. They found that, consistent with a negative impact of TEs, the strains with a larger number of insertions have a lower fitness.

In *Drosophila*, early surveys of insertion site polymorphisms in both natural populations and among strains revealed that the majority of TE insertions are at low frequency in populations and that fixed elements are rare [99–104]. This pattern suggests that in *Drosophila* TEs go through a rapid turn-over of elements, in which the insertion of new elements is offset by the selective loss of element-containing loci [100, 105–107]. In humans, population genetics studies have shown that the majority of *L1* elements behave as neutral alleles and accumulate readily in the genome of their host. This does not mean that *L1* activity is fully neutral. A fitness cost related to the length of *L1* elements has been demonstrated, yet it is insufficient to prevent the fixation of most elements, hence the extremely large number of copies in mammals [108–110]. The genomic distribution of TEs in *Drosophila* and human is also consistent with a deleterious effect of TE insertions. TEs tend to be more abundant in regions of low recombination or no recombination [109–112] because low-recombining regions tend to accumulate deleterious mutations due to Hill-Robertson interactions [113] or because elements in those low-recombining regions are less likely to be deleterious by mediating ectopic recombination events (see below).

Although it is widely accepted that TEs are indeed deleterious to their host, the basis for the deleterious effect of TEs has long been a matter of debate. Three non-exclusive deleterious effects of TEs have been proposed: (1) the direct effect of where elements insert (e.g. gene inactivation); (2) the effect of genetic rearrangements caused by ectopic (non-allelic) recombination between copies; (3) the effect of the transposition process per se. There is no doubt that TE insertions can indeed be deleterious when inserted into genes or even in introns, as suggested by the more than 60 disease-causing insertions reported in human (reviewed in [114]). It is also unquestionable that recombination between non-allelic TE insertions can create chromosomal rearrangements and large genomic deletions [115] which are very likely to be deleterious. Furthermore, the transposition process per se can be deleterious, for instance when the endonuclease of *L1* retrotransposons makes double strand breaks in the host genome [116]. The question remains: which one of these mechanisms affects the dynamics of TEs in natural populations the most?

Although the issue is still debated, some recent population genetics results suggest an important role of ectopic recombination. As longer elements are more likely to be involved in ectopic recombination than shorter ones, they are more likely to be deleterious and eliminated by purifying selection. In addition, elements that belong to large families are more likely to find a partner for ectopic recombination than families with small copy numbers, thus larger families should be more deleterious than smaller ones. This is exactly what Petrov et al. [117, 118] tested using a population genetics approach. They found that, as predicted by the ectopic recombination model, selection against insertions was length-dependent and copy number-dependent. Long elements segregated in *Drosophila* populations at lower frequency than short elements and many elements belonging to small copy number families reached fixation in the *Drosophila* genome. Using the same approach, it was shown that the longest *L1* elements in the human genome were at lower frequency in human populations than the more truncated ones [108]. This result suggests that full-length elements are imposing a genetic load on their human host, but that short, truncated elements behaved like neutral alleles, thus supporting the ectopic exchange model.

A second line of evidence in favor of the ectopic recombination model comes from the distribution of TEs across the genome. In *Drosophila* and in human, TEs accumulate in low-recombining regions of the genome, possibly because insertions in low- or non-recombining regions are less likely to be involved in non-allelic recombination. However, deleterious mutations are expected to accumulate in low-recombining regions of the genome whatever the nature of their deleterious effect is because of Hill-Robertson interactions. Simulation studies show that the accumulation of deleterious TEs caused by Hill-Robertson interactions would, however, occur only in populations of very small size and in regions of very low recombination [119]. These conditions are very restrictive and suggest that the ectopic exchange model provides a more general and likely explanation to the biased genomic distribution of TEs. In addition, a careful examination of the human genome revealed that the distribution bias of *L1* elements was also length-dependent. Short, truncated elements (<1.2 kb) are more homogeneously distributed across the genome than elements longer than 1.2 kb, which tend to accumulate in low-recombining regions [109, 110].

Although selection is a powerful force limiting the spread of TEs, positive selection could also be acting on some insertions, thus increasing their chance of fixation and their copy number. TEs can be an important source of evolutionary novelties, either by affecting the expression of host genes, by participating in regulatory networks, by incorporating into coding sequences or by creating new genes [120–123]. Some of these domestication events have had a very significant impact on the evolutionary fate of their hosts. For instance, a large number of regulatory sequences in vertebrates are derived from TEs and certainly had a profound impact on the evolution of this group [124, 125]. In *Drosophila*, some insertions have undoubtedly been positively selected as they confer resistance to insecticide [126]. However, one might wonder if positive selection in favor of TE insertions is strong enough and occurs often enough to affect the abundance and diversity of TEs in eukaryote genomes. A recent screen of *Drosophila* populations outside of Africa identified at least 13 insertions that could have been under positive selection [127], although the actual number of adaptive insertions is probably between 25 and 50. This is a remarkably large number considering that *D. melanogaster* has left Africa only 10 to 16,000 years ago. Despite this high level of TE-related adaptation the *Drosophila* genome contains, as described earlier, mostly low-frequency insertions and the number of fixed insertions is relatively small. Thus, adaptive insertions, as crucial as they might be as a potent source of evolutionary novelty, seem to be too rare to have a significant effect on copy number. In fact, for positive selection to significantly increase copy number, it would require a massive amount of favorable alleles or a general advantageous function provided by TEs. To our knowledge there are only 2 cases that would fit this model. Some elements in *Drosophila* have an important role in the maintenance of telomeres and in fact are the main constituents of *Drosophila* telomeres. *Drosophila* lacks telomerase and their telomeres are actually composed of 3 non-LTR retrotransposons, named *HeT-A, TART* and *TAHRE*, that specifically insert at the tips of chromosomes and reach significant copy number because of their function at maintaining chromosome integrity [128]. Second, selection seems to be favoring the fixation of *L1* elements on the eutherian X chromosome. It was proposed that *L1* can act as a booster element that would facilitate the inactivation of the X chromosome [129]. Indeed, the X chromosomes of all mammals, except the opossum that lacks a *Xist* homolog [124], are enriched in *L1* elements. It is possible that this accumulation of *L1* on the X is due to a favorable effect of *L1* insertion, related to a functional role of these elements in X inactivation.

The Impact of Host Demography and Life History on Transposable Elements

As TEs are obligatory parasites, their dynamics in the genome is affected by the evolution and natural history of their host. In particular, any factor that affects the effective population size *(N_e)* of the host will modify the equilibrium between drift and selection. When N_e is large, selection dominates over drift, but any factor that decreases N_e (e.g. bottleneck, mating system) will strengthen drift. Thus, one would expect that in populations with large N_e, selection would be more efficient at removing deleterious TE insertions than in small populations in which the rate of fixation will be higher. This can have far-reaching consequences in terms of genomic evolution because the long-term accumulation of insertions will lead to an increase in genome size, whereas one expects species with large population size to have smaller genomes. This hypothesis was supported by a study by Lynch and Conery [1], who compared N_e and the genome size of a wide range of organisms. They found that species with very large population size tend to have smaller genomes than species with small population size, thus supporting an important role of non-adaptive demographic factors in shaping genome size evolution. However, a more recent analysis limited to plants [130] failed to find such a correlation and more comparative analysis will probably be required to settle this issue.

An effect of drift on the rate of fixation of TEs has also been examined using a population genetics approach. In *Drosophila subobscura* [131] and in *Arabidopsis lyrata* [132], TE insertions are found at higher frequency in bottlenecked populations than in populations that have retained a large long-term population size, an observation consistent with a reduced efficacy of purifying selection. In fact, the frequency distribution of TEs in strongly bottlenecked *A. lyrata* populations is consistent with neutrality [132]. Similarly, it was shown that the strength of selection against TE insertions was strongly reduced in populations of *D. melanogaster* that emigrated out of Africa [133]. Thus, populations subjected to strong genetic drift tend to accumulate TE insertions whereas large populations will eliminate them.

Another factor that affects N_e is the mating system. For instance, inbred species face a reduced N_e relative to outbred species; consequently, the efficacy of selection should also be reduced in these taxa. In addition, inbreeding will result in an increase in homozygosity. As an element is more likely to be involved in ectopic recombination in the heterozygous state, selection against insertions will be more efficient in outcrossing species than in inbred ones. Thus, it is predicted that inbred populations will carry more fixed TEs than outcrossing ones. This was tested by comparing the selfing worm species *Caenorhabditis elegans* with its outcrossing relative *C. remanei* [134] and the self-fertilizing *A. thaliana* with the outcrossing *A. lyrata* [135, 136]. In both comparisons, TE insertions were at higher frequency in the selfing species, consistent with a reduced efficiency of selection. The mating system also affected the patterns of genomic distribution. Unlike what was found in *Drosophila* and human, TEs are not more abundant in low-recombining regions in self-fertilizing *A. thaliana* [137] and *C. elegans* [138], suggesting a lack of genome-wide purifying selection. This is not to say that TEs can be considered purely neutral. The density of TEs is lower in or near host genes, suggesting a deleterious effect of TE insertions related to disruption of normal gene function.

Post-Insertional Control of Transposon Copy Number

Eukaryote genomes have hosted TEs since their origin and we can observe numerous TEs that are fixed in their host species. Since it is unlikely that any eukaryotic lineage has continuously evolved under demographic conditions consistently conducive to the elimination of TEs by selection (such as large, stable population size and random mating), there exists a cumulative effect over time in which complex demographic histories give TEs a chance to become fixed in the genome. Thus, we should expect eukaryote genomes to be much larger than they actually are. The reason we don't is that, once inserted and fixed, TEs decay over time and accumulate deletions until eventually the elements are no longer present. This DNA loss will prevent the unlimited accumulation of TEs, counteracting the expansion of the genome through transposition.

In *Arabidopsis* and in rice, LTR retrotransposons decay rapidly and it was estimated that the half-life of an LTR retrotransposon sequence in rice is less than 6 Myr [139, 140]. Thus, the abundance of young elements in plants is as much a reflection of recent expansion as it is a reflection of a rapid rate of DNA loss. In plants, the main mechanism responsible for DNA loss is illegitimate recombination, resulting in severely truncated TEs [140]. A similar mechanism might be at play in lizards [17] where TEs seem to decay very rapidly due to large deletions that encompass the ends of insertions. An interesting observation made in plants and insects is that the rate of DNA loss varies among species and could account for difference in the abundance of TEs among genomes. The rate of deletion in *Drosophila* is twice as high as the rate in the Hawaiian crickets of the genus *Laupala* which are characterized by large genomes, and the size of these deletions is on average 4 times larger in *Drosophila* [141]. Thus, in addition to the strong effect of purifying selection against TE insertions, the *Drosophila* genome is subject to a high rate of DNA loss. Together these 2 processes account for the relatively small genome of *Drosophila*. Surprisingly, significant differences in the rate of DNA loss can occur even within the same genus as demonstrated in the plant genus *Gossypium* [142]. Thus genome size differences seem to result from interspecific differences in both the rate of fixation of TEs and the rate of elimination of DNA.

Conclusion

In this chapter we have reviewed the impact of population genetic forces, drift and selection on the dynamics of TEs. One aspect we have not examined here is the variation in the rate of transposition. It is well known that the rate at which TEs transpose varies considerably through time and among lineages and consequently has a profound impact on the TE profile. However, this aspect of TE dynamics is the least understood. In the past decade, a number of mechanisms controlling the rate of transposition have been uncovered, including DNA methylation, RNA interference, repeat induced point mutation and post-translational gene silencing. In addition, some TEs have evolved the ability to regulate their own transposition, possibly to limit their deleterious impact on the host. Even demography can indirectly act on the rate of transposition, if selection fails to remove the most active element from the populations. All these factors certainly affect the rate of transposition, but how the regulation of transposition impacts the dynamics of transposition in natural populations remains one of the least understood aspects of TE biology and will require further studies.

References

1 Lynch M, Conery JS: The origins of genome complexity. Science 2003;302:1401–1404.

2 Arkhipova IR: Distribution and phylogeny of Penelope-like elements in eukaryotes. Syst Biol 2006;55:875–885.

3 Kapitonov VV, Tempel S, Jurka J: Simple and fast classification of non-LTR retrotransposons based on phylogeny of their RT domain protein sequences. Gene 2009;448:207–213.

4 Malik HS, Burke WD, Eickbush TH: The age and evolution of non-LTR retrotransposable elements. Mol Biol Evol 1999;16:793–805.

5 Luan DD, Korman MH, Jakubczak JL, Eickbush TH: Reverse transcription of R2Bm RNA is primed by a nick at the chromosomal target site: a mechanism for non-LTR retrotransposition. Cell 1993;72:595–605.

6 Khazina E, Weichenrieder O: Non-LTR retrotransposons encode noncanonical RRM domains in their first open reading frame. Proc Natl Acad Sci USA 2009;106:731–736.

7 Boissinot S, Furano AV: Adaptive evolution in LINE-1 retrotransposons. Mol Biol Evol 2001;18:2186–2194.

8 Martin SL, Branciforte D, Keller D, Bain DL: Trimeric structure for an essential protein in L1 retrotransposition. Proc Natl Acad Sci USA 2003;100:13815–13820.

9 Martin SL: Ribonucleoprotein particles with LINE-1 RNA in mouse embryonal carcinoma cells. Mol Cell Biol 1991;11:4804–4807.

10 Martin SL, Bushman FD: Nucleic acid chaperone activity of the ORF1 protein from the mouse LINE-1 retrotransposon. Mol Cell Biol 2001;21:467–475.

11 Swergold GD: Identification, characterization, and cell specificity of a human LINE-1 promoter. Mol Cell Biol 1990;10:6718–6729.

12 Cost GJ, Feng Q, Jacquier A, Boeke JD: Human L1 element target-primed reverse transcription in vitro. EMBO J 2002;21:5899–5910.

13 Martin SL, Li WL, Furano AV, Boissinot S: The structures of mouse and human L1 elements reflect their insertion mechanism. Cytogenet Genome Res 2005;110:223–228.

14 Ichiyanagi K, Okada N: Mobility pathways for vertebrate L1, L2, CR1, and RTE clade retrotransposons. Mol Biol Evol 2008;25:1148–1157.

15 Duvernell DD, Pryor SR, Adams SM: Teleost fish genomes contain a diverse array of L1 retrotransposon lineages that exhibit a low copy number and high rate of turnover. J Mol Evol 2004;59:298–308.

16 Furano AV, Duvernell D, Boissinot: L1 (LINE-1) retrotransposon diversity differs dramatically between mammals and fish. Trends Genet 2004;20:9–14.

17 Novick PA, Basta H, Floumanhaft M, McClure MA, Boissinot S: The evolutionary dynamics of autonomous non-LTR retrotransposons in the lizard *Anolis carolinensis* shows more similarity to fish than mammals. Mol Biol Evol 2009;26:1811–1822.

18 Lander ES, Linton LM, Birren B, Nusbaum C, Zody MC: Initial sequencing and analysis of the human genome. Nature 2001;409:860–921.

19 Furano AV: The biological properties and evolutionary dynamics of mammalian LINE-1 retrotransposons. Prog Nucleic Acid Res Mol Biol 2000;64:255–294.

20 Adey NB, Schichman SA, Graham DK, Peterson SN, Edgell MH, Hutchison CA 3rd: Rodent L1 evolution has been driven by a single dominant lineage that has repeatedly acquired new transcriptional regulatory sequences. Mol Biol Evol 1994;11:778–789.

21 Boissinot S, Chevret P, Furano AV: L1 (LINE-1) retrotransposon evolution and amplification in recent human history. Mol Biol Evol 2000;17:915–928.

22 Khan H, Smit A, Boissinot S: Molecular evolution and tempo of amplification of human LINE-1 retrotransposons since the origin of primates. Genome Res 2006;16:78–87.
23 Warren WC, Hillier LW, Marshall Graves JA, Birney E, Ponting CP, et al: Genome analysis of the platypus reveals unique signatures of evolution. Nature 2008;453:175–183.
24 Wei W, Gilbert N, Ooi SL, Lawler JF, Ostertag EM, et al: Human L1 retrotransposition: *cis* preference versus *trans* complementation. Mol Cell Biol 2001; 21:1429–1439.
25 Dewannieux M, Esnault C, Heidmann T: LINE-mediated retrotransposition of marked Alu sequences. Nat Genet 2003;35:41–48.
26 Ohshima K, Hamada M, Terai Y, Okada N: The 3′ ends of tRNA-derived short interspersed repetitive elements are derived from the 3′ ends of long interspersed repetitive elements. Mol Cell Biol 1996;16: 3756–3764.
27 Piskurek O, Nishihara H, Okada N: The evolution of two partner LINE/SINE families and a full-length chromodomain-containing Ty3/Gypsy LTR element in the first reptilian genome of *Anolis carolinensis.* Gene 2009;441:111–118.
28 Piskurek O, Austin CC, Okada N: Sauria SINEs: Novel short interspersed retroposable elements that are widespread in reptile genomes. J Mol Evol 2006; 62:630–644.
29 Gentles AJ, Wakefield MJ, Kohany O, Gu W, Batzer MA, et al: Evolutionary dynamics of transposable elements in the short-tailed opossum *Monodelphis domestica.* Genome Res 2007;17:992–1004.
30 Eickbush TH, Jamburuthugoda VK: The diversity of retrotransposons and the properties of their reverse transcriptases. Virus Res 2008;134:221–234.
31 Havecker ER, Gao X, Voytas DF: The diversity of LTR retrotransposons. Genome Biol 2004;5:225.
32 Malik HS, Eickbush TH: Phylogenetic analysis of ribonuclease H domains suggests a late, chimeric origin of LTR retrotransposable elements and retroviruses. Genome Res 2001;11:1187–1197.
33 Kim A, Terzian C, Santamaria P, Pélisson A, Purd'homme M, Bucheton A: Retroviruses in invertebrates: the gypsy retrotransposon is apparently an infectious retrovirus of *Drosophila melanogaster.* Proc Natl Acad Sci USA 1994;91:1285–1289.
34 Song SU, Gerasimova T, Kurkulos M, Boeke JD, Corces VG: An env-like protein encoded by a *Drosophila* retroelement: evidence that gypsy is an infectious retrovirus. Genes Dev 1994;8:2046–2057.
35 Ribet D, Harper F, Dupressoir A, Dewannieux M, Pierron G, Heidmann T: An infectious progenitor for the murine IAP retrotransposon: emergence of an intracellular genetic parasite from an ancient retrovirus. Genome Res 2008;18:597–609.
36 Gifford R, Tristem M: The evolution, distribution and diversity of endogenous retroviruses. Virus Genes 2003;26:291–315.
37 Andersson ML, Lindeskog M, Medstrand P, Westley B, May F, Blomberg J: Diversity of human endogenous retrovirus class II-like sequences. J Gen Virol 1999;80(Pt 1):255–260.
38 Stuart-Rogers C, Flavell AJ: The evolution of Ty1-*copia* group retrotransposons in gymnosperms. Mol Biol Evol 2001;18:155–163.
39 Goodwin TJ, Poulter RT: The DIRS1 group of retrotransposons. Mol Biol Evol 2001;18:2067–2082.
40 Jiang N, Bao Z, Temnykh S, Cheng Z, Jiang J, et al: *Dasheng*: a recently amplified nonautonomous long terminal repeat element that is a major component of pericentromeric regions in rice. Genetics 2002; 161:1293–1305.
41 Feschotte C, Pritham EJ: DNA transposons and the evolution of eukaryotic genomes. Annu Rev Genet 2007;41:331–368.
42 Kapitonov VV, Jurka J: *Helitrons* on a roll: eukaryotic rolling-circle transposons. Trends Genet 2007;23:521–529.
43 Kapitonov VV, Jurka J: Rolling-circle transposons in eukaryotes. Proc Natl Acad Sci USA 2001;98:8714–8719.
44 Kapitonov VV, Jurka J: Self-synthesizing DNA transposons in eukaryotes. Proc Natl Acad Sci USA 2006;103:4540–4545.
45 Pritham EJ, Putliwala T, Feschotte C: *Mavericks*, a novel class of giant transposable elements widespread in eukaryotes and related to DNA viruses. Gene 2007;390:3–17.
46 Fischer MG, Suttle CA: A virophage at the origin of large DNA transposons. Science 2011;332:231–234.
47 Hartl DL, Lozovskaya ER, Lawrence JG: Nonautonomous transposable elements in prokaryotes and eukaryotes. Genetica 1992;86:47–53.
48 Novick PA, Smith JD, Floumanhaft M, Ray DA, Boissinot S: The evolution and diversity of DNA transposons in the genome of the lizard *Anolis carolinensis.* Genome Biol Evol 2011;3:1–14.
49 Yang G, Nagel DH, Feschotte C, Hancock CN, Wessler SR: Tuned for transposition: molecular determinants underlying the hyperactivity of a *Stowaway* MITE. Science 2009;325:1391–1394.
50 Kordis D, Lovsin N, Gubensek F: Phylogenomic analysis of the L1 retrotransposons in Deuterostomia. Syst Biol 2006;55:886–901.

51 Waters PD, Dobigny G, Waddell PJ, Robinson TJ: Evolutionary history of LINE-1 in the major clades of placental mammals. PLoS One 2007;2:e158.

52 Kordis D, Gubensek F: Unusual horizontal transfer of a long interspersed nuclear element between distant vertebrate classes. Proc Natl Acad Sci USA 1998;95:10704–10709.

53 Schaack S, Gilbert C, Feschotte C: Promiscuous DNA: horizontal transfer of transposable elements and why it matters for eukaryotic evolution. Trends Ecol Evol 2010;25:537–546.

54 Bartolome C, Bello X, Maside X: Widespread evidence for horizontal transfer of transposable elements across *Drosophila* genomes. Genome Biol 2009;10:R22.

55 Thomas J, Schaack S, Pritham EJ: Pervasive horizontal transfer of rolling-circle transposons among animals. Genome Biol Evol 2010;2:656–664.

56 Pace JK 2nd, Gilbert C, Clark MS, Feschotte C: Repeated horizontal transfer of a DNA transposon in mammals and other tetrapods. Proc Natl Acad Sci USA 2008;105:17023–17028.

57 Gilbert C, Hernandez SS, Flores-Benabib J, Smith EN, Feschotte C: Rampant horizontal transfer of *SPIN* transposons in squamate reptiles. Mol Biol Evol 2011 [Epub ahead of print].

58 Novick P, Smith J, Ray D, Boissinot S: Independent and parallel lateral transfer of DNA transposons in tetrapod genomes. Gene 2010;449:85–94.

59 Gilbert C, Schaack S, Pace JK 2nd, Brindley PJ, Feschotte C: A role for host-parasite interactions in the horizontal transfer of transposons across phyla. Nature 2010;464:1347–1350.

60 Piskurek O, Okada N: Poxviruses as possible vectors for horizontal transfer of retroposons from reptiles to mammals. Proc Natl Acad Sci USA 2007;104: 12046–12051.

61 Carlton JM, Hirt RP, Silva JC, Delcher AL, Schatz M, et al: Draft genome sequence of the sexually transmitted pathogen *Trichomonas vaginalis.* Science 2007;315:207–212.

62 Gardner MJ, Hall N, Fung E, White O, Berriman M, et al: Genome sequence of the human malaria parasite *Plasmodium falciparum.* Nature 2002;419:498–511.

63 Bringaud F, Ghedin E, El-Sayed NM, Papadopoulou B: Role of transposable elements in trypanosomatids. Microbes Infect 2008;10:575–581.

64 Wickstead B, Ersfeld K, Gull K: Repetitive elements in genomes of parasitic protozoa. Microbiol Mol Biol Rev 2003;67:360–375.

65 Bringaud F, Ghedin E, Blandin G, Bartholomeu DC, Caler E, et al: Evolution of non-LTR retrotransposons in the trypanosomatid genomes: *Leishmania major* has lost the active elements. Mol Biochem Parasitol 2006;145:158–170.

66 Lorenzi H, Thiagarajan M, Haas B, Wortman J, Hall N, Caler E: Genome wide survey, discovery and evolution of repetitive elements in three *Entamoeba* species. BMC Genomics 2008;9:595.

67 Pritham EJ, Feschotte C, Wessler SR: Unexpected diversity and differential success of DNA transposons in four species of *Entamoeba* protozoans. Mol Biol Evol 2005;22:1751–1763.

68 Schnable PS, Ware D, Fulton RS, Stein JC, Wei F, et al: The B73 maize genome: complexity, diversity, and dynamics. Science 2009;326:1112–1115.

69 Baucom RS, Estill JC, Chaparro C, Upshaw N, Jogi A, et al: Exceptional diversity, non-random distribution, and rapid evolution of retroelements in the B73 maize genome. PLoS Genet 2009;5:e1000732.

70 Piegu B, Guyot R, Picault N, Roulin A, Sanyal A, et al: Doubling genome size without polyploidization: dynamics of retrotransposition-driven genomic expansions in *Oryza australiensis*, a wild relative of rice. Genome Res 2006;16:1262–1269.

71 Zuccolo A, Sebastian A, Talag J, Yu Y, Kim H, et al: Transposable element distribution, abundance and role in genome size variation in the genus *Oryza.* BMC Evol Biol 2007;7:152.

72 Arabidopsis Genome Initiative: Analysis of the genome sequence of the flowering plant *Arabidopsis thaliana.* Nature 2000;408:796–815.

73 Kaminker JS, Bergman CM, Kronmiller B, Carlson J, Svirskas R, et al: The transposable elements of the *Drosophila melanogaster* euchromatin: a genomics perspective. Genome Biol 2002;3:RESEARCH0084.

74 Lerat E, Burlet N, Biémont C, Vieira C: Comparative analysis of transposable elements in the melanogaster subgroup sequenced genomes. Gene 2011; 473:100–109.

75 Tribolium Genome Sequencing Consortium, Richards S, Gibbs RA, Weinstock GM, Brown SJ, et al: The genome of the model beetle and pest *Tribolium castaneum.* Nature 2008;452:949–955.

76 Honeybee Genome Sequencing Consortium: Insights into social insects from the genome of the honeybee *Apis mellifera.* Nature 2006;443:931–949.

77 Xia Q, Zhou Z, Lu C, Cheng D, Dai F, et al: A draft sequence for the genome of the domesticated silkworm *(Bombyx mori).* Science 2004;306:1937–1340.

78 Osanai-Futahashi M, Suetsugu Y, Mita K, Fujiwara H: Genome-wide screening and characterization of transposable elements and their distribution analysis in the silkworm, *Bombyx mori*. Insect Biochem Mol Biol 2008;38:1046–1057.

79 Volff JN: Genome evolution and biodiversity in teleost fish. Heredity 2005;94:280–294.

80 Roest Crollius H, Jaillon O, Dasilva C, Ozouf-Costaz C, Fizames C, et al: Characterization and repeat analysis of the compact genome of the freshwater pufferfish *Tetraodon nigroviridis*. Genome Res 2000; 10:939–949.

81 Volff JN, Bouneau L, Ozouf-Costaz C, Fischer C: Diversity of retrotransposable elements in compact pufferfish genomes. Trends Genet 2003;19:674–678.

82 Basta HA, Buzak AJ, McClure MA: Identification of novel retroid agents in *Danio rerio, Oryzias latipes, Gasterosteus aculeatus* and *Tetraodon nigroviridis*. Evol Bioinform Online 2007;3:179–195.

83 Hellsten U, Harland RM, Gilchrist MJ, Hendrix D, Jurka J, et al: The genome of the Western clawed frog *Xenopus tropicalis*. Science 2010;328:633–636.

84 Alföldi J, Di Palma F, Grabherr M, Williams C, Kong L, et al: The genome of the green anole lizard and a comparative analysis with birds and mammals. Nature 2011;477:587–591.

85 Kordis D: Transposable elements in reptilian and avian (Sauropsida) genomes. Cytogenet Genome Res 2009;127:94–111.

86 Castoe TA, Hall KT, Guibotsy Mboulas ML, Gu W, de Koning AP, et al: Discovery of highly divergent repeat landscapes in snake genomes using high throughput sequencing. Genome Biol Evol 2011;3: 641–653.

87 International Chicken Genome Sequencing Consortium: Sequence and comparative analysis of the chicken genome provide unique perspectives on vertebrate evolution. Nature 2004;432:695–716.

88 Mouse Genome Sequencing Consortium, Waterston RH, Lindblad-Toh K, Birney E, Rogers J, et al: Initial sequencing and comparative analysis of the mouse genome. Nature 2002;420:520–562.

89 Pace JK 2nd, Feschotte C: The evolutionary history of human DNA transposons: evidence for intense activity in the primate lineage. Genome Res 2007; 17:422–432.

90 Barbulescu M, Turner G, Seaman MI, Deinard AS, Kidd KK, Lenz J: Many human endogenous retrovirus K (HERV-K) proviruses are unique to humans. Curr Biol 1999;9:861–868.

91 Turner G, Barbulescu M, Su M, Jensen-Seaman MI, Kidd KK, Lenz J: Insertional polymorphisms of full-length endogenous retroviruses in humans. Curr Biol 2001;11:1531–1535.

92 Cantrell MA, Ederer MM, Erickson IK, Swier VJ, Baker RJ, Wichman HA: MysTR: an endogenous retrovirus family in mammals that is undergoing recent amplifications to unprecedented copy numbers. J Virol 2005;79:14698–14707.

93 Ray DA, Feschotte C, Pagan HJ, Smith JD, Pritham EJ, et al: Multiple waves of recent DNA transposon activity in the bat, *Myotis lucifugus*. Genome Res 2008;18:717–728.

94 Casavant NC, Scott L, Cantrell MA, Wiggins LE, Baker RJ, Wichman HA: The end of the LINE?: lack of recent L1 activity in a group of South American rodents. Genetics 2000;154:1809–1817.

95 Cantrell MA, Scott L, Brown CJ, Martinez AR, Wichman HA: Loss of LINE-1 activity in the megabats. Genetics 2008;178:393–404.

96 Bingham PM, Kidwell MG, Rubin GM: The molecular basis of P-M hybrid dysgenesis: the role of the P element, a P-strain-specific transposon family. Cell 1982;29:995–1004.

97 Schaefer RE, Kidwell MG, Fausto-Sterling A: Hybrid dysgenesis in *Drosophila melanogaster:* morphological and cytological studies of ovarian dysgenesis. Genetics 1979;92:1141–1152.

98 Pasyukova EG, Nuzhdin SV, Morozova TV, Mackay TF: Accumulation of transposable elements in the genome of *Drosophila melanogaster* is associated with a decrease in fitness. J Hered 2004;95:284–290.

99 Biemont C, Lemeunier F, Garcia Guerreiro MP, Brookfield JF, Gautier C, et al: Population dynamics of the copia, mdg1, mdg3, gypsy, and P transposable elements in a natural population of *Drosophila melanogaster*. Genet Res 1994;63:197–212.

100 Charlesworth B, Charlesworth D: The population dynamics of transposable elements. Genet Res 1983; 42:1–27.

101 Charlesworth B, Langley CH: The population genetics of *Drosophila* transposable elements. Annu Rev Genet 1989;23:251–287.

102 Charlesworth B, Lapid A, Canada D: The distribution of transposable elements within and between chromosomes in a population of *Drosophila melanogaster*. II. Inferences on the nature of selection against elements. Genet Res 1992;60:115–130.

103 Nuzhdin SV, Mackay TF: The genomic rate of transposable element movement in *Drosophila melanogaster*. Mol Biol Evol 1995;12:180–181.

104 Biemont C, Vieira C, Hoogland C, Cizeron G, Loevenbruck C, et al: Maintenance of transposable element copy number in natural populations of *Drosophila melanogaster* and *D. simulans*. Genetica 1997;100:161–166.

105 Montgomery E, Charlesworth B, Langley CH: A test for the role of natural selection in the stabilization of transposable element copy number in a population of *Drosophila melanogaster.* Genet Res 1987; 49:31–41.

106 Kaplan NL, Brookfield JF: Transposable elements in mendelian populations. III. Statistical results. Genetics 1983;104:485–495.

107 Langley CH, Brookfield JF, Kaplan NL: Transposable elements in mendelian populations. I. A theory. Genetics 1983;104:457–471.

108 Boissinot S, Davis J, Entezam A, Petrov D, Furano AV: Fitness cost of LINE-1 (L1) activity in humans. Proc Natl Acad Sci USA 2006;103:9590–9594.

109 Boissinot S, Entezam A, Furano AV: Selection against deleterious LINE-1-containing loci in the human lineage. Mol Biol Evol 2001;18:926–935.

110 Song M, Boissinot S: Selection against LINE-1 retrotransposons results principally from their ability to mediate ectopic recombination. Gene 2007;390: 206–213.

111 Bartolome C, Maside X, Charlesworth B: On the abundance and distribution of transposable elements in the genome of *Drosophila melanogaster.* Mol Biol Evol 2002;19:926–937.

112 Rizzon C, Marais G, Gouy M, Biémont C: Recombination rate and the distribution of transposable elements in the *Drosophila melanogaster* genome. Genome Res 2002;12:400–407.

113 Hill WG, Robertson A: The effect of linkage on the limit to artificial selection. Genet Res 1966;8:269–294.

114 Callinan PA, Batzer MA: Retrotransposable elements and human disease. Genome Dyn 2006;1: 104–115.

115 Han K, Lee J, Meyer TJ, Remedios P, Goodwin L, Batzer MA: L1 recombination-associated deletions generate human genomic variation. Proc Natl Acad Sci USA 2008;105:19366–19371.

116 Gasior SL, Wakeman TP, Xu B, Deininger PL: The human LINE-1 retrotransposon creates DNA double-strand breaks. J Mol Biol 2006;357:1383–1393.

117 Petrov D, Aminetzach YT, Davis JC, Bensasson D, Hirsh AE: Size matters: non-LTR retrotransposable elements and ectopic recombination in *Drosophila.* Mol Biol Evol 2003;20:880–892.

118 Petrov DA, Fiston-Lavier AS, Lipatov M, Lenkov K, González J: Population genomics of transposable elements in *Drosophila melanogaster.* Mol Biol Evol 2011;28:1633–1644.

119 Dolgin ES, Charlesworth B: The effects of recombination rate on the distribution and abundance of transposable elements. Genetics 2008;178:2169–2177.

120 Goodier JL, Kazazian HH Jr: Retrotransposons revisited: the restraint and rehabilitation of parasites. Cell 2008;135:23–35.

121 Hua-Van A, Le Rouzic A, Boutin TS, Filée J, Capy P: The struggle for life of the genome's selfish architects. Biol Direct 2011;6:19.

122 Muotri AR, Marchetto MC, Coufal NG, Gage FH: The necessary junk: new functions for transposable elements. Hum Mol Genet 2007;16(Spec No. 2): R159–R167.

123 Oliver KR, Greene WK: Transposable elements: powerful facilitators of evolution. Bioessays 2009; 31:703–714.

124 Mikkelsen TS, Wakefield MJ, Aken B, Amemiya CT, Chang JL, et al: Genome of the marsupial *Monodelphis domestica* reveals innovation in non-coding sequences. Nature 2007;447:167–177.

125 Bejerano G, Lowe CB, Ahituv N, King B, Siepel A, et al: A distal enhancer and an ultraconserved exon are derived from a novel retroposon. Nature 2006;441:87–90.

126 Aminetzach YT, Macpherson JM, Petrov DA: Pesticide resistance via transposition-mediated adaptive gene truncation in *Drosophila.* Science 2005;309:764–767.

127 Gonzalez J, Lenkov K, Lipatov M, Macpherson JM, Petrov DA: High rate of recent transposable element-induced adaptation in *Drosophila melanogaster.* PLoS Biol 2008;6:e251.

128 Pardue ML, Debaryshe PG: Retrotransposons that maintain chromosome ends. Proc Natl Acad Sci USA 2011;108:20317–20324.

129 Lyon MF: X-chromosome inactivation: a repeat hypothesis. Cytogenet Cell Genet 1998;80:133–137.

130 Whitney KD, Baack EJ, Hamrick JL, Godt MJ, Barringer BC, et al: A role for nonadaptive processes in plant genome size evolution? Evolution 2010; 64:2097–2109.

131 Garcia Guerreiro MP, Chávez-Sandoval BE, Balanyà J, Serra L, Fontdevila A: Distribution of the transposable elements *bilbo* and *gypsy* in original and colonizing populations of *Drosophila subobscura.* BMC Evol Biol 2008;8:234.

132 Lockton S, Ross-Ibarra J, Gaut BS: Demography and weak selection drive patterns of transposable element diversity in natural populations of *Arabidopsis lyrata.* Proc Natl Acad Sci USA 2008;105:13965–13970.

133 Gonzalez J, Macpherson JM, Messer PW, Petrov DA: Inferring the strength of selection in *Drosophila* under complex demographic models. Mol Biol Evol 2009;26:513–526.

134 Dolgin ES, Charlesworth B, Cutter AD: Population frequencies of transposable elements in selfing and outcrossing *Caenorhabditis* nematodes. Genet Res (Camb) 2008;90:317–329.
135 Lockton S, Gaut BS: The evolution of transposable elements in natural populations of self-fertilizing *Arabidopsis thaliana* and its outcrossing relative *Arabidopsis lyrata*. BMC Evol Biol 2010;10:10.
136 Wright SI, Le QH, Schoen DJ, Bureau TE: Population dynamics of an Ac-like transposable element in self- and cross-pollinating *Arabidopsis*. Genetics 2001; 158:1279–1288.
137 Wright SI, Agrawal N, Bureau TE: Effects of recombination rate and gene density on transposable element distributions in *Arabidopsis thaliana*. Genome Res 2003;13:1897–1903.
138 Duret L, Marais G, Biemont C: Transposons but not retrotransposons are located preferentially in regions of high recombination rate in *Caenorhabditis elegans*. Genetics 2000;156:1661–1669.
139 Ma J, Devos KM, Bennetzen JL: Analyses of LTR-retrotransposon structures reveal recent and rapid genomic DNA loss in rice. Genome Res 2004; 14:860–869.
140 Devos KM, Brown JK, Bennetzen JL: Genome size reduction through illegitimate recombination counteracts genome expansion in *Arabidopsis*. Genome Res 2002;12:1075–1079.
141 Petrov DA, Sangster TA, Johnston JS, Hartl DL, Shaw KL: Evidence for DNA loss as a determinant of genome size. Science 2000;287:1060–1062.
142 Hawkins JS, Proulx SR, Rapp RA, Wendel JF: Rapid DNA loss as a counterbalance to genome expansion through retrotransposon proliferation in plants. Proc Natl Acad Sci USA 2009;106:17811–17816.

Stéphane Boissinot, PhD
Department of Biology
Queens College, The City University of New York
65–30 Kissena Boulevard, Flushing, NY 11367–1597 (USA)
Tel. +1 718 997 3437, E-Mail stephane.boissinot@qc.cuny.edu

Garrido-Ramos MA (ed): Repetitive DNA.
Genome Dyn. Basel, Karger, 2012, vol 7, pp 92–107

SINEs as Driving Forces in Genome Evolution

J. Schmitz

Institute of Experimental Pathology, University of Münster, Münster, Germany

Abstract

SINEs are short interspersed elements derived from cellular RNAs that repetitively retropose via RNA intermediates and integrate more or less randomly back into the genome. SINEs propagate almost entirely vertically within their host cells and, once established in the germline, are passed on from generation to generation. As non-autonomous elements, their reverse transcription (from RNA to cDNA) and genomic integration depends on the activity of the enzymatic machinery of autonomous retrotransposons, such as long interspersed elements (LINEs). SINEs are widely distributed in eukaryotes, but are especially effectively propagated in mammalian species. For example, more than a million Alu-SINE copies populate the human genome (approximately 13% of genomic space), and few master copies of them are still active. In the organisms where they occur, SINEs are a challenge to genomic integrity, but in the long term also can serve as beneficial building blocks for evolution, contributing to phenotypic heterogeneity and modifying gene regulatory networks. They substantially expand the genomic space and introduce structural variation to the genome. SINEs have the potential to mutate genes, to alter gene expression, and to generate new parts of genes. A balanced distribution and controlled activity of such properties is crucial to maintaining the organism's dynamic and thriving evolution.

Copyright © 2012 S. Karger AG, Basel

SINEs are short interspersed elements whose master copies repetitively retropose via RNA intermediates. The molecular origin of SINEs can be traced back to an organism's own cellular small RNAs, especially highly expressed tRNAs and parts of the 7SL RNA, and less frequently 5S rRNA, that are equipped with their own internal RNA polymerase III (Pol III) promoter. Actively transcribed RNAs are coincidently recognized at their 3′ ends by contemporary autonomous retrotransposon-derived enzymes (e.g. from long interspersed elements, LINEs), reverse transcribed, and more or less randomly inserted into the genome [1]. The new genomic location is crucial to determining whether the new insertion becomes an active SINE (rarely) or an inactive retropseudosequence like millions of other elements. Effective subsequent transcription requires additional sequence motifs upstream of the insertion site [2]. The transcription of, for example, new tRNA- or 7SL-derived SINEs starts 10–12 nucleotides (nt) upstream of their internal Pol III promoter box A, extends beyond their characteristic oligo(A)-tail, and occasionally continues further downstream to a random (T) stretch (TTTT or more complex terminator motifs) [3]. Continuous and frequent transcription is a first requirement towards a prospective functional master SINE element. As non-autonomous elements, their reverse transcription (from RNA to cDNA) and genomic integration depends on the activity of the enzymatic machinery of autonomous retrotransposons, such as LINEs or the LINE-derived retroposon-like transposable elements (RTEs). Thus, the second requirement is the proficient exploitation of the retrotranspositional LINE system. To 'mimic' a suitable LINE RNA target (the genuine template for LINE retrotransposition), the 3′ ends of SINEs are generally similar or even derived from the 3′ ends of the LINE mRNAs. But organisms also have ways to hinder the proliferation of such elements, so the final path to success lies in escaping the organism's epigenetic or other defense systems. Because active SINEs do not directly contribute to the organism's fitness, they usually do not fall under natural selection that might prevent the decay of the essential transcriptional recognition sequences. This and possible protective effects of neighboring gene regions influence the limited lifetime of such elements. Furthermore, SINEs live and die with their associated autonomous retrotransposons. For example, more than 140 million years ago (Mya) the mammalian LINE2/3 activity terminated; consequently, the associated mammalian-wide interspersed repeat (MIR) SINEs suffered the same fate [4].

SINEs were first reported in rodents [5] and primates [6], but their presence is well documented in many eukaryotic lineages, including mammals, reptiles, birds, fishes, insects, molluscs, and plants (reviewed in [7]). In monotremes, which represent the first mammalian divergence, the LINE2-dependent Mon1 SINEs are the predominant mobile elements. Moving along the evolutionary tree, on the lineage leading to therians (marsupials and placentals) an especially effective autonomous retrotransposon association evolved, the LINE1-SINE system, which is particularly active in placentals. This more recently evolved LINE1 retrotransposon machinery is rather unspecific in its recognition and retrotransposition of *any* available oligo(A)-tailed RNA, such as (1) mRNAs, (2) tRNA derivates, (3) the primate-specific *Alu*-SINEs derived from the 7SL signal recognition particle, or (4) more complex composite retrotransposons, such as the SVA elements in apes (composed of a $(CCCTCT)_n$ multimer – an *Alu*-like part – a variable number of tandem repeats (VNTR) – a SINE-R derived from an LTR element of the human endogenous retrovirus (HERV) – and an oligo(A) tail) [8, 9]. Thus, in the LINE1-SINE system, the retrotranspositional frequency corresponds to the quantity of available transcribed oligo(A)-tailed templates.

Although similar SINEs might emerge de novo independently in different lineages, specific SINEs or SINE families are usually restricted to respective orders, rarely crossing order boundaries. SINEs often are divided into different subfamilies with successive, partially overlapping waves of activity [10]. Usually, however, there is just 1 LINE/SINE system active at a given time, as is the case for the LINE1-*Alu*-SINE association in higher primates. At present, ~100 or much less *Alu*-SINE loci are possibly retrotranspositionally active in primates [11, 12], which roughly corresponds to the number of retrotranspositionally competent LINEs [13]. In human, it is estimated that active, retrotransposed SINE and LINE sequences lead to new germline insertions in at least every 30th and 50th birth, respectively [13]. However, in marsupials there is strong indication that 3 different LINE/SINE systems were active at the same time [14].

Interestingly, it has been shown for humans that LINE1s (containing AT-rich sequences) and *Alu*-SINEs (GC-rich sequences) are correspondingly distributed in AT-rich heterochromatic or GC-rich euchromatic genomic regions [15]. This is surprising because the insertion mechanisms for both are identical and possess the same slight preference for 5′-TT/AAAA-3′ genomic motifs [16]. It is speculated that the resultant lack of LINEs in gene-rich regions may be due to negative selection against such large transposable elements (TEs) that carry many internal transcription factor binding sites [17]; therefore, they might interfere with the transcription of adjacent genes. Negative selection against TE insertion in euchromatic regions may also counteract the deleterious effects triggered by TE-induced ectopic recombination [18]. Recent investigations show that, for example, the insertion preference of *Alu*-SINEs differs between old (insertion in GC-rich regions) and young (insertion in AT-rich regions) elements [19].

The effects of SINEs on genomes are wide-ranging, and the present review summarizes some of the most important aspects of SINEs as substrates for evolution.

SINEs Substantially Expand the Genomic Space

Because retroposition involves the copying of sequences and insertion of the copy back into the genome, SINEs and other TEs have the potential to substantially increase the size of genomes. One of the most impressive examples of genomic expansion is the doubling of the maize genome in just ~5 million years via proliferation of TEs [20]. Also in mammals the contribution of TEs to genome size is substantial; up to half of the mammalian genome is derived from recognizable TEs. Fortunately, only a small number of such insertions remain active. Most SINE copies and other retroposed RNAs as well, remain as inactive parts of the genome and are never transcribed directly. For example, only a few of the million *Alu*-SINE genomic copies are still actively transcribed [11] and contribute further to the continuous increase in genomic size. Below are 2 of the more interesting examples of SINEs or SINE-like elements recently discovered that have contributed to increases in genome sizes.

Example of Genome Extension by Tailless Retropseudogenes

As indicated above, the LINE system is not only responsible for its own proliferation, but also that of the SINEs and other small nuclear RNAs. In particular, over the last 140 million years, the special LINE1 retrotransposon has produced an enormous number of fragmented RNA genomic insertions known as tailless retropseudogenes [21]. Such tailless RNA copies are specifically truncated at structural loop regions (e.g. in tRNA loops or in other single stranded regions), and the insertion of the cDNAs is directed by sequence complementarity of their 3′-terminal 2–18 nucleotides and corresponding genomic loci. However, this variant insertion mechanism does not necessarily exclude the initial recognition of an oligo(A)-tail by LINE1 and retroposition via internal priming. Because tailless retropseudogenes were so far only found in therian mammals, their distribution by the LINE1 retrotranspositional system is obvious.

Example of Genome Extension by RTE-snoRNA Retroposition

The retrotranspositional process can be very 'inventive' in expanding the genome size, as demonstrated by the procedure that is called RTE-snoRNA retrotransposition [22]. In platypus, an intronic small nucleolar RNA (snoRNA) housekeeping gene (that is normally co-transcribed and processed with its host protein-coding gene to modify ribosomal RNAs) has been fused with the 3′-tail of a BovB_Plat RTE (bovine B platypus retrotransposon-like non-LTR transposable element). Although snoRNA distributions are usually limited to a single or few copies due to their distribution via rare duplication events and the adjacent RTE fragment can no longer be independently transcribed, together they form a formidable, genome-expanding element. The snoRNA endows the RTE fragment with the Pol II transcription of the host gene and the subsequent snoRNA-specific processing, and the RTE fragment provides the processed snoRNA with the RTE-tail necessary for co-opting the retrotranspositional machinery encoded by the active BovB_Plat autonomous element. Combined these 2 elements constitute an extremely efficient cooperation that produced more than 40,000 genomic copies in platypus [23]. Several of them are still actively expressed and at least 1 of them performs the housekeeping functions of the snoRNA, suggesting the possible functionality of many of the others also. This chimera is also present in echidna that diverged from the platypus lineage 17 Mya (unpublished data), providing strong evidence for its evolutionary conservation.

In mammals, SINEs or SINE-like elements expand the genome size significantly. They occupy ~13% of the human, ~8% of the mouse, ~10% of the opossum, and ~22% of the platypus genome (reviewed in [24]). While to date no chicken-specific SINEs have been detected, the zebra finch genome contains a few thousand CR1 LINE-mobilized SINEs (~0.03% of the genome). In reptiles, BovB LINE-mobilized SINEs are widespread and more than 100,000 copies are estimated to be present in the lizard genome (~2–5% of the genome) [25, 26]. SINEs were also detected in many fishes, where the copy number can vary substantially among different taxa. The fugu genome contains just a few thousand copies [24], while ~10% genomic SINE coverage

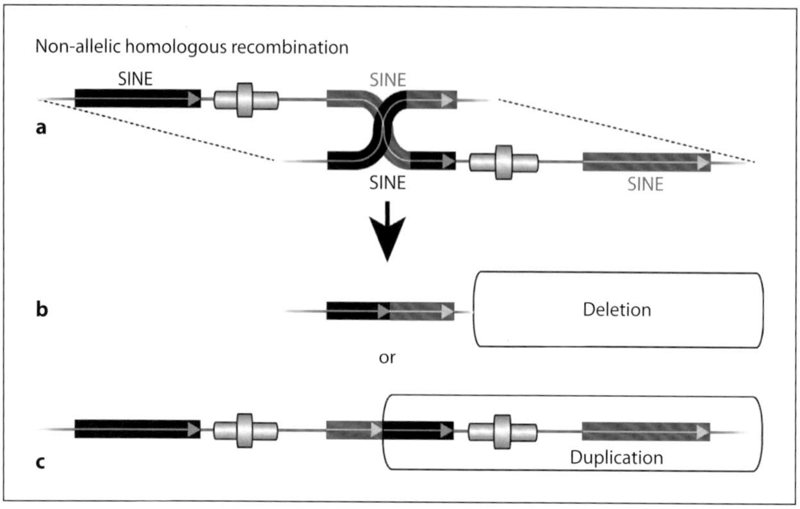

Fig. 1. SINE-induced non-allelic homologous recombination. **a** The illustration shows 2 different SINEs (black and grey bars) with high sequence similarities and embedded arrows indicating their orientations. The homologous allelic regions are shifted (dotted lines), and non-homologous (non-allelic) SINEs lead to recombination. The similar SINEs match, break, and rejoin, resulting in either a deletion of the SINE enclosed sequence region including the gene (**b**) or a larger recombined sequence carrying 2 versions of the SINE enclosed sequence which leads to a duplication of the enclosed gene (indicated by the single-exon (large grey cylinder) flanked by 2 untranslated regions (small grey cylinders) (**c**). The duplicated/deleted sequence information is framed.

is estimated for the zebrafish genome [27]. SINEs are also known from diverse invertebrate phyla, including Tunicata, Mollusca, Plathelminthes, Echinodermata, Arthropoda, and Nematoda, and plant genomes (for an overview, see [7]).

SINEs Introduce Structural Variants of the Genome

When retroposed SINEs are frequent inhabitants of the genome, they can cause sequence rearrangements. Sequence rearrangements induce variation by duplicating, deleting, or inverting sequence information. Duplicated genes can adopt the original function or acquire variant and subsequently novel functions, thus generating novel genes or multigene families. Furthermore, duplicated sequence regions constitute hot spots for subsequent segmental duplications. Because such rearrangements occur randomly, their outcome is not necessarily advantageous and they frequently lead to serious genetic disorders [28] that will eventually be selected against at the individual or population level. Any repetitive sequence block is a potential initiation point for segmental duplication. Because of their high abundance, TEs such as SINEs are frequently involved in structural variations via non-allelic homologous recombination (fig. 1).

Example TE-Associated Human Intraspecific Structural Variation
From a comparison of 2 human individuals, Xing et al. [29] determined that 706 of 8,000 intraspecies structural variants detected were induced by transposed elements, resulting in 305 kb of additional and 126 kb of removed genomic sequences. It has been shown that many segmental duplications are flanked by *Alu* elements, a strong indication for their impact on the duplication process [30]. Compared to lemurs, the human genome is expanded by ~15–20%, and 90% of the expansion is due to retroposition [31]. Some of the recent retrotransposon induced rearrangements led to numerous human diseases [28, 32]. From this analysis, Xing et al. [29] described 4 different processes of mobile element-mediated rearrangements: canonical retrotransposon insertions, non-canonical insertions associated with double-strand break repair, TE-mediated non-allelic homologous recombination leading to insertions or deletions, and non-homologous end-joining-mediated deletion via 1–7-nt 'microhomology' between TE internal breakpoints. However, structural variation is different in hetero- versus euchromatic regions. As mentioned above, TEs are often excluded (counterselected at the individual or population level) from recombination-sensitive functional (gene-rich) regions and accumulate more frequently in low-recombination-rate genomic areas, thereby maintaining genomic integrity [33]. Thus, while structural variants are selected against in euchromatic regions, because many of them cause disease, the few that are established in euchromatic regions (that do not cause disease) have a high potential to lead to novelties. The low abundance of LINEs and young *Alu*-SINEs in gene-rich regions is an example of potential selection against deleterious effects of such elements [19].

SINEs Influence or Regulate Gene Expression

Jordan et al. [34] showed that ~25% of investigated human promoter regions (~500 bp upstream of annotated transcription start sites) contain TEs, and they proposed that these elements are frequently involved in the regulatory network of transcription. SINEs and other transposed sequences control gene expression through multiple ways: by integrating into transcriptional control regions of genes, providing binding sites for transcription factors, and serving directly as enhancers or silencers of gene expression (fig. 2a). In so doing, they modulate the genomic landscape by contributing regulatory motifs. SINEs directly influence transcription and/or post-transcriptional modification in several ways. Pre-transcriptionally, they have the ability to temporally and spatially lock or unlock gene transcription during developmental stages by changing the chromatin status from a heterochromatic to a gene-activating, euchromatic structure [35]. When TEs are transcribed from the opposite strand of a gene [36], they interfere with the expression of the complementary hnRNA due to the collision of convergent RNA polymerases (fig. 2b). Post-transcriptionally, SINEs influence modification by RNA interference (RNAi). After endonucleolytic cleavage and

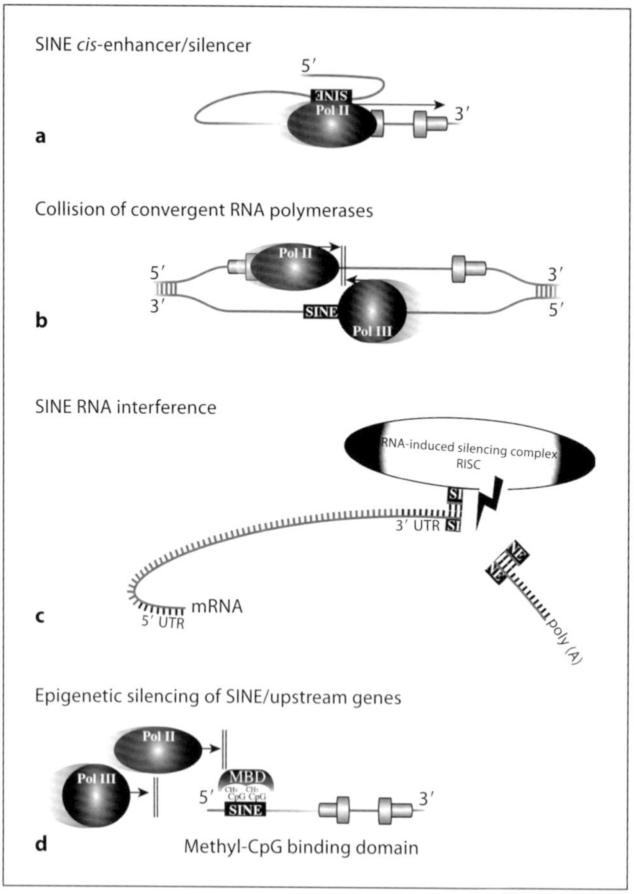

Fig. 2. SINE-induced variation of gene regulation. **a** SINE *cis*-enhancer/silencer. The upstream-located SINE (black box, written in reverse) interacts directly with the polymerase II (Pol II) promoter to enhance or silence the transcription of the associated gene. **b** SINE-induced collision of convergently moving RNA polymerases. The protein-coding gene and the SINE are located on opposite DNA strands and are transcribed in parallel by Pol II and Pol III, respectively. Due to steric inhibition of the 2 polymerases (collision), the transcription of the gene and SINE are interrupted (indicated by a double line). **c** SINE-induced RNA interference. A transcribed SINE RNA binds to a complementary sequence (e.g. a similar SINE located in the 3′ UTR of an mRNA). The double-stranded RNA region attracts the host-specific RNA-induced silencing complex (RISC) and the mRNA is fragmented (flash symbol). **d** Epigenetic silencing of a SINE and the upstream-located gene. The methyl-CpG binding domain (MBD) binds to methyl groups (CH_3) on CpG dinucleotides within the SINE sequence. The SINE-specific transcription by Pol III and the downstream-located Pol II genes are blocked (indicated by double lines). Genes in the 5′ to 3′ orientation are indicated by large cylinders (exons) and small cylinders (untranslated regions).

removal of the 3′ poly(A) tail protection, the mRNA is exposed to the cellular mRNA degradation (fig. 2c). Upon stress-induced up-regulation of SINE transcription (e.g. increased Pol III transcription after viral infection or stress-induced demethylation of CpGs), SINE RNAs can directly bind to and repress RNA Pol II transcription of mRNAs (*trans*-regulation [37]). SINEs can also serve as competitive/supportive promoters of RNA Pol II transcription [38]. Finally, the TE-targeted methylation of CpG dinucleotides, a part of the epigenetic defensive system used to silence retrotranspositional products, can also influence the transcriptional activity of adjacent genes (fig. 2d) [39]. Since repetitive elements in the human genome contain more than 50% of these CpG dinucleotides [40], this effect can be substantial.

One should note that predominantly old elements and element families are involved in regulatory networks [41], which agrees with our own observation that predominantly old SINE insertions (see below) became persistently exonized. The following are a few examples of ultraconserved SINE modules that became exonized and acquired important cellular tasks.

Example LF-SINEs

It was recently shown that ultraconserved LF (living fossil)-SINEs, which were established ~410 Mya, eventually contribute a variety of different protein-coding cassettes and functionally conserved regulatory elements. One example is an ultraconserved enhancer derived from an LF-SINE that is located ~0.5 Mb upstream of the neurodevelopmental insulin gene enhancer gene (*ISL1*), a gene that is expressed in developing motor neurons and well conserved in tetrapodes [42]. These LF-SINEs are highly conserved non-protein-coding elements that were active in the distant past; however, some still active LF-SINEs were detected in the coelacanth, the well-known 'living fossil' that is name-giving for these elements.

Example AmnSINE1s

AmnSINE1s originated 300 Mya in the common ancestor of amniotes (reptiles, birds, and mammals) and, similar to LF-SINEs, are thought to provide building blocks involved in ancient morphological innovations [43]. They were thought to have been effectively selected by genetic drift following drastic population bottlenecks induced by the rapid change in atmospheric oxygen concentration during the Permian-Triassic era. This change probably led to the mass extinction that occurred 250 Mya [44], with few survivors carrying possibly critical genomic changes from ultraconserved SINE insertions to adapt to the altered atmospheric oxygen concentration.

More than 100 AmnSINE1 loci are conserved in mammalian genomes and are thought to convey essential mammalian-specific features mainly by influencing brain-specific transcription. That SINEs are able to influence brain-specific expression and permit functional or behavioral changes was first demonstrated with BC1, a rodent-specific tRNA-derived SINE that was exapted to brain-specific tasks [45]. Furthermore, BC1 is thought to be the master gene of thousands of related ID SINEs in rodents [46].

Example CORE-SINEs

In addition to LF-SINEs and AmnSINE1s, another ancient group of SINEs, the so-called CORE-SINEs (including MIRs, the monotreme-specific Mon1, placental Ther1,

Ther2, and the marsupial-specific, recently active MAR1s [47]) evolved important mammalian features. Approximately 10 kb downstream of the proopiomelanocortin gene (*POMC*), Santangelo et al. [48] found a specific CORE-SINE that regulates the gene activity for producing important hormone precursors in all mammalian species. The brain-specific CORE-SINE enhancer is ultraconserved in all mammals but absent in other vertebrates where regulation is thought to be performed by different regulatory elements [48].

SINEs Evolve into Protein-Coding Sequences

Sela et al. [49] found that, although retrotransposon insertion is random, ~60% of retroposed sequences in human and mouse are located in intronic regions that comprise just ~24% of the genome. Some of these well-localized retrotransposons (close to protein-coding sequences) evolve into protein-coding sequences via exonization and might subsequently acquire function in a process called exaptation [50]. The original non-protein-coding modules are usually short (up to a few hundred nucleotides) and can be included in adjacent exons via low degrees of alternative splicing without introducing shifts in open reading frames. Comparable to duplicated genes, such alternative splice products evolve mainly unconstrained because the original, dominant variant ensures functionality. If by chance they evolve an advantageous feature, natural selection improves the initially cryptic alternative splice site by selecting advantageously mutated splice sites, thereby increasing the proportion of the alternative products. In some cases, the optimization leads to the complete replacement of the original variant and constitutive expression of the new form. Evolutionary time is crucial for such a process, and we showed that from the insertion of a new SINE to its functional exonization and exaptation additional mutagenic steps are necessary, and tens or even hundreds of millions of years can elapse [51]. In principle, any suitably positioned stretch of DNA might get exonized, but reverse-oriented SINEs, such as *Alu*s [52] or MIRs [51], appear to be almost predestined for this because they are also coincidentally equipped with splice sites, comprising a 3′ AG-cryptic splice site and a preceding oligo(T) stretch – the complement of an oligo(A) tail that serves as poly-pyrimidine tract in splicing (fig. 3). There is no common codon pattern for exonized SINE cassettes. For example, the 19 exonized LF-SINEs so far analyzed used all 3 possible reading frames [42].

We think that the actual sequence of additional amino acids introduced into the protein-coding sequence by the exonized SINEs is not as significant for the evolution of the gene as is the space that they introduce, possibly optimizing protein structures by separating active protein domains to more advantageous distances from one another. An example might be the constitutively expressed MIR-cassette within the zinc finger protein 639 gene (*ZNF639*) [51].

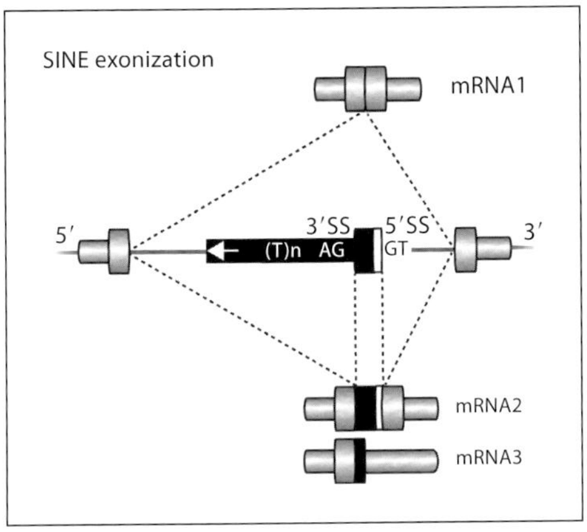

Fig. 3. SINE-induced exonization. The illustration in the middle represents the gene after an intronic reverse insertion of a SINE element. The original mRNA (without exonization) is shown at the top and the inclusion forms at the bottom. A SINE (black bar) inserts in the reverse orientation (white arrow) into an intron (thin grey line) of a gene (small cylinders = 5′ and 3′ UTRs, large cylinders = exons). The cellular splicing system is attracted by an internal 3′ cryptic AG splice site in the reverse SINE and by the poly-pyrimidine-like branch site (reverse of the SINE A-rich tail; $(T)_n$). The necessary 5′ GT splice site is randomly selected from the adjacent intron. The new, alternatively spliced exon comprises a part of the SINE (thick black box) as well as sequences of the exonized intronic region (white bar) and yields the processed mRNA2. Initially, the original mRNA1 (without the SINE exon) is predominantly processed, ensuring functionality. The new splice variant is further shaped by random changes and natural selection. If a new advantageous splice variant results, it can completely replace the original splicing product and lead to constitutive expression of the SINE cassette. However, in some instances, the new exon cassette might interrupt the open reading frame and lead to a truncated mRNA3 that is exposed to nonsense-mediated mRNA decay. The dotted lines indicate the alternatively skipped intronic regions of the mature mRNAs.

Example ZNF639

ZNF639 binds DNA and may function as a transcriptional repressor. More than 160 Mya, before the monotremes diverged from the mammalian lineage, a LINE2-mobilized MIR element (CORE-SINE) inserted into intron 5 of *ZNF639* and was fixed in the germline of the common ancestor of all mammals. The exact time of insertion is difficult to determine but occurred somewhere between the divergence of sauropsids and mammals and the first split of mammals (~300–160 Mya). However, before the mammals diversified, the element was already exonized and contributed a new exon 6 to *ZNF639*. Today, this exon is present and constitutively expressed in all mammalian species. We showed that the selection pressure on the 45 additional amino acids is moderate (K_a/K_s value 0.19) compared to the 9 zinc finger domains of the adjacent exon 7 (K_a/K_s value 0.03) [51]. This indicates that the sequence of the new amino acids is possibly not the crucial improvement, but rather, that the spacer function of the new exonized cassette, separating the protein domains of the 5th and 7th exon, is the more decisive innovation. Comparing the mammalian *ZNF639* to other vertebrate homologs provided further support for this idea. As the classical MIRs are mammalian-specific, frogs and all other non-mammalian *ZNF639*s lack the additional MIR-coding cassette. Surprisingly, in all bird clades we also found another additional 6th exon. Moreover, the number of additional amino acids in the birds' exonized sequences is identical to that in the exonized MIR-coding cassette in mammals, but the amino acid composition is completely different, and in the case of the birds can not be assigned to any known TE. This suggests that a selectively advantageous spacer evolved independently in the 2 amniote lineages. And this is not the only example in which an exonized sequence seems to position flanking protein domains at more functionally favorable distances.

Example ADAR2

Double-strand RNA-specific editase 1 (*ADAR2/RED1*) is involved in the editing of precursor mRNAs by site-specific conversion of adenosine to inosine. Two splice variants are known for the human gene (*hADAR2*), one of which includes an *Alu*J exonization (40 amino acids) in the center of the deaminase domain (active core of the protein). The additional amino acids contribute an extra loop within 2 ß-strands of the adjacent amino acids and increase the distance between the 2nd and 3rd putative Zn^{2+}-chelating amino acids in the deaminase domain. Both variants have the same substrate specificity, but the catalytic activity of the *Alu*J inclusion variant is different, and additional potential protein interactions and regulatory functions for the enlarged splice variant have been discussed [53].

Usually it is a long obstacle course from insertion to the evolution of a novel advantageous exon function. Most exonized genes lose their additional protein-coding sequences (splice variants) by randomly accumulated mutations before an improving or novel function with a selective advantage is established. The long path to a new exon can be very versatile and may include editing processes and changes at both the DNA and the RNA levels [54].

Example NARF – Exonization via Editing

More than 43 Mya, in the common ancestor of higher primates 2 independent tail-to-tail *Alu*Sx element insertions occurred in intron 7 of the gene encoding the nuclear prelamin A recognition factor (*NARF*), which binds and processes the carboxyl-terminal tail of prelamin A. The reverse-oriented, 3′-located element already carried a C-to-GT change that later provided the 5′ splice site of a new exon 8. At the latest, sometime in the common ancestor of Great Apes, several additional changes were introduced into the precursor mRNA by adenosine-to-inosine (A-to-I) RNA editing that generated a functional 3′ splice site (AA-to-AI editing; AI functions as an AG splice site). Such editing also converted an internal UAG stop codon to a UIG codon (functions as UGG tryptophan codon). These changes were facilitated by the back-folding of the 2 adjacent *Alu*Sx elements in the unspliced pre-mRNA, building a partial double-stranded RNA structure which is a signal for the editing enzymes to introduce inosine at adenosine positions. Finally, in the common ancestor of chimpanzee and human a TGA-to-CGA mutation at the DNA level facilitated the uninterrupted open reading frame of the new alternatively spliced exon 8 [55]. Although RNA editing is rarely involved in exonization, this example shows how evolution can act at the DNA and RNA levels to provide variations for natural selection. The process of editing is known to not only facilitate new splice variants, but is also involved in preventing additional aberrant exons [56].

In a review of the data, Sorek [57] summarized that more than 90% of new human exons were derived from repetitive sequences (~3,400 cases), with a clear dominance of *Alu*-SINE cassettes (~62%). The preexisting cryptic 3′-AG splice sites and the oligo-pyrimidine tracts of reverse SINEs, such as *Alu*s and MIRs, provide 'prefabricated' functional modules for exonization. Thus, more than 80% of exonized *Alu* elements and 60% of MIR exonizations have occurred in elements integrated in the reverse orientation [51]. In these cases, the second necessary splice sites (5′-GT) were randomly selected within the retrotransposon or taken from the adjacent intronic region, thereby including intronic sequence stretches in the new exon (fig. 3).

Down-Regulation of SINE Activity

It has been shown that SINEs, as well as other repetitive sequences, play an extraordinary role in genome evolution. Today we can trace the successes of and innovations brought about by SINEs that have already been selected, but we have only an incomplete imagination of all the past blind alleys and individual disasters caused by SINE activity. Similarly to managing a wild animal in the zoo, organisms must permanently control the activity of TEs and suppress their uncontrolled spread. This goal cannot always be met and may then lead to retrotransposon-induced diseases; whereupon, the regulatory challenge is met by purifying selection at the individual or population level. Some of the most important regulatory mechanisms to tame TE activity are described in the following section.

Example DNA Methylation (Transcriptional Gene Silencing)

In eukaryotic cells epigenetic silencing mechanisms regulate the activity of transposable elements by labeling (methylation via methyltransferase) cytosine residues at CpG dinucleotide sites and subsequently binding methyl-CpG proteins that block transcription. As mentioned previously, more than 50% of all human CpG sites are associated with transposed elements, and *Alu*-SINEs contain about 30% of the total genomic CpG sites [58].

This methylation labeling co-evolved very efficiently with the reviving activity of LINE1s in therian mammals and is also well distributed in plants. The methylation status of a genomic sequence can be copied and thus inherited upon genome replication. Because retroposition is assumed to occur predominantly in germ and embryonic cells, it is epigenetically silenced by CpG methylation in most somatic cells (reviewed in [59]). Ongoing genomic analyses of somatic cells will soon provide a more comprehensive understanding of the retrotransposon insertion frequency in gametes vs. somatic cells.

Example RNAi (Post-Transcriptional Gene Silencing)
In general, RNAi is involved in controlling the activities of genes. The ~20-nt-long microRNAs and small interfering RNAs are the principle components for silencing mRNA and other RNAs, and especially in defending the host cell from intruding viral or other transposed elements. Expressed LINE1 sense and antisense promoters that are 'co-inserted' with the LINE in the 5′ UTR region of genes (similar to the SINE in fig. 2c), or transcribed independently from a host gene, lead to double-stranded RNA that is processed by the RNAse III homolog DICER (an endoribonuclease that cleaves double-stranded RNA) to generate short interfering RNAs. Such RNAs are induced into the RNAi-induced silencing complex (RISC), which leads to the endonucleolytic cleavage of, for example, LINE1 mRNA [60] and consequently to the silencing of LINE activities and their LINE-dependent SINE mobilization.

Example Piwi Protein and piRNA-Interacting RNA Silencing
In vertebrates, Piwi silencing activity is restricted to the germline and is mediated by an animal-specific subclass of Argonaute proteins (Piwi proteins). In zebrafish, expression was detected in both male and female gonads, but in mammals, Piwi proteins are only effective in the male germline and are associated with the ~29-nt-long piRNAs that guide the silencing machinery to bind and cleave homologous RNAs. Because many piRNAs are derived from mobile elements, it is expected that they are mainly involved in post-transcriptional gene silencing of retrotransposons and other TEs [61].

Summary

SINEs are only one component of transposed genomic elements but in many organisms, such as human, at least numerically, they predominate. A significant genomic presence of SINEs provides a pool of evolutionary building blocks that might contribute directly or indirectly to an organism's fate. The reshuffling of genomic regions induced by repetitive modules is one way of influencing affected genes by partial or complete duplication or deletion. Such significant genomic interference exposes the genome to strong purifying selection and for the organism or its next generation can mean anything from perdition to innovation. Novel properties might be suitable for adaptive survivability, especially if the environment changes drastically. Also, some indirect regulatory effects of SINEs influence gene regulatory networks by contributing expression enhancers or silencers, which evolved especially efficiently in the extraordinarily retrotranspositionally active mammalian genomes.

When located close to protein-coding genes, parts of SINEs frequently evolve into protein-coding sequences. SINE-internal sequence stretches in the reverse orientation resemble splice-like components and promote the conversion to protein-coding sequences, perhaps contributing to the optimization of special structural properties in the derived proteins. Time is obviously a crucial factor in generating and establishing evolutionary novelties. Many young exonizations or regulatory changes have not had enough time to evolve advantages to be significantly favored by natural selection and are often lost in certain populations, lineages, or species. A good example are the primate-specific *Alu*-SINE exonizations that were present in some older primate lineages but subsequently lost in younger ones. By contrast, more ancient exonizations, such as those derived from the more than 160-million-year-old MIR elements, had enough time to lead to persistent exonization or even constitutive expression of new exons in all mammalian lineages. Similarly, ultraconserved SINE sequences that have survived nearly unchanged for millions of years facilitated the evolution of essential regulatory modules. Especially we primates are much influenced by SINEs for better or worse because of the extraordinarily efficient activity and distribution of *Alu*s, which presents both a challenge for an operable genome and, at the same time, a chance to evolve novelties in the struggle of species survival.

Acknowledgements

Comparative surveys are only possible with the help of the enormous efforts of the many genome sequencing groups and colleagues that provide the tools to understand the informative content of the genome needed to unravel the secrets of mobile elements. I would like to thank Jürgen Brosius, Carsten A. Raabe and Gennady Churakov for valuable comments relating to the manuscript and Marsha Bundman for assistance in editing the information into a hopefully interesting context. Most of our own contributions to understanding the dynamics and influences of mobile elements were supported by the Deutsche Forschungsgemeinschaft (SCHM1469) and the Medical Faculty of the University of Münster.

References

1 Singer MF: SINEs and LINEs: highly repeated short and long interspersed sequences in mammalian genomes. Cell 1982;28:433–434.

2 Ludwig A, Rozhdestvensky TS, Kuryshev VY, Schmitz J, Brosius J: An unusual primate locus that attracted two independent Alu insertions and facilitates their transcription. J Mol Biol 2005;350:200–214.

3 Kramerov DA, Vassetzky NS: SINEs. Wiley Interdiscip Rev RNA 2011;2:772–786.

4 Smit AF, Riggs AD: MIRs are classic, tRNA-derived SINEs that amplified before the mammalian radiation. Nucleic Acids Res 1995;23:98–102.

5 Kramerov DA, Grigoryan AA, Ryskov AP, Georgiev GP: Long double-stranded sequences (dsRNA-B) of nuclear pre-mRNA consist of a few highly abundant classes of sequences: evidence from DNA cloning experiments. Nucleic Acids Res 1979;6:697–713.

6 Houck CM, Rinehart FP, Schmid CW: A ubiquitous family of repeated DNA sequences in the human genome. J Mol Biol 1979;132:289–306.

7 Kramerov DA, Vassetzky NS: Short retroposons in eukaryotic genomes. Int Rev Cytol 2005;247:165–221.

8 Hancks DC, Goodier JL, Mandal PK, Cheung LE, Kazazian HH Jr: Retrotransposition of marked SVA elements by human L1s in cultured cells. Hum Mol Genet 2011;20:3386–3400.

9 Wang H, Xing J, Grover D, Hedges DJ, Han K, et al: SVA elements: a hominid-specific retroposon family. J Mol Biol 2005;354:994–1007.

10 Quentin Y: The Alu family developed through successive waves of fixation closely connected with primate lineage history. J Mol Evol 1988;27:194–202.

11 Kapitonov VV, Pavlicek A, Jurka J: Anthology of human repetitive DNA; in Weinheim RAM (ed): Encyclopedia of Molecular Cell Biology and Molecular Medicine. Wiley-VCH, 2004, pp 251–305.

12 Comeaux MS, Roy-Engel AM, Hedges DJ, Deininger PL: Diverse *cis* factors controlling *Alu* retrotransposition: what causes *Alu* elements to die? Genome Res 2009;19:545–555.

13 Kazazian HH Jr: Mobile elements: drivers of genome evolution. Science 2004;303:1626–1632.

14 Nilsson MA, Churakov G, Sommer M, Tran NV, Zemann A, et al: Tracking marsupial evolution using archaic genomic retroposon insertions. PLoS Biol 2010;8:e1000436.

15 Smit AF: Interspersed repeats and other mementos of transposable elements in mammalian genomes. Curr Opin Genet Dev 1999;9:657–663.

16 Jurka J: Sequence patterns indicate an enzymatic involvement in integration of mammalian retroposons. Proc Natl Acad Sci USA 1997;94:1872–1877.

17 Nigumann P, Redik K, Matlik K, Speek M: Many human genes are transcribed from the antisense promoter of L1 retrotransposon. Genomics 2002; 79:628–634.

18 Graham T, Boissinot S: The genomic distribution of L1 elements: the role of insertion bias and natural selection. J Biomed Biotech 2006;2006:75327.

19 Belle EM, Webster MT, Eyre-Walker A: Why are young and old repetitive elements distributed differently in the human genome? J Mol Evol 2005; 60:290–296.

20 SanMiguel P, Tikhonov A, Jin YK, Motchoulskaia N, Zakharov D, et al: Nested retrotransposons in the intergenic regions of the maize genome. Science 1996;274:765–768.

21 Schmitz J, Churakov G, Zischler H, Brosius J: A novel class of mammalian-specific tailless retropseudogenes. Genome Res 2004;14:1911–1915.

22 Schmitz J, Zemann A, Churakov G, Kuhl H, Grutzner F, et al: Retroposed SNOfall–a mammalian-wide comparison of platypus snoRNAs. Genome Res 2008;18:1005–1010.

23 Warren WC, Hillier LW, Marshall Graves JA, Birney E, Ponting CP, et al: Genome analysis of the platypus reveals unique signatures of evolution. Nature 2008;453:175–183.

24 Mandal PK, Kazazian HH Jr: SnapShot: vertebrate transposons. Cell 2008;135:192–192.e1.

25 Piskurek O, Austin CC, Okada N: Sauria SINEs: novel short interspersed retroposable elements that are widespread in reptile genomes. J Mol Evol 2006;62:630–644.

26 Alfoldi J, Di Palma F, Grabherr M, Williams C, Kong L, et al: The genome of the green anole lizard and a comparative analysis with birds and mammals. Nature 2011;477:587–591.

27 Izsvak Z, Ivics Z, Garcia-Estefania D, Fahrenkrug SC, Hackett PB: DANA elements: a family of composite, tRNA-derived short interspersed DNA elements associated with mutational activities in zebrafish *(Danio rerio)*. Proc Natl Acad Sci USA 1996;93:1077–1081.

28 Callinan PA, Batzer MA: Retrotransposable elements and human disease. Genome Dyn 2006;1: 104–115.

29 Xing J, Zhang Y, Han K, Salem AH, Sen SK, et al: Mobile elements create structural variation: analysis of a complete human genome. Genome Res 2009; 19:1516–1526.

30 Bailey JA, Liu G, Eichler EE: An *Alu* transposition model for the origin and expansion of human segmental duplications. Am J Hum Genet 2003;73:823–834.

31 Liu G, Zhao S, Bailey JA, Sahinalp SC, Alkan C, et al: Analysis of primate genomic variation reveals a repeat-driven expansion of the human genome. Genome Res 2003;13:358–368.

32 Kazazian HH Jr: Mobile elements and disease. Curr Opin Genet Dev 1998;8:343–350.

33 Bartolome C, Maside X, Charlesworth B: On the abundance and distribution of transposable elements in the genome of *Drosophila melanogaster*. Mol Biol Evol 2002;19:926–937.

34 Jordan IK, Rogozin IB, Glazko GV, Koonin EV: Origin of a substantial fraction of human regulatory sequences from transposable elements. Trends Genet 2003;19:68–72.

35 Lunyak VV, Prefontaine GG, Nunez E, Cramer T, Ju BG, et al: Developmentally regulated activation of a SINE B2 repeat as a domain boundary in organogenesis. Science 2007;317:248–251.

36 Conley AB, Miller WJ, Jordan IK: Human cis natural antisense transcripts initiated by transposable elements. Trends Genet 2008;24:53–56.

37 Mariner PD, Walters RD, Espinoza CA, Drullinger LF, Wagner SD, et al: Human Alu RNA is a modular transacting repressor of mRNA transcription during heat shock. Mol Cell 2008;29:499–509.
38 Ferrigno O, Virolle T, Djabari Z, Ortonne JP, White RJ, Aberdam D: Transposable B2 SINE elements can provide mobile RNA polymerase II promoters. Nat Genet 2001;28:77–81.
39 Slotkin RK, Martienssen R: Transposable elements and the epigenetic regulation of the genome. Nat Rev Genet 2007;8:272–285.
40 Xie H, Wang M, Bonaldo Mde F, Rajaram V, Stellpflug W, et al: Epigenomic analysis of Alu repeats in human ependymomas. Proc Natl Acad Sci USA 2010;107:6952–6957.
41 Gentles AJ, Wakefield MJ, Kohany O, Gu W, Batzer MA, et al: Evolutionary dynamics of transposable elements in the short-tailed opossum *Monodelphis domestica*. Genome Res 2007;17:992–1004.
42 Bejerano G, Lowe CB, Ahituv N, King B, Siepel A, et al: A distal enhancer and an ultraconserved exon are derived from a novel retroposon. Nature 2006;441:87–90.
43 Nishihara H, Smit AF, Okada N: Functional noncoding sequences derived from SINEs in the mammalian genome. Genome Res 2006;16:864–874.
44 Okada N, Sasaki T, Shimogori T, Nishihara H: Emergence of mammals by emergency: exaptation. Genes Cells 2010;15:801–812.
45 DeChiara TM, Brosius J: Neural BC1 RNA: cDNA clones reveal nonrepetitive sequence content. Proc Natl Acad Sci USA 1987;84:2624–2628.
46 Kim J, Martignetti JA, Shen MR, Brosius J, Deininger P: Rodent BC1 RNA gene as a master gene for ID element amplification. Proc Natl Acad Sci USA 1994; 91:3607–3611.
47 Gilbert N, Labuda D: CORE-SINEs: eukaryotic short interspersed retroposing elements with common sequence motifs. Proc Natl Acad Sci USA 1999;96:2869–2874.
48 Santangelo AM, de Souza FS, Franchini LF, Bumaschny VF, Low MJ, Rubinstein M: Ancient exaptation of a CORE-SINE retroposon into a highly conserved mammalian neuronal enhancer of the proopiomelanocortin gene. PLoS Genet 2007; 3:1813–1826.
49 Sela N, Mersch B, Gal-Mark N, Lev-Maor G, Hotz-Wagenblatt A, Ast G: Comparative analysis of transposed element insertion within human and mouse genomes reveals Alu's unique role in shaping the human transcriptome. Genome Biol 2007;8:R127.
50 Brosius J, Gould SJ: On 'genomenclature': a comprehensive (and respectful) taxonomy for pseudogenes and other 'junk DNA'. Proc Natl Acad Sci USA 1992;89:10706–10710.
51 Krull M, Petrusma M, Makalowski W, Brosius J, Schmitz J: Functional persistence of exonized mammalian-wide interspersed repeat elements (MIRs). Genome Res 2007;17:1139–1145.
52 Lev-Maor G, Sorek R, Shomron N, Ast G: The birth of an alternatively spliced exon: 3′ splice-site selection in Alu exons. Science 2003;300:1288–1291.
53 Gerber A, O'Connell MA, Keller W: Two forms of human double-stranded RNA-specific editase 1 (hRED1) generated by the insertion of an Alu cassette. RNA 1997;3:453–463.
54 Lev-Maor G, Sorek R, Levanon EY, Paz N, Eisenberg E, Ast G: RNA-editing-mediated exon evolution. Genome Biol 2007;8:R29.
55 Möller-Krull M, Zemann A, Roos C, Brosius J, Schmitz J: Beyond DNA: RNA editing and steps toward *Alu* exonization in primates. J Mol Biol 2008;382:601–609.
56 Sakurai M, Yano T, Kawabata H, Ueda H, Suzuki T: Inosine cyanoethylation identifies A-to-I RNA editing sites in the human transcriptome. Nat Chem Biol 2010;6:733–740.
57 Sorek R: The birth of new exons: mechanisms and evolutionary consequences. RNA 2007;13:1603–1608.
58 Rollins RA, Haghighi F, Edwards JR, Das R, Zhang MQ, et al: Large-scale structure of genomic methylation patterns. Genome Res 2006;16:157–163.
59 Beauregard A, Curcio MJ, Belfort M: The take and give between retrotransposable elements and their hosts. Ann Rev Genet 2008;42:587–617.
60 Soifer HS, Zaragoza A, Peyvan M, Behlke MA, Rossi JJ: A potential role for RNA interference in controlling the activity of the human LINE-1 retrotransposon. Nucleic Acids Res 2005;33:846–856.
61 Houwing S, Kamminga LM, Berezikov E, Cronembold D, Girard A, et al: A role for Piwi and piRNAs in germ cell maintenance and transposon silencing in zebrafish. Cell 2007;129:69–82.

Jürgen Schmitz
Institute of Experimental Pathology (ZMBE)
University of Münster, Von-Esmarch-Str. 56
DE–48149 Münster (Germany)
Tel. +49 251 835 2133, E-Mail jueschm@uni-muenster.de

Garrido-Ramos MA (ed): Repetitive DNA.
Genome Dyn. Basel, Karger, 2012, vol 7, pp 108–125

Unstable Microsatellite Repeats Facilitate Rapid Evolution of Coding and Regulatory Sequences

A. Jansen[a,b,c] · R. Gemayel[a,b] · K.J. Verstrepen[a,b]

[a]Laboratory for Systems Biology, VIB and [b]Laboratory for Genetics and Genomics, Centre of Microbial and Plant Genetics (CMPG), KU Leuven, Heverlee, [c]Human Genome Laboratory, Department for Molecular and Developmental Genetics, VIB, Leuven, Belgium

Abstract

Tandem repeats are intrinsically highly variable sequences since repeat units are often lost or gained during replication or following unequal recombination events. Because of their low complexity and their instability, these repeats, which are also called satellite repeats, are often considered to be useless 'junk' DNA. However, recent findings show that tandem repeats are frequently found within promoters of stress-induced genes and within the coding regions of genes encoding cell-surface and regulatory proteins. Interestingly, frequent changes in these repeats often confer phenotypic variability. Examples include variation in the microbial cell surface, rapid tuning of internal molecular clocks in flies, and enhanced morphological plasticity in mammals. This suggests that instead of being useless junk DNA, some variable tandem repeats are useful functional elements that confer 'evolvability', facilitating swift evolution and rapid adaptation to changing environments. Since changes in repeats are frequent and reversible, repeats provide a unique type of mutation that bridges the gap between rare genetic mutations, such as single nucleotide polymorphisms, and highly unstable but reversible epigenetic inheritance.

Copyright © 2012 S. Karger AG, Basel

Repeats, Repeats, Repeats

Repetitive DNA sequences are important components of genomes. In eukaryotes, the presence of repeated DNA sequences was uncovered in the 1960s, when researchers sought to resolve the observation that genome size does not correlate with the phenotypic complexity of an organism [1]. DNA hybridization experiments revealed that this discrepancy is the result of large variations in non-coding repetitive DNA that

A.J. and R.G. contributed equally to this work.

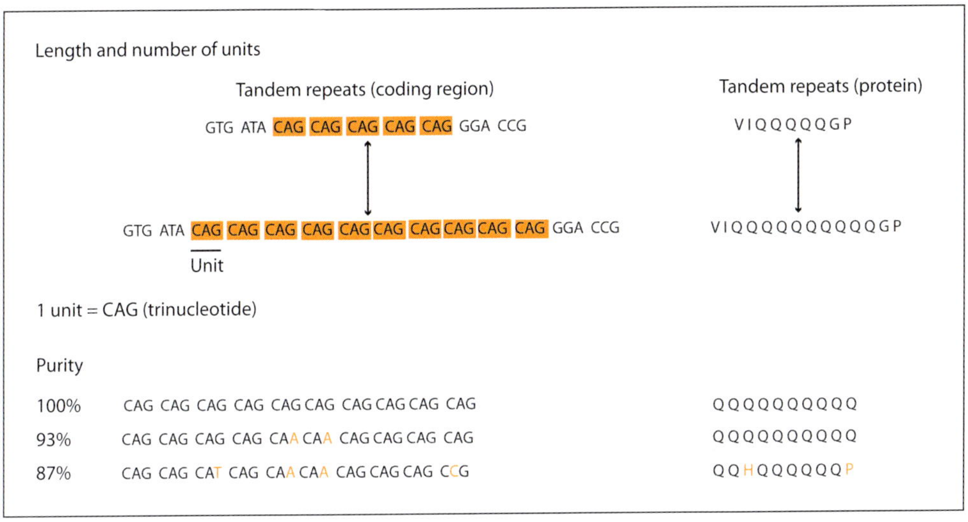

Fig. 1. Characteristics of TRs. The length and composition of the repeat unit, the number of repeat units and the purity of the repeat tract define a TR sequence. Microsatellite polymorphisms arise due to changes in the number of repeat units. The frequency of TR mutations depends on the repeat tract; long and pure TR sequences are more unstable than short tracts containing multiple point mutations [4].

makes up large portions of almost any genome. In humans, for example, no less than 46% of the DNA consists of repeated sequences. Unexpectedly, in the mid 1990s, large amounts of repetitive sequences were also discovered in prokaryotes. Indeed, despite their small genome sizes, prokaryotes contain a wide variety of repetitive sequences which can account for 10% or more of the total genome [2].

Repeated sequences in the genome are divided into 2 main categories. Most abundant are the interspersed repeats, where the repeated units are transposable elements that are dispersed throughout the genome. Variations in genome size are mostly generated by alterations in the number of interspersed repeated sequences. Tandem repeats (TRs) form the second category of repetitive DNA (fig. 1). Here, each repeat unit is located directly adjacent to the other, in other words, they occur in tandem. TRs are also referred to as satellite DNA, because they were originally identified as the sequences making up a second or 'satellite' band occurring when genomic DNA is separated in density-gradient centrifugation.

TRs are further classified based on the size of the repeated unit. Repeats with units equal to or greater than 10 nucleotides (nt) in length are generally known as minisatellites, those with unit lengths above 135 nt as megasatellites. Repetitions of shorter sequences (up to 9 nt) are termed microsatellites, simple sequence repeats (SSRs) or short tandem repeats (STRs). In this chapter, we will focus on studies involving microsatellites (for a review covering minisatellite repeats, see [3]). Here we will use the terms 'TRs' and 'microsatellites' interchangeably throughout the text.

TRs are particularly interesting from an evolutionary point of view because they are extremely unstable, mutating at higher frequencies than other sequences in the genome. Mutation rates of TRs vary from 10^{-7} to 10^{-3} per generation, in other words 1 to 10 orders of magnitude higher than the rate of point mutations (which varies between organisms) [4]. Whereas point mutations do occur, most mutations in microsatellites involve addition or removal of one or more units, resulting in a new repeat with the same unit sequence, but a different total length (fig. 1). There are 2 major models explaining how TRs can undergo expansions and contractions, namely strand-slippage replication and unequal recombination (reviewed in [3]).

Historically, hypervariable TRs were considered to be nonfunctional 'junk' DNA [5]. While it is true that TRs mainly occur in gene deserts, the recent availability of whole genome sequences revealed that microsatellites are also present in coding regions, promoters and other regulatory regions [6]. In some instances, TR polymorphisms in genes have been linked to functional changes in the corresponding proteins, suggesting that repeat sequences fulfill specific biological roles. This would imply that, because of their instability, TRs may allow swift adaptive evolution of genes and their associated phenotypes.

In this chapter, we discuss examples of microsatellites with diverse biological functions in a wide range of organisms. The focus will be on those studies that reveal the mechanisms for TR-mediated phenotypic changes. Special attention will be devoted to the few studies that link hypervariable TRs to the rapid evolution of genes. The first part of the chapter summarizes notable examples of variable microsatellites within coding regions. In the second part, the effects of TRs located in promoters and other regulatory regions will be discussed.

Tandem Repeats in Gene Coding Regions

The wealth of information gathered by whole genome sequencing projects and our increased knowledge about genome biology indicate that, in addition to being present in gene deserts (non-functional parts of the genome), TRs are also present in functional parts of the genome (coding regions and regulatory regions) [4]. This section will be devoted to microsatellite repeats found in gene coding regions. We will highlight their occurrence among a large spectrum of living organisms and discuss the implications of their variability.

Tandem Repeats in Gene Coding Regions: Functional Enrichment Rather than Random Occurrence

The percentage of genes with TRs in their coding sequence is relatively high in many organisms. In the human genome, for example, 17% of genes contain repeats in their open reading frames; similar percentages are found in other species [3]. When we consider the distribution of these intragenic repeats, it is becoming increasingly clear that it is not random. TRs with long repeat units (>10 nt, i.e. minisatellite repeats) are mostly found in genes encoding extracellular or cell-surface proteins. By contrast, repeats with short units (<10 nt, i.e. microsatellite repeats) are enriched in regulatory genes, including genes encoding transcription factors, DNA- and RNA-binding proteins, and regulatory hubs [4, 7].

Intragenic repeats mostly have repeat units that contain a multiple of 3 nt, presumably because of selection against frameshift mutations that would occur with unstable repeats of units that do not contain a multiple of 3 nt [4]. Exceptions (mostly in prokaryotes) do occur, and their outcome is beneficial as exemplified by studies on phase variation in bacteria (see further). On the protein level, certain amino acids are enriched in repeats. In eukaryotes, glutamine, arginine, glutamate, alanine, and serine are mostly represented [Verstrepen K.J., unpublished results], whereas hydrophobic amino acids such as isoleucine, methionine, and tryptophan are largely absent.

Taken together, this functional and structural enrichment indicates a non-random distribution of repeats within coding sequences. Is this merely a consequence of some gene categories being more tolerant of 'internal junk DNA', or could the enrichment indicate a functional role for these repeats? To answer this central question, we need to turn to studies that have investigated the functional consequences of repeat variation.

Repetitive DNA in Human Diseases: The Harmful Consequences of Repeat Instability

The earliest reports linking unstable TRs to functional variation came from the search into the origins of human neurodegenerative diseases. In the early 1990s, 3 landmark papers were published describing nucleotide repeat expansions as the causative mutations in spinobulbar muscular atrophy [8], Fragile X syndrome [9] and Huntington disease [10]. Since then, around 20 diseases caused by unstable repeat expansions have been documented (for a review, see [11]). Although the pathologies of these diseases are quite different, they all present a number of common features. For most of the genes in question the repeats are variable within the unaffected population, and symptoms only appear when a certain critical threshold is exceeded. Individuals carrying greater repeat expansions have earlier disease onset and more severe symptoms, demonstrating the dynamic nature of these mutations.

Disease-causing repeat expansions can be located in gene coding regions (exons) as well as in regulatory regions (introns or 3′ and 5′ untranslated regions, UTRs). Extensive research, using various model organisms, revealed that different mechanisms underlie the many diseases originating from repeat expansions. Briefly, these could be protein loss-of-function (transcription/translation deficiency), protein gain-of-function (novel properties gained by the mutant protein) or RNA-mediated gain-of-function (downstream toxic effects of the mutated RNAs) [11].

In this section, we will discuss Huntington disease as an example of a disorder caused by the expansion of unstable repeats in a gene coding region, with emphasis on the role of the expanded repeats in pathogenesis. In the second section of this chapter, we will discuss Fragile X syndrome, a disease caused by the expansion of repeats in the 5′ UTR of a gene. These 2 examples have been chosen to provide insights into how unstable repeats could become detrimental. Since diseases of repeat expansions are not in the scope of this chapter, we refer the readers to the following more comprehensive review covering the genetics, pathologies and molecular mechanisms of diseases caused by repeat expansions [11].

Huntington Disease (HD)
The most notable of the so-called polyglutamine diseases, is caused by the expansion of a CAG repeat (coding for a stretch of glutamine residues) in exon 1 of the *IT15* gene [10]. It is dominantly inherited and symptoms typically begin in mid-life. The symptoms of this devastating neurological disorder include chorea, cognitive decline, and psychiatric disturbances. Death usually occurs 10–15 years after disease onset [11].

The CAG repeat number in *IT15* is variable in healthy individuals and ranges from 6 to 35. Alleles with 36–39 repeats are associated with an increased risk of developing HD, and 40 repeats or more cause HD [12]. The age of onset and the severity of the disease correlate with the length of the CAG tract in *IT15*. Long repeats cause early disease onset and more severe symptoms. Alleles of ≥70 repeats, for example, are found in the juvenile form of HD [11].

The *IT15* gene encodes a large protein named huntingtin (Htt) with the polyglutamine tract situated at its amino-terminus. Huntingtin is a 348-kDa multifunctional scaffold protein with a large network of interactors and binding partners. Through these numerous interactions, Htt is involved in many cellular functions and pathways, mainly transcription, transport, and signaling. The precise pathological mechanism of HD is still not entirely elucidated but, owing to the numerous functions of Htt, strong evidence supports the idea of multiple pathogenic pathways being at work in HD. These are transcriptional dysregulation, transport defects, and mitochondrial dysfunction [11]. Whereas older studies mainly hypothesized that the expanded repeats gave rise to toxic protein aggregates, recent evidence suggests that the pathogenesis of expanded Htt repeats is rather a consequence of changes in Htt conformation [11]. The conformational changes triggered by the expanded polyglutamine tract likely interfere with the interaction of Htt and its numerous partners in the nucleus, such as the global transcriptional activator CREB binding protein (CBP), resulting in aberrant expression of target genes [13].

The Positive Side of Instability: Tandem Repeats as a Source of Useful Genomic Variation and Facilitators of Evolution
The plethora of repeat-linked diseases show that having an aberrant number of repeats in certain genes can have devastating consequences, hinting at a (positive) biological role for 'normal-sized' repeats. However, as a result of the 'junk DNA' dogma, TRs received little attention in the early days of molecular genetics. They

were believed to be neutral sequences with no functional value. Their high instability was only linked to negative consequences, as can be argued from their role in human neurodegenerative diseases. The advent of whole genome sequencing, however, revealed that TRs might play a functional role. Several lines of evidence argue for the utility of repetitive DNA sequences in genomes despite their high mutation rates and apparent low information content. They are relatively abundant in functional regions of the genomes, suggesting a selective advantage in having a higher mutability rate in certain genomic locations. Instead of being removed, some repeats have been kept over long evolutionary distances. For example, the same microsatellite repeats are present in the *SIS2* coding region in the common brewer's yeast *Saccharomyces cerevisiae* and its evolutionarily distant homolog, the yeast *Kluyveromyces lactis* [Verstrepen K.J., unpublished results]. The date of divergence of these 2 species was estimated to be 150 million years ago [14]. Amino acid repeats seem to be mostly encoded by pure DNA repeats (identical codons), with no evidence of selective pressure against unstable repeats [6]. In fact, most amino acids can be encoded by various codons and only a few mutations (impurities) in repeated codons can dramatically reduce their propensity to mutate [4]. Taken together, these observations seem to indicate that repeats might play a more beneficial role in genomes than previously believed.

In this section, we review the growing evidence supporting the role of TRs as facilitators of rapid evolutionary adaptation. It all begins with the observation that pathogenic bacteria can, in the span of few generations, evolve into more resistant variants capable of surviving host defenses. Through the variation of just a few repeat units, flies can tune their internal molecular clocks to changes in their environments. The remarkable variation in dog morphology can be explained through the functional variation of a repeat-containing developmental gene. It is important to mention that, due to space restrictions, we are only summarizing evidence from a few seminal studies (for a more elaborate overview, see [3]).

Phase Variation in Pathogenic Bacteria

Studies investigating microsatellite repeat variation in bacterial genes can be credited for being the earliest reports that documented beneficial roles for these unstable sequences. Through a process called phase variation, bacteria can rapidly switch between phenotypes allowing them to adapt, at an accelerated rate, to changes in their environment. This property is mainly observed in pathogenic bacteria that are in constant evolutionary struggle with their host's defense system. Phase variation is basically a reversible and random loss or gain of a phenotype that occurs at a high frequency. It is mediated by changes in expression of one or multiple genes. In *Neisseria gonorrhoeae*, for example, varying the outer membrane structure can be used as a strategy to evade attack by the host immune response. Members of the P.II family of cell surface genes contain a CTCTT microsatellite repeat in the region coding for the membrane signal peptide [15]. Spontaneous variation in repeat number can cause frameshifts, and consequently the protein is either correctly translated or not correctly translated in different cells of the population. This switching of phenotype occurs during infection and leads to the creation of new, more resistant variants capable of evading the immune system [15]. A similar strategy is employed by *Haemophilus influenzae.* The *lic1* gene, responsible for the addition of phosphorylcholine moieties to the membrane lipopolysaccharides, contains an intragenic CAAT repeat. In this case too, variation in repeat number results in different membrane structures depending on whether the Lic1 protein is correctly translated or not [16]. Frequent stochastic changes in the repeat number therefore give rise to a mixed population of Lic1+ and Lic1– cells, which in turn generates diversity at the cell surface, helping the pathogen to evade the host's immune system [16].

Genes encoding outer-membrane proteins are only a portion of the total phase variable genes in pathogenic bacteria. Recent genome-wide sequencing has uncovered that regulatory genes also have the potential to be phase variable. The DNA-methyltransferase *modA* in *Neisseria* and *Haemophilus* species contains intragenic AGTC or AGCC microsatellite repeats whose unit number directly influences the rate of phase variation [17]. In this example also, the outcome of repeat variation can be a fully functional or a non-functional protein since the *modA* gene has 2 potential start codons. As a consequence of this variation, there is a change in the expression of a subset of genes (due to differential methylation), such as those encoding outer membrane proteins, iron transporters, and heat shock proteins. Some of the *modA* targets are even predicted to be phase variable themselves. Here again, this multi-level phase variation increases the species' fitness under various stress conditions (e.g. heat shock, antimicrobial agent). In particular, when the ModA protein is not active, *N. gonorrhoeae* cells can survive better in human cervical epithelial cells [18].

These examples illustrate the immense adaptive capacity TRs can confer to pathogenic prokaryotes. Recent genome-wide studies have shown that in virulent prokaryotes the number of TRs is particularly high, suggesting that variable repeats may serve similar biological roles in both prokaryotic and eukaryotic pathogens [19].

Keeping Internal Clocks Synchronized with External Conditions

Living organisms possess timing mechanisms, called circadian clocks, that maintain the regular rhythm of different biological processes when changes in environmental cues (light intensity, temperature) occur. Therefore, having a circadian clock in tune with the environment provides a competitive advantage in fitness and survival. The duration of the circadian cycle, also known as period, is typically 24 h and different molecular mechanisms responsible for keeping this period in phase with the environment have been identified. Several studies have elegantly illustrated a role for TRs in genes involved in maintaining the circadian rhythm. In the fungus *Neurospora crassa*, for example, intragenic CAG repeats found in the transcription factor white collar-1 (WC-1) play a key functional role. WC-1 controls the expression of a central regulator of the circadian clock [20]. The WC-1 repeats are variable between different

N. crassa strains and this variability correlates dynamically with the length of the circadian period in these strains. Longer CAG repeats correlate with shorter circadian periods in strains collected from low latitudes [21]. These initial observations were validated by experimental data from a cross between 2 *N. crassa* strains with different CAG repeat numbers. In the resulting progeny, repeat length cosegregated with the length of the circadian period, arguing in favor of a role for these CAG repeats in maintaining the rhythm of the biological clock [21].

Another case for a role of variable repeats in adjusting the circadian period came from a study on the *period* (*per*) gene in the fly *Drosophila melanogaster*. The *per* gene is an essential regulatory component of the circadian clock. In natural populations, this gene has a variable hexanucleotide repeat (coding for a threonine-glycine repeat), the 2 most common alleles having 17 and 20 repeats. Natural isolates and transgenic flies with 17 repeats are more suited for warm climates since their 24 h circadian period gets shorter at cooler temperatures. The flies with 20 repeats, however, show better temperature compensation (i.e. less thermal sensitivity) at low temperatures and are thus more favored in colder environments [22]. Here it seems that selection might be favoring long repeats for cold environments and short repeats for warmer ones. This is supported by the findings that in the warmer parts of Europe and North Africa the 17-repeat allele is predominant over the 20-repeat allele in natural populations, whereas the reverse is observed in populations from the colder regions of these continents [22], although distinguishing between neutral variation that segregates with geographic isolation and true adaptation can sometimes be difficult.

Again, contrary to the idea that unstable repeats can only be detrimental, the high mutation rates of microsatellites in genes involved in the circadian clock can be beneficial to organisms, allowing them to robustly adjust their internal clocks to fluctuations in external temperature.

Gross Morphological Variation over Short Evolutionary Timescales

Point mutations in *cis*-regulatory sequences are believed to be the predominant source of the genetic diversity that underlies morphological variation [23]. Fondon and Garner [24] have presented an alternative hypothesis, supported by strong evidence, implicating TR variation as another source of phenotypic diversity. The authors compared genomic and morphological data from different dog breeds, specifically examining repeat polymorphisms in coding regions of developmental genes. Most of these repeats were variable among different breeds and 2 polymorphisms had striking consequences. Dogs with a 51-nt deletion in the repeat region (coding for a proline-glutamine tract) of the *Alx-4* gene have an additional rear claw (polydactyly, a signature feature of the Great Pyrenees breed), whereas dogs with the full-length repeat do not present this feature. Interestingly, the complete loss of *Alx-4* function in knockout mice results in a similar polydactyly. The *Runx-2* gene (a transcription factor controlling osteoblast development) in dogs contains 2 polymorphic repeats, one coding for polyglutamine and the other for polyalanine. The ratio of glutamine over alanine repeats strongly correlates with the degree of dorsoventral nose bend and midface length in different dog breeds [24]. Transcriptional activation of a *Runx-2* target gene was shown to correlate with the polyglutamine over polyalanine ratio in transgenic *Runx-2* constructs, arguing in favor of repeat variation being directly responsible for functional variation in *Runx-2* [25].

The striking aspect about these findings is the pace at which these morphological changes happened. Significant evolution in skull morphology is evident in as little as 50 years. This is even more remarkable if we consider the strong selection against genetic diversity as a result of domestication and inbreeding. The rate of single point mutations can probably not sustain this rapid evolution. These facts, in addition to the high occurrence of microsatellite repeats in genes controlling development and body morphology, provide compelling evidence for TRs as facilitators of rapid morphological evolution in higher eukaryotes [24].

Quantitative Variation in Gene Transcription

A number of studies have shown that intragenic repeat variation in transcription factors directly influences the rate at which target genes are transcribed. The study on *Runx-2* (see above) measured expression of its target gene [25], whereas in a study by Gerber et al. [26] the open reading frame of the *GAL4* transcription factor was fused to different numbers of CAG (codes for glutamine) or CCG (codes for proline) repeats. Here, the rate of mRNA transcription from a reporter gene increased quantitatively with repeat number both in vitro and in vivo, with an optimal number of repeats beyond which no further increase was observed [26]. These studies indicate that variable TRs can provide small quantitative adjustments to gene expression and ultimately small variations in the corresponding phenotype. Although they do not provide a mechanistic explanation for the changes in transcriptional activity as a result of changes in repeat number, it is striking to see that these phenotypes are gradual and quantitative, similar to some examples described so far (see also the section on TRs in promoters). Could the effects of intragenic repeat variation be due to structural variation in the protein sequence that leads to changes in protein-protein or protein-DNA interactions? TR domains in proteins are indeed predicted to be mostly flexible and unstructured loops that could serve as protein interaction domains [27].

Tandem Repeats in Regulatory Sequences

Polymorphic TR sequences do not only occur in coding regions but also within regulatory sequences of genes such as promoters, introns and the 5′ and 3′ UTRs of transcripts. As in coding regions, microsatellites in gene regulatory regions fulfill functional roles. An increasing number of studies link microsatellite polymorphisms to variation in gene expression (see below). In this section, we will first explore how unstable microsatellites in regulatory regions of genes cause prevalent human neurodegenerative diseases by modulating gene expression. The second part of this section will be devoted to the various mechanisms that unstable TRs in gene regulatory regions use to impact gene expression. Finally, we discuss how microsatellites in regulatory regions are associated with the rapid evolution of gene expression.

Tandem Repeats and Human Neurodegenerative Diseases: the Case of the FMR1 Gene
As mentioned in the first section of this chapter, polymorphic TRs in the human genome are most notably known for underlying several neurodegenerative and neuromuscular diseases. Whereas many disease-linked repeats are located within coding regions (see above), several other diseases are caused by repeat expansions in introns and UTRs. Here we discuss the case of the *FMR1* gene where expansion of CGG repeats in the 5′ UTR can result in 2 discrete pathologies. The fragile X syndrome (FXS), one of the most common forms of inherited mental retardation, occurs when the trinucleotide repeat is massively expanded (>200 repeats) [9]. Such a 'full mutation' leads to increased methylation of the CpG island in the 5′ UTR as well as to decreased histone acetylation at the 5′ end of *FMR1*. These epigenetic changes result in transcriptional silencing of the *FMR1* gene and loss of FMRP expression through the inhibition of transcription factor binding. The FMRP protein is an RNA-binding protein that plays a crucial role in intracellular RNA transport and in the regulation of translation of target mRNAs. The absence of FMRP leads to an uncontrolled synthesis of proteins involved in cytoskeletal structure, synaptic transmission, and neuronal maturation [28].

A different pathology emerges when a so-called 'premutation' allele of the *FMR1* gene is present (55–200 repeats). The fragile X-associated tremor/ataxia syndrome (FXTAS), causing tremors, balance problems and dementia which progressively worsen over time, appears to be an RNA-mediated gain-of-function phenotype. Here, the formation of a premutation mRNA that impairs protein synthesis results in reduced levels of FMRP. This in turn leads to elevated production of the *FMR1* mRNA, and the consequent 2–10-fold higher levels of the transcript presumably trigger an over-interaction with trinucleotide-binding proteins and translational factors. As a consequence, the amount of these proteins and factors may be reduced in the cell pool, thus impairing several cellular processes [29].

Tandem Repeats Modulating Gene Expression: a Plethora of Mechanisms
An increasing number of studies reveal a link between microsatellite polymorphisms in regulatory sequence elements and variation in gene expression, function, or both. In most cases, more research needs to be conducted to confirm that changes in TR copy number lead to altered gene expression levels. However, a number of studies have uncovered a wide range of mechanisms that link microsatellite variation in gene regulatory regions to changes in gene expression. In this section, we will take an in-depth look at these mechanisms by focusing on some notable examples.

Polymorphic Tandem Repeats Overlapping with Regulatory Protein Binding Sites Modulate Gene Expression

TRs can overlap with binding sites for regulatory proteins such as transcription factors. Hence, variations in the number of repeat units can modulate gene expression by altering the number of transcription factor binding sites. For example, in the bacterial pathogen *Neisseria meningitidis*, the promoter of the gene encoding the NadA adhesion protein contains TAAA repeats. The stochastic loss or gain of repeat units alters the binding of the IHF transcription factor, resulting in phase-variable transcription (see also above) [30]. Similarly, in humans, the number of TCC repeats in the promoter of the epidermal growth factor (EGF) modulates binding of the Sp1 transcriptional regulator to the promoter, thereby affecting EGF expression [31].

Microsatellites that contain transcription factor binding sites have also been shown to play a role in human cancer development. The *PIG3* gene is activated by the tumor suppressor p53 which interacts with a pentanucleotide TR sequence in the *PIG3* promoter. The number of repeat units directly correlates with the extent of transcriptional activation by p53, raising the intriguing possibility that an individual's susceptibility to cancer might show some dependence on the number of repeat units [32]. Another example occurs in Ewing's sarcoma, a malignant bone tumor that affects children. Here, a chromosomal translocation causes the formation of the EWS/FLI fusion oncoprotein. EWS/FLI functions as an aberrant transcription factor that modulates expression of several target genes through a length-dependent interaction with a GGAA microsatellite in the promoter of these genes [33].

TR polymorphisms affecting transcription factor binding are not only at play in promoter regions but also in other regulatory regions of the genome. For example, the tyrosine hydroxylase (*TH*) gene contains a TCAT repeat in the first intron. Here, variation in repeat copy number correlates with changes in the binding of the transcription factor ZNF191 [34]. These and other examples illustrate that microsatellites overlapping with regulatory protein binding sites may cause dynamic changes in gene expression.

Tandem Repeat Polymorphisms Affect Spacing between Functional Elements in Promoters

A number of studies show that polymorphic TRs located between 2 regulatory elements in a promoter can affect transcription rates by changing the distance between these regulatory sequences. Non-optimal distances between functional elements often affect transcriptional initiation. This mechanism is employed by bacteria to regulate phase variation (see also above). For example, in *Mycoplasma hyorhinis*, antigenic diversity is generated by combinatorial expression and phase variation of multiple surface lipoproteins (Vlps). Rapid on-off switching of the *vlp* genes is achieved by length variation of a poly(A) tract located between the –10 and –35 consensus sequences in their promoter region, presumably affecting the position of the RNA polymerase [35]. The high variability in TR size allows for the rapid production of variants in a pathogenic bacterial population, which increases the chances of evading the host immune system. Variations in the length of different microsatellites between the –10 and the –35 domains have also been reported for genes encoding adhesive and immunogenic surface proteins in other pathogens, including the *hifA* and *hifB* genes in *H. influenzae* (AT repeat) [36].

Tandem Repeats Influence Chromatin Structure

Intergenic TRs have been shown to affect gene expression through their effects on the chromatin structure of gene promoters. Chromatin is the complex of DNA and associated proteins in eukaryotes and archaeans. The basic unit of chromatin is the nucleosome which is created when a stretch of DNA is bound by a complex of histone proteins. Nucleosome formation has a large impact on many processes involving DNA, including transcription, because DNA that is part of a nucleosome is less accessible to other regulatory proteins. The underlying DNA sequence determines in large part where nucleosomes are formed. Most notably, homopolymeric TRs, primarily poly(A) and poly(T) stretches, have been identified as strong nucleosome deterring sequences [37]. A recent study by Vinces et al. [38] showed that nucleosome-free regions of promoters in yeast and humans are enriched for TRs, pointing towards a common role for TRs in nucleosome positioning. Moreover, in the same study, dinucleotide AT-rich repeats in the promoters of the yeast *YHB1, MET3* and *SDT1* genes were mutated, resulting in altered gene expression and nucleosome positioning [38]. Thus, microsatellites are common functional elements in promoters. They affect the local nucleosome organization, and frequent mutations in the unstable repeat tracts cause changes in the expression levels of the corresponding gene, thereby making these promoters more 'evolvable' (see also below) [38]. It is important to note that although nucleosome-deterring AT-rich repeats are extremely common in promoters, not all TRs act as nucleosome deterring sequences. In fact, some GC-rich repeats may stimulate nucleosome formation. For example, CTG triplet repeats promote the positioning of nucleosomes, and a $(CTG)_{12}$ repeat has been shown to promote gene expression in a reporter system [39].

Tandem Repeats and RNA Structure

Long microsatellites can form hairpin RNA structures which in turn can affect RNA processing, stability and translation through various mechanisms. For example, the expanded CUG repeats in the UTR of the myotonic dystrophy protein kinase (*DMPK*) gene transcript can form very stable hairpin structures, which may be able to recruit and activate PKR, a double-stranded RNA-binding and pro-apoptotic kinase [40]. In addition, the hairpin structures in the *DMPK* gene sequester the muscleblind-like 1 (MBNL1) splicing factor and cause misregulation of the alternative splicing of multiple genes [41]. RNA splicing is also affected by expansion of a TG repeat in the cystic fibrosis transmembrane conductance regulator (*CFTR*) gene. Expanded alleles of this TR form stable secondary structures that can result in decreased RNA splicing efficiency and cause non-classic cystic fibrosis [42].

Repeats in mRNA may also affect translation. An example is the accumulation of toxic, homopolymeric proteins that may contribute to pathogenicity in several trinucleotide expansion diseases. These proteins form when expanded, hairpin-forming CUG and CTG trinucleotide repeats undergo translation in the absence of an ATG start codon through a process termed repeat-associated non-ATG (RAN) translation [43].

Tandem Repeats Affect DNA Structure

The most common form of DNA is the right-handed double helix known as B-DNA. Microsatellite sequences are capable of generating the rather unusual Z-DNA, a left-handed double helical structure of DNA. The formation of Z-DNA is generally not favored but is promoted under certain conditions, including the presence of alternating purines and pyrimidines in TR sequences. The uncommon structure of Z-DNA influences gene expression and Z-DNA by itself promotes transcription [44]. In addition, the drastically different structure of Z-DNA affects the binding of proteins to the DNA. For example, a CA repeat upstream of the rat prolactin gene forms Z-DNA and inhibits gene transcription, presumably by inhibiting the transcriptional efficiency of RNA polymerase II [45]. The formation of Z-DNA may also influence gene expression by recruiting Z-DNA-binding proteins. An example is the double-stranded RNA adenosine deaminase (ADAR1) protein that binds promoter Z-DNA and activates gene expression [44].

Regulatory Tandem Repeats Are Associated with Rapid Evolution of Gene Expression

The examples discussed in the section above convincingly illustrate how microsatellites, by means of their inherent instabilities, act as hypervariable sequence elements that control levels of gene expression. Unstable TRs may impact the evolution of gene expression by creating diversity in populations and subsequently allowing quick Darwinian evolution and adaptation. A study by Vinces et al. [38] confirmed that TRs in promoters can support rapid evolution of gene expression (fig. 2). Here, reporter constructs allowing selection for different levels of gene expression were subjected to several rounds of selection. As a result, promoter constructs containing TRs yielded many variants with higher expression levels, and expression level changes were linked to changes in TR number (fig. 2), demonstrating how unstable repeats can stimulate 'evolvability' [38].

Another example is the promoter of the human matrix metalloproteinase-3 (*MMP3*) gene which contains homopolymeric poly(T) sequences. Polymorphisms in one of these poly(T) stretches have been well-documented and are associated with heart disease. It appears that in the European population there has been positive selection for a shorter allele that is associated with higher levels of *MMP3* expression and myocardial infarction and aneurysm. Moreover, the polymorphic tracts of *MMP3* evolve rapidly among several primates, suggesting that they are mutational

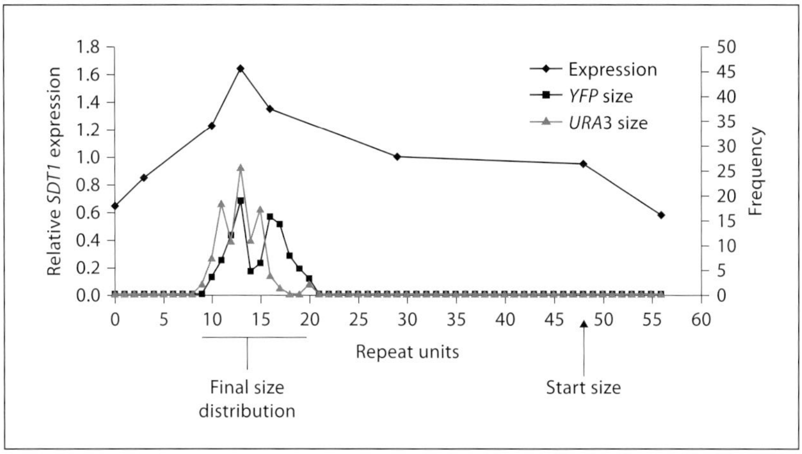

Fig. 2. Experimental evolution of gene expression mediated by TRs. The *S. cerevisiae SDT1* promoter was inserted upstream of the *URA3* and *YFP* reporter genes and the resulting mutant strains were subjected to selection for higher expression (using the reporter gene functions as selectable traits). After just a few rounds of selection, this resulted in changes in the number of TR units in the *SDT1* promoter, of which many were linked to higher expression levels. The starting strain (48 repeat units, indicated by arrow) has relatively low expression (diamonds, left axis). The *URA3* (triangles) and *YFP* (squares) lines represent the final size distributions of TRs after selection for higher expression by using *URA3* or *YFP* reporters, respectively. TR size distribution without selection remains mostly at the initial 48 units (see [38] for details).

hotspots that drive rapid evolution of *MMP3* gene expression and its associated phenotypes [46].

It is important to note that the examples in this text might only represent the tip of the iceberg. Unstable microsatellite repeats might impact evolution of gene expression on a much larger scale than what has been documented to date. A genome-wide study among different yeast strains and species revealed that promoters containing TRs evolved faster, resulting in more divergent gene expression patterns [38]. In humans, analysis of polymorphic regions revealed their overrepresentation in regulatory regions where they may have a significant impact on expression variation. This suggests that variable TRs may be important sources of genetic variation that drive the evolution of gene expression [47, 48].

Conclusions

Here, we have focused on the biological roles of TRs, namely in the evolution of gene function and expression. The studies described in this chapter demonstrate the benefits of unstable TRs as facilitators of rapid functional changes allowing for increased fitness and accelerated adaptation to novel conditions. The ability of TRs to facilitate

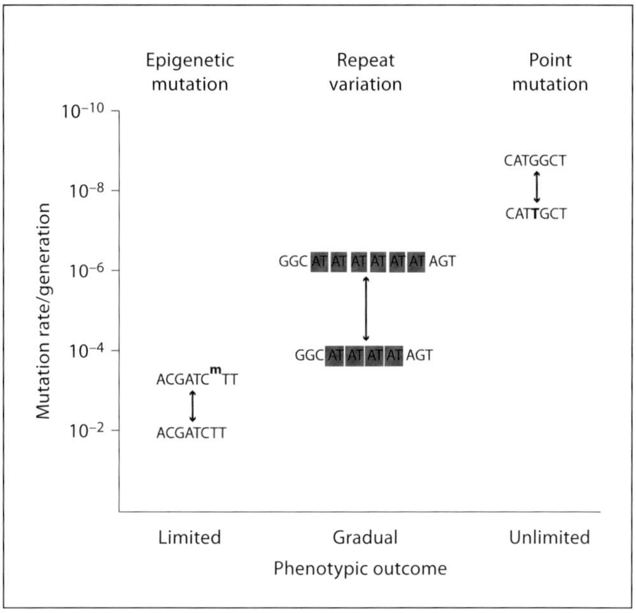

Fig. 3. The time scale of different mutations and their underlying phenotypic changes. The rate of TR mutations lies between that of epigenetic changes, such as cytosine methylation, and point mutations (e.g. G to T mutation). The phenotypic outcome of repeat variation is more subtle and gradual than the on-off changes resulting from epigenetic changes, but more limited than the phenotypes that can result from point mutations. As such, variable TRs help bridge the gap between rapidly-evolving hypervariable epigenetic traits, and slow-evolving, robust genetic variation.

phenotypic evolution is mostly due to their instability. With mutation rates of 10^{-2} and 10^{-5} per generation, repeat instability lies between that of point mutations (10^{-8} to 10^{-9}) and the more frequent epigenetic mutations (10^{-1} to 10^{-2}) (fig. 3). Like epigenetic changes, variations in repeats are reversible and probably only have a limited biological potential (e.g. fine-tuning or regulating a specific existing function, rather than generating completely novel functions as is possible with mutations in non-repetitive DNA). Therefore, in line with being a major source of genetic variation in genomes [49], TRs may provide genomes with a type of mutational capacity and dynamics that bridges the gap between the highly variable, but functionally limited, epigenetic mutations and the rare, but functionally unlimited, point mutations.

It is interesting to note that, in contrast to most epigenetic traits which often act as switches, variable repeats can change in small incremental steps, which may in turn lead to small, incremental changes in quantitative (continuously changing) phenotypes. Such quantitative phenotypes are often believed to depend on complex genetic interactions and multiple genes dubbed 'quantitative trait loci'. However, things do not always need to be complex, since variation in 1 repeat located within 1 gene or regulatory region could also underlie a quantitative trait. In the case of most intragenic repeats, for example, variation will only result in a shorter or longer protein and consequently a gradual quantitative adjustment in gene function, as illustrated by the study on dog skeletal morphology [24]. Similarly, gradual changes in gene expression can also result from repeat variation in promoters [38]. The examples provided here show that TRs can confer quantitative changes in phenotypes through a simple monogenic mechanism. We should point out that some repeat changes can also mediate a

Table 1. Key studies implicating TRs as modulators of rapid and efficient functional diversity

Keyword	Impact and interest	Reference
Tuning knobs	One of the earliest reviews highlighting the benefits of variable tandem repeats and proposing the role of evolutionary significant, genetic 'tuning knobs'.	[50]
Morphological evolution	A significant study implicating tandem repeats as a source of phenotypic diversity in domestic dogs.	[24]
Phase variation	Illustrates how variable repeats in promoters alter transcription factor binding.	[30]
Functional variability	Experimentally validates the role of tandem repeats in mediating gradual quantitative phenotypic changes.	[6]
Transcriptional evolvability	Provides experimental evidence that variable tandem repeats in promoters can mediate changes in gene expression and promote transcriptional evolvability.	[38]

binary switch (on- or off-switching between phenotypes) as seen in bacterial phase variation.

The question remains whether TRs are selected for or just form through polymerase slippage and are mostly neutral unless they expand to a critical number. The multiple examples described here suggest that some TRs are useful to genomes and therefore may be selected for, allowing them to spread through the population, reach fixation and even become conserved among various species. Variable TRs may provide genomes with a degree of evolutionary flexibility with minimum risk, while other parts of the genome remain stable and robust [50]. Some repeats may form in parts of the genome where instability is deleterious. These will most likely be selected against. Finally, it is not implausible that many repeats may be (virtually) neutral, or in other words, true 'junk DNA'.

In Depth Readings

In table 1, we highlight a number of landmark papers that contributed to our understanding of variable TRs as functionally relevant DNA sequences with an impact on the evolution of gene function or expression.

Acknowledgements

The authors would like to apologize for the omission of multiple relevant studies due to space limitations. Research in the lab of K.J.V. is supported by Human Frontier Science Program HFSP RGY79/2007, ERC Young Investigator Grant 241426, the EMBO YIP program, VIB, KU Leuven, IWT, the FWO-Odysseus program and the AB InBev Baillet-Latour foundation. R.G. acknowledges an F+ fellowship from the KU Leuven.

References

1 Hartl DL: Molecular melodies in high and low C. Nat Rev Genet 2000;1:145–149.

2 van Belkum A, Scherer S, van Alphen L, Verbrugh H: Short-sequence DNA repeats in prokaryotic genomes. Microbiol Mol Biol Rev 1998;62:275–293.

3 Gemayel R, Vinces MD, Legendre M, Verstrepen KJ: Variable tandem repeats accelerate evolution of coding and regulatory sequences. Ann Rev Genet 2010;44:445–477.

4 Legendre M, Pochet N, Pak T, Verstrepen KJ: Sequence-based estimation of minisatellite and microsatellite repeat variability. Genome Res 2007;17:1787–1796.

5 Orgel LE, Crick FH: Selfish DNA: the ultimate parasite. Nature 1980;284:604–607.

6 Verstrepen KJ, Jansen A, Lewitter F, Fink GR: Intragenic tandem repeats generate functional variability. Nat Genet 2005;37:986–990.

7 Young ET, Sloan JS, Van Riper K: Trinucleotide repeats are clustered in regulatory genes in *Saccharomyces cerevisiae*. Genetics 2000;154:1053–1068.

8 Laspada AR, Wilson EM, Lubahn DB, Harding AE, Fischbeck KH: Androgen receptor gene-mutations in X-linked spinal and bulbar muscular atrophy. Nature 1991;352:77–79.

9 Verkerk A, Pieretti M, Sutcliffe JS, Fu YH, Kuhl DPA, et al: Identification of a gene (FMR-1) containing a CGG repeat coincident with a breakpoint cluster region exhibiting length variation in fragile-X syndrome. Cell 1991;65:905–914.

10 Macdonald ME, Ambrose CM, Duyao MP, Myers RH, Lin C, et al: A novel gene containing a trinucleotide repeat that is expanded and unstable on Huntingtons-disease chromosomes. Cell 1993;72:971–983.

11 Orr H, Zoghbi H: Trinucleotide repeat disorders. Ann Rev Neurosci 2007;30:575–621.

12 Rubinsztein DC, Leggo J, Coles R, Almqvist E, Biancalana V, et al: Phenotypic characterization of individuals with 30–40 CAG repeats in the Huntington disease (HD) gene reveals HD cases with 36 repeats and apparently normal elderly individuals with 36–39 repeats. Am J Hum Genet 1996;59:16–22.

13 Nucifora FC Jr, Sasaki M, Peters MF, Huang H, Cooper JK, et al: Interference by huntingtin and atrophin-1 with CBP-mediated transcription leading to cellular toxicity. Science 2001;291:2423–2428.

14 Wolfe KH, Shields DC: Molecular evidence for an ancient duplication of the entire yeast genome. Nature 1997;387:708–713.

15 Stern A, Brown M, Nickel P, Meyer TF: Opacity genes in *Neisseria-gonorrhoeae* – control of phase and antigenic variation. Cell 1986;47:61–71.

16 Weiser JN, Love JM, Moxon ER: The molecular mechanism of phase variation of *H. influenzae* lipopolysaccharide. Cell 1989;59:657–665.

17 De Bolle X, Bayliss CD, Field D, van de Ven T, Saunders NJ, et al: The length of a tetranucleotide repeat tract in *Haemophilus influenzae* determines the phase variation rate of a gene with homology to type III DNA methyltransferases. Mol Microbiol 2000;35:211–222.

18 Srikhanta YN, Dowideit SJ, Edwards JL, Falsetta ML, Wu HJ, et al: Phasevarions mediate random switching of gene expression in pathogenic *Neisseria*. PLoS Pathog 2009;5:e1000400.

19 Mrazek J, Guo XX, Shah A: Simple sequence repeats in prokaryotic genomes. Proc Natl Acad Sci USA 2007;104:8472–8477.

20 Froehlich AC, Liu Y, Loros JJ, Dunlap JC: White collar-1, a circadian blue light photoreceptor, binding to the frequency promoter. Science 2002;297:815–819.

21 Michael TP, Park S, Kim TS, Booth J, Byer A, et al: Simple sequence repeats provide a substrate for phenotypic variation in the *Neurospora crassa* circadian clock. PLoS One 2007;2:e795.

22 Sawyer LA, Hennessy JM, Peixoto AA, Rosato E, Parkinson H, et al: Natural variation in a *Drosophila* clock gene and temperature compensation. Science 1997;278:2117–2120.

23 Carroll SB: Endless forms: the evolution of gene regulation and morphological diversity. Cell 2000;101:577–580.

24 Fondon JW, Garner HR: Molecular origins of rapid and continuous morphological evolution. Proc Natl Acad Sci USA 2004;101:18058–18063.

25 Sears KE, Goswami A, Flynn JJ, Niswander LA: The correlated evolution of *Runx2* tandem repeats, transcriptional activity, and facial length in Carnivora. Evol Dev 2007;9:555–565.

26 Gerber HP, Seipel K, Georgiev O, Hofferer M, Hug M, et al: Transcriptional activation modulated by homopolymeric glutamine and proline stretches. Science 1994;263:808–811.

27 Simon M, Hancock JM: Tandem and cryptic amino acid repeats accumulate in disordered regions of proteins. Genome Biol 2009;10:R59.

28 Brown V, Jin P, Ceman S, Darnell JC, O'Donnell WT, et al: Microarray identification of FMRP-associated brain mRNAs and altered mRNA translational profiles in fragile X syndrome. Cell 2001;107:477–487.

29 Tassone F, Iwahashi C, Hagerman PJ: FMR1 RNA within the intranuclear inclusions of fragile X-associated tremor/ataxia syndrome (FXTAS). RNA Biol 2004;1:103–105.
30 Martin P, Makepeace K, Hill SA, Hood DW, Moxon ER: Microsatellite instability regulates transcription factor binding and gene expression. Proc Natl Acad Sci USA 2005;102:3800–3804.
31 Johnson AC, Jinno Y, Merlino GT: Modulation of epidermal growth factor receptor proto-oncogene transcription by a promoter site sensitive to S1 nuclease. Mol Cell Biol 1988;8:4174–4184.
32 Contente A, Dittmer A, Koch MC, Roth J, Dobbelstein M: A polymorphic microsatellite that mediates induction of *PIG3* by p53. Nat Genet 2002;30:315–320.
33 Gangwal K, Sankar S, Hollenhorst PC, Kinsey M, Haroldsen SC, et al: Microsatellites as EWS/FLI response elements in Ewing's sarcoma. Proc Natl Acad Sci USA 2008;105:10149–10154.
34 Albanese V, Biguet NF, Kiefer H, Bayard E, Mallet J, Meloni R: Quantitative effects on gene silencing by allelic variation at a tetranucleotide microsatellite. Hum Mol Genet 2001;10:1785–1792.
35 Yogev D, Rosengarten R, Watson-McKown R, Wise KS: Molecular basis of *Mycoplasma* surface antigenic variation: a novel set of divergent genes undergo spontaneous mutation of periodic coding regions and 5′ regulatory sequences. EMBO J 1991; 10:4069–4079.
36 van Ham SM, van Alphen L, Mooi FR, van Putten JP: Phase variation of *H. influenzae* fimbriae: transcriptional control of two divergent genes through a variable combined promoter region. Cell 1993;73: 1187–1196.
37 Iyer V, Struhl K: Poly(dA:dT), a ubiquitous promoter element that stimulates transcription via its intrinsic DNA structure. EMBO J 1995;14:2570–2579.
38 Vinces MD, Legendre M, Caldara M, Hagihara M, Verstrepen KJ: Unstable tandem repeats in promoters confer transcriptional evolvability. Science 2009;324:1213–1216.
39 Tomita N, Fujita R, Kurihara D, Shindo H, Wells RD, Shimizu M: Effects of triplet repeat sequences on nucleosome positioning and gene expression in yeast minichromosomes. Nucleic Acids Res Suppl 2002;231–232.
40 Tian B, White RJ, Xia T, Welle S, Turner DH, et al: Expanded CUG repeat RNAs form hairpins that activate the double-stranded RNA-dependent protein kinase PKR. RNA 2000;6:79–87.
41 Mykowska A, Sobczak K, Wojciechowska M, Kozlowski P, Krzyzosiak WJ: CAG repeats mimic CUG repeats in the misregulation of alternative splicing. Nucleic Acids Res 2011;39:8938–8951.
42 Hefferon TW, Groman JD, Yurk CE, Cutting GR: A variable dinucleotide repeat in the *CFTR* gene contributes to phenotype diversity by forming RNA secondary structures that alter splicing. Proc Natl Acad Sci USA 2004;101:3504–3509.
43 Zu T, Gibbens B, Doty NS, Gomes-Pereira M, Huguet A, et al: Non-ATG-initiated translation directed by microsatellite expansions. Proc Natl Acad Sci USA 2011;108:260–265.
44 Oh DB, Kim YG, Rich A: Z-DNA-binding proteins can act as potent effectors of gene expression in vivo. Proc Natl Acad Sci USA 2002;99:16666–16671.
45 Naylor LH, Clark EM: d(TG)n.d(CA)n sequences upstream of the rat prolactin gene form Z-DNA and inhibit gene transcription. Nucleic Acids Res 1990; 18:1595–1601.
46 Rockman MV, Hahn MW, Soranzo N, Loisel DA, Goldstein DB, Wray GA: Positive selection on *MMP3* regulation has shaped heart disease risk. Curr Biol 2004;14:1531–1539.
47 Rockman MV, Wray GA: Abundant raw material for *cis*-regulatory evolution in humans. Mol Biol Evol 2002;19:1991–2004.
48 Tirosh I, Barkai N, Verstrepen KJ: Promoter architecture and the evolvability of gene expression. J Biol 2009;8:95.
49 Tautz D, Trick M, Dover GA: Cryptic simplicity in DNA is a major source of genetic variation. Nature 1986;322:652–656.
50 King DG, Soller M, Kashi Y: Evolutionary tuning knobs. Endeavour 1997;21:36–40.

Kevin J. Verstrepen
Laboratory for Systems Biology, VIB
Gaston Geenslaan 1
BE–3001 Heverlee (Belgium)
E-Mail Kevin.Verstrepen@biw.vib-kuleuven.be

Garrido-Ramos MA (ed): Repetitive DNA.
Genome Dyn. Basel, Karger, 2012, vol 7, pp 126–152

Satellite DNA Evolution

M. Plohl · N. Meštrović · B. Mravinac

Ruđer Bošković Institute, Zagreb, Croatia

Abstract

Satellite DNAs represent the most abundant fraction of repetitive sequences in genomes of almost all eukaryotic species. Long arrays of satellite DNA monomers form densely packed heterochromatic genome compartments and also span over the functionally important centromere locus. Many specific features can be ascribed to the evolution of tandemly repeated genomic components. This chapter focuses on the structural and evolutionary dynamics of satellite DNAs and the potential molecular mechanisms responsible for rapid changes of the genomic areas they constitute. Monomer sequences of a satellite DNA evolve concertedly through a process of molecular drive in which mutations are homogenized in a genome and fixed in a population. This process results in divergence of satellite sequences in reproductively isolated groups of organisms. However, some satellite DNA sequences are conserved over long evolutionary periods. Since many satellite DNAs exist in a genome, the evolution of species-specific satellite DNA composition can be directed by copy number changes within a library of satellite sequences common for a group of species. There are 2 important features of these satellite DNAs: long time sequence conservation and, at the same time, proneness to rapid changes through copy number alterations. Sequence conservation may be enhanced by constraints such as those imposed on functional motifs and/or architectural features of a satellite DNA molecule. Such features can limit the selection of sequences able to persist in a genome, and can direct the evolutionary course of satellite DNAs spanning the functional centromeres.

Copyright © 2012 S. Karger AG, Basel

Satellite DNAs are repetitive DNA sequences arranged as arrays of highly abundant, head-to-tail tandem repetitions that commonly outmatch 50% of the genome [1, 2]. The term satellite DNA has its origin in early gradient centrifugation experiments, in which genomic fragments composed of tandem sequences were isolated from auxiliary or satellite bands because of the different buoyant density of satellite sequences compared to the main fraction of genomic DNA.

Cytogenetic mapping experiments defined satellite DNA sequences as the main component of heterochromatic genome compartments, transcriptionally suppressed, gene-poor chromosomal regions that remain condensed throughout the cell cycle. In this regard, satellite DNAs span centromeric and pericentromeric areas, but can also be located at subtelomeric chromosomal positions. In some species, satellite DNAs and heterochromatin are also placed at interstitial chromosomal locations. Many satellite DNAs exist in a genome and among species. Satellite DNAs differ in nucleotide sequence, sequence complexity, repeat unit length and abundance, with only 2 characteristics shared: ability to build long arrays of tandem head-to-tail arranged repeats and ability to form heterochromatic regions.

Despite decades of intensive research, our knowledge concerning functional significance of satellite DNAs is limited, and a decades-long debate evaluating the functionality of these sequences still goes on. Established in the early 1970's, the 'junk DNA' hypothesis treated satellite DNAs as a useless genomic portion, deposited in heterochromatin as in a kind of a genomic junkyard [3]. Absence of coding potential, extreme diversity of satellite DNAs and lack of direct evidence for any possible function(s) made this hypothesis attractive for a long period of time. Consequently, satellite DNAs were often recognized as monotonous and useless material, able to accumulate until they become a too heavy load for a genome. However, based on the opposite view, efforts were done in order to understand the functional potential of this genomic fraction [2]. These studies were inspired by an elementary question: if satellite DNAs and heterochromatin do not bring any evolutionary benefit, why are so large genomic fractions not lost or diminished, particularly in the light of mechanisms of efficient (hetero)chromatin diminution in somatic cell lines of different organisms (e.g. [4]). Recent comprehensive studies fostered by advances in methodological approaches started to accumulate experimental data pointing to the functional roles. In the first instance, it is now clearly evident that satellite DNAs have impact on genomic functions at the higher level, such as chromosome organization and pairing. A new, challenging field addresses the impact of satellite DNA transcripts, particularly on formation and maintenance of heterochromatin structure [5]. Among other possible roles, it has recently been evidenced that species specificity of DNA sequences in *Drosophila* heterochromatin affects chromosome segregation in hybrids [6]. This result supports the assumption that evolutionary dynamics of satellite DNAs can raise reproductive barriers among groups of organisms and trigger speciation [7, 8].

Satellite DNAs and heterochromatin are still the least understood genomic compartments, underrepresented and neglected in outputs of every genome project. In addition to a possible lack of interest caused by historical arguments explained above, the major reason lies in the specificities of evolution of DNA sequences repeated in tandem, which is governed by different rules and mechanisms compared to those affecting any other genomic component. As a result of this particular pattern, satellite DNAs characterize arrays composed of nearly identical repeating units, making long fragments hardly accessible for current sequencing, assembly and mapping techniques. This obstacle is well illustrated by the genome sequencing project output of the insect model organism *Tribolium castaneum* in which the major satellite DNA, comprising 17% of the genome, is represented by only 0.3% in the assembled sequence

[9]. A broad consequence of the assembly problem is that long-range sequential organization of satellite repeats has been studied in heterochromatin of very few species in which compositions and variability of DNA sequences allow overlapping of sequenced segments into contigs (e.g. [10, 11]). On the contrary, on the level of basic repeating units or satellite DNA monomers, hundreds of satellite sequences were characterized in organisms belonging to all major taxonomic groups of plant and animal species and at different levels of genome complexity, from simple organisms such as nematodes to the complex ones as humans. Novel satellite sequences are normally detected by ladder-like bands of electrophoretically separated genomic DNA degraded by restriction endonucleases. A characteristic ladder pattern results from tandem organization of satellite monomers. It is obtained by partial degradation of consecutive repeats as a consequence of sequence variability within the restriction site. Cloning of DNA material eluted from ladder-like bands enables further, sequence-detailed analysis of monomeric satellite DNA units. Even in the genomic era, available structural and evolutionary information has mostly been obtained by analysis of individual monomers and short multimers detected and cloned in this simple way.

Extreme diversity of satellite DNAs and limited view on long-range organization and functional significance raise major difficulties in deriving general conclusions about origins and evolution of satellite repeats. In addition, despite a large number of characterized satellite sequences, studies focused on specific evolutionary questions are scarce and done on a hardly representative number of species and species groups. From the most general point of view, these studies revealed high complexity of satellite DNA evolution and specificities of particular systems. Based on experimental data and postulated models, the time has come to integrate views on principles that define evolution of satellite DNAs in particular species as well as in general. This review addresses some of the most intriguing questions of satellite DNA evolution, most of them still only partially understood. For example, are there requirements on satellite DNA monomers or can any sequence persist amplified as a satellite DNA? What determines sequence dynamics of a satellite DNA? Why do some satellite DNAs remain conserved during tens of millions years (Myr), whereas some accumulate species-specific mutations in a short evolutionary period? What mechanisms are exactly involved in spread and maintenance of the particular satellite DNA in a particular species? Why do satellite repeats remain chromosome-limited in some species, while distributed over the whole chromosome complement in others? In addition to the interpretation of basic concepts in satellite DNA evolution, these and other related questions will be addressed in the following overview.

Satellite DNA Sequence Features

Satellite DNA monomers differ in length and nucleotide sequence, and many satellite DNAs exist or can be formed by amplifying sequences of different structural features, complexity and evolutionary origin. Monomer length between 150–180 bp and 300–360 bp was observed in many satellite DNAs and can be considered as evolutionarily favored (e.g. in insects [12]). Nevertheless, it is difficult to propose a generalized assumption about possible constraints affecting this feature because satellites with repeat unit lengths in the whole range, from only few nucleotides to over 1 kb, are far from being exceptional. The current hypothesis links preferred monomer length and the length of DNA wrapped around 1 or 2 nucleosomes as a requirement that may facilitate regular phasing of nucleosomes in the heterochromatin [13]. So far there has been no efficient experimental assay to verify this assumption, but rare length-affecting mutations observed in sequence analyses of many satellites support the idea about significance of monomer length.

Functional features of some satellite repeats can be encoded by short nucleotide sequence elements residing within a satellite monomer, and monomers may appear as simple carriers of a functional motif. It is therefore assumed that functional segments should evolve under constraints, while monomer sequence outside the motif can accumulate variability in a more rapid way. Insertions and deletions are scarce, even in variable regions, and the monomer length often remains fixed, what can be further explained as the need for proper spacing of interaction sites. Sequence comparisons revealed alternations of mutation-rich and mutation-poor sequence segments within satellite DNA monomers of various animal and plant species such as human and *Arabidopsis* [14, 15]. Conserved sequence segments may represent yet uncharacterized motifs involved in interactions with structural components of heterochromatin, or may participate in homologous recombination as regions of increased similarity [16]. An exceptionally complex pattern of sequence variability was found in a family of satellite DNAs of root-knot nematode species from the genus *Meloidogyne*, stressing the role of selective constraints in formation of novel satellites [17]. Short conserved motifs detected between centromeric satellite DNAs of rice and maize may represent functional elements originating from the ancestral sequence, arising about 50–70 Myr ago [18]. Sequence motifs residing within satellite DNAs may also represent sequence determinants in epigenetic modifications. Methylation-sensitive sites in clustered satellite monomer variants direct epigenetic modifications and help to differentiate pericentromeric heterochromatin from centromeric chromatin (centrochromatin) in *Arabidopsis* and maize [19].

Among potential sequence motifs, the CENP-B box is the best explored example. The CENP-B box is the 17-bp-long sequence segment embedded within 171-bp-long monomers of alpha satellite DNA of higher primates, the diverse family of centromeric satellites present on all chromosomes except on Y [20]. The motif probably facilitates kinetochore formation through binding the centromere-associated protein CENP-B [21, 22]. In the conserved functional form, it is present in approximately 23% of satellite monomers existing in the human genome, while changes in the nucleotide sequence make interaction inefficient in the rest. Monomers bearing the functional motif are dispersed among those with non-functional sites, thus providing a spatial arrangement considered to be needed for functionality of the CENP-B protein [23]. Similarity of the CENP-B protein to transposase of the mobile element *Tigger* opens a speculative possibility that this regularly distributed DNA-protein complex may promote recombination processes involved in maintenance of the satellite DNA array [24]. Sequence motifs resembling the CENP-B box were recognized in unrelated satellite DNAs of evolutionarily distant species, for example in the mouse [25] and Antarctic scallop *Adamussium colbecki* [26], and were hypothesized to bear similar functional roles. Other not so well defined sequence segments of alpha and other satellites may have potential to recognize different proteins involved in centromere formation and maintenance, such as CENP-C [27].

Despite differences in the nucleotide sequence of satellite monomers, secondary and tertiary structures of the DNA molecule, namely dyad structures and sequence-induced bent helix axis, respectively, may represent characteristics needed for certain functional interactions. Different combinations of nucleotides can build similar structures which might be evolving under constraints, while the sequence can be altered as long as the structure itself is not impaired. Bent helix axis of the satellite monomer and a resulting structure of the DNA molecule composed of tandemly repeated monomers are induced by periodic distribution of nucleotides, particularly by distribution of short tracts of As and/or Ts phased with a turn of double helix [28]. Approximately 50% of satellite DNA sequences are able to build tertiary structures prominent enough to be of structural and/or functional significance [29]. Structures induced by bent helix axis may be involved in specific recognition of DNA-binding protein components of the heterochromatin, since intercalary compounds releasing these structures also impair the binding affinity [30]. Evolutionary constraints imposed both on the monomer length and on the structure induced by bent DNA may explain evolution of satellite DNA from the beetle *Palorus subdepressus*. In this species, two 72-bp-long repeat variants were amplified together as subrepeats in a new, composite satellite DNA monomer conforming the repeat unit length of 142–144 bp and the tertiary structure, features shared with other, non-homologous satellites of related congeneric species [31].

Inversely duplicated sequence segments with a potential to form dyad structures are components of many satellite DNAs. It was proposed that such structures can be involved in heterochromatin and/or centromere formation [32], although the true range of their putative functional role(s) is difficult to assess. In addition, inversely repeated segments of different length are quite frequent in diverse satellite monomers, making functional significance of each of them questionable. Much less frequent are complex satellite monomers composed of inversely duplicated subrepeats, sometimes several hundred nucleotides long and able to provide regions of homology which can build large and potentially stable dyad structures. Evolutionary history of these complex monomers can be clearly tracked through inverse duplications of primordial repeating units [33]. Sequence analysis indicates that such structures may be evolutionarily favored [34]. Inversely duplicated segments, either those within a satellite monomer or based on whole inverted subrepeats, are probably a simple outcome of stochastic processes in sequence dynamics, favoring segment inversions. Once emerged, dyad structures could provide enhanced proneness of a particular sequence element to interactions in the heterochromatin genomic environment, and even if the benefit is subordinate, it might put constraints on sequence evolution. Inversely duplicated sequence segments can be also important for the sake of their own dispersal, since these structures can be recognized by transposition-related mechanisms. Structural similarity to putative miniature inverted-repeat transposable elements (MITE) is probably responsible for broad distribution, diversity and persistence of the over 500 Myr old satellite DNA family of mollusks [35].

Concerted Evolution

Homogenization and Fixation of Tandem Repeats

Each satellite DNA is characterized by a monomer sequence, and thousands of monomers build homogeneous, megabase-long genomic segments. Normally, any DNA sequence accumulates mutations in time, and copies of a sequence scattered throughout the genome diverge with a rate which is inversely correlated to constraints imposed on the sequence. Although neutral evolution was first suggested as a mode of satellite DNA evolution, more recent works assume that at least some satellite DNA sequences evolve under low constraints which may be imposed on a part of the monomer or on some structural features beyond the nucleotide sequence level, as discussed in the above paragraph. Opposite to what may be expected, nucleotide sequence divergence among monomers within satellite DNA arrays is usually quite low, not exceeding few percents. For the purpose of sequence analysis it is therefore often convenient to manipulate with the satellite DNA consensus sequence, which is derived by taking the most frequent nucleotide at each position of the monomer repeat as a representative.

Sequence homogeneity of satellite DNAs is a result of non-independent evolution of repeating units. This means that mutations do not accumulate in a single monomer sequence; instead, they either spread among repetitive units of a satellite DNA or they become eliminated. This particular mode of evolution, known as concerted evolution, is consequence of a 2-level process called molecular drive, consisting of sequence homogenization and fixation [36, 37] (fig. 1a). At the first level, within the genome mutations are homogenized among all repeats of the satellite DNA by mechanisms of non-reciprocal sequence transfer. On the population level they become fixed among individuals as a result of random assortment of genetic material in meiosis and chromosome segregation. Persistent low sequence variability of monomers in the satellite DNA is therefore the net effect of 2 opposite processes, accumulation of mutations on one side and the rate of their spread or elimination on the other. In reproductively isolated organisms, accumulation of mutations homogenized

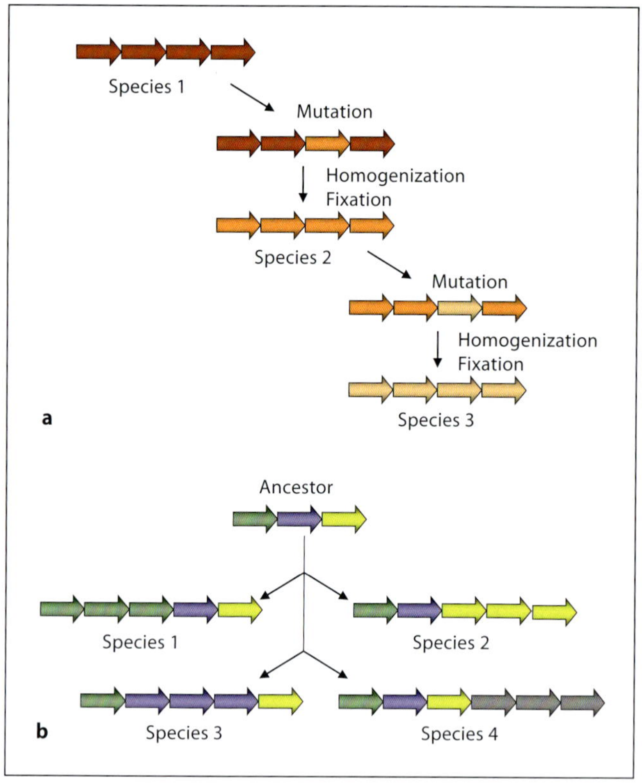

Fig. 1. Schematic representation of satellite DNA evolutionary concepts. **a** Concerted evolution. Satellite DNA is changed due to gradual accumulation of sequence divergence. **b** Satellite DNA library concept. Variation in satellite profiles is obtained by changes in copy number.

within the satellite DNA would rapidly lead to the higher homogeneity of repeats in the genome and within reproductively linked individuals than between 2 separated groups of organisms. This causes divergent evolution of satellite sequences in groups of organisms, while sequence homogeneity among monomers remains unaltered in each of them. According to the model of concerted evolution, mutations accumulate and spread within satellite DNAs gradually, and in this case satellite DNAs can be used as phylogenetically informative markers [38] (fig. 1a). A special case of concerted evolution is observed in species with asexual reproduction. In these organisms mutations are only homogenized, because fixation is dependent on population dynamics in sexual reproduction. In agreement with the model, the consequence of disabled fixation in parthenogenetic organisms is sequence variability of satellite monomers, which is comparable on all taxonomic levels of explored organisms [39]. In some cases sequence homogenization can be also suppressed, and this is explained by peculiar biological traits that lead to non-concerted evolution of satellite repeats in which mutations accumulate without possibility to spread or get eliminated. This pattern was observed in satellite DNAs of organisms with a limited number of reproducers, such as termites [40]. In addition, sequence dynamics of some satellite DNAs conserved during long evolutionary periods indicate predominant accumulation of

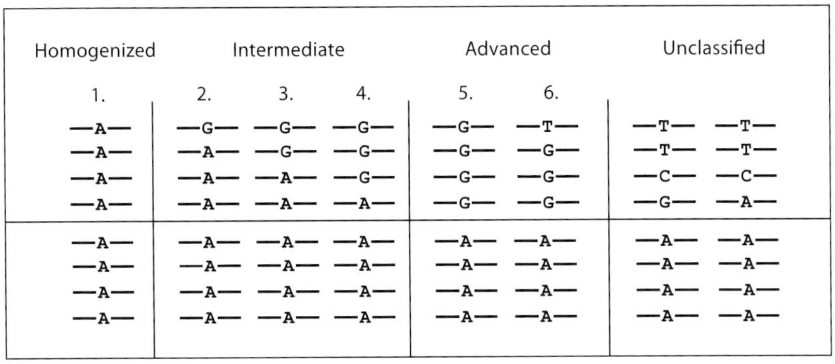

Fig. 2. Graphic representation of transition stages during the spread of new mutations between 2 groups of satellite monomers. Detailed explanation is given in section 'Homogenization and Fixation of Tandem Repeats'.

mutations with their limited spread and essentially non-concerted pattern of satellite DNA evolution (see also the section about satellite DNA sequence dynamics).

In the course of sequence homogenization, spread of mutations at each nucleotide position can be followed by comparing 2 sets of satellite DNA monomer sequence variants according to the method of Strachan et al. [41] (fig. 2). This method has been used in many studies of satellite sequences (e.g. [42, 43]), and it greatly improved our views on long-time preserved satellites and their mode of evolution [17, 44]. Briefly, variations at each position are classified into 6 classes according to appearance and spread of the mutated nucleotide in one set of monomers, compared to the homogeneous position in the other set. Classes can be further grouped into 3 transition stages [17]. Homogenized stage (class 1) represents identity of a nucleotide position in both sets. In the intermediate stage (classes 2, 3 and 4), the gradual spread of a new mutation in one set of monomers is followed, while in the advanced stage (classes 5 and 6) the mutated position becomes completely homogeneous, and ultimately a subsequent mutation is introduced (fig. 2). As unclassified are considered those nucleotide positions that are heterogeneous in both sets. Among them, identical mutations at the same nucleotide position in both sets of variants are thought to result from a mutation event that occurred in the ancestral set of sequences, rather than being a result of independent events.

Mechanisms of Sequence Homogenization

Homogenization of satellite repeats is driven by molecular mechanisms of non-reciprocal sequence transfer, such as unequal crossover, gene conversion, rolling circle replication, transposition and maybe some other, not yet disclosed mechanism(s) [37, 45, 46]. In general, efficiency of recombinational processes drops with increased sequence divergence. Initial homogeneity of satellite arrays is therefore necessary for efficient homogenization of mutated variants by mechanisms such as unequal crossover. It has been reported that recombination mechanisms can tolerate up to 30% sequence divergence of alpha satellite DNA [47]; however, the threshold level can differ in different systems. Alternatively, conserved sequence segments in divergent monomers may be sufficient as recombinational hot-spots, as commented in the previous section [16, 45, 48]. The rate of sequence change by unequal crossover depends on sequence divergence, array length and chromosomal position, and these limitations lead to differentiation of isolated groups of satellite DNA monomers in a genome, as it will be explained in the following section. In addition to unequal crossover, gene conversion is a mechanism of homologous recombination considered to be involved in homogenization of tandem repeats. Gene conversion can be identified in many satellite sequences as tracts of mutations shared by a group of monomer variants, often of submonomer length [33].

Mechanisms that can contribute to efficient dispersal of satellite sequences throughout the genome are mechanisms related to transposition. Sequence similarity points to segments of mobile elements as a source of satellite repeats in both animals and plants, such as in *Drosophila* [49] and in the cycas *Zamia paucijuga* [50]. For instance, the hypervariable 3′ region of plant retrotransposons frequently embeds short arrays of tandemly repeated segments, variable in DNA sequence and in copy number. These repeats can be further amplified into novel satellite DNAs [51]. In addition, fragments of non-autonomous mobile elements or fragments of sequences that only resemble non-autonomous mobile elements, such as those containing inversely oriented sequence motifs, can be amplified into satellite DNA. Sequences resembling non-autonomous MITEs embed short arrays of 2–6 tandem repeats in a part of their sequence, and elements related in sequence and repeat unit length are monomers of the BIV160 family of satellite DNAs, broadly distributed in mollusks [35, 52]. Although sequence similarity is usually evident, the true nature of mechanism(s) that expand fragments of mobile or mobile-like elements into long arrays of satellite DNAs is not known. While evolution of sequence segments from mobile elements to satellite DNAs seems to be a logical scenario, the possibility that satellite DNA fragments were simply captured by mobile elements is also open [53]. At this point it may be speculated that both pathways are possible and that a sequence can be repeatedly reverted from one organizational form to the other.

Segments of satellite DNA sequences may be a common constituent of extrachromosomal circular DNA (eccDNA), as shown by hybridization experiments. These results strongly support the idea that excision, rolling circle replication and reinsertion of eccDNA play a significant role in the evolution of satellite repeats [54]. In addition to transposition-related mechanisms, this pathway can also be important for efficient dispersion of satellite arrays on genomic locations with limited sequence homology, such as heterologous chromosomes. In conclusion, it is difficult to explore the exact role(s) of this and other mechanisms proposed to be involved in satellite DNA evolution, and the experimental evidence concerning molecular details of any of them is still missing. However, in the nucleotide sequence of satellite DNAs characteristic marks are left, enabling tracking of evolutionary history and understanding contribution of each particular mechanism by extensive sequence analyses.

Dynamics of Satellite DNA Evolution

Intragenomic Diversification of Satellite Arrays

An important outcome of mechanisms involved in homogenization is a higher degree of sequence similarity observed among adjacent repeats than among those retrieved at random. Monomers can often, but not always (see also below), be clustered into groups or subfamilies based on homogenized variant nucleotide positions and chromosome-specific distribution [37, 55]. Sequence divergences accumulate because of higher homogenization efficiency among adjacent monomers than among those positioned in different arrays on the same chromosome, and progressively, on homologous and heterologous chromosomes [37]. According to the model of concerted evolution, the result is divergent evolution of satellite DNA arrays within the genome and formation of chromosome-specific monomer variants and satellite subfamilies. Effects of limited homogenization efficiency among distal genomic locations can be interpreted as a primary outcome of unequal crossover [45]. Accumulated divergence of chromosome-specific subfamilies further reduces efficiency of interchromosomal sequence homogenization and ultimately reaches the level when homogenization becomes completely suppressed. Formation of subfamilies within the genome and across the species was studied in details in alpha satellite DNA of primates, and these data are presented in the last section of this review.

Theoretical models predict enhanced variability of satellite DNA monomers at ends of arrays homogenized predominantly by unequal crossover [45, 56]. Although studies on this issue are rare and done on a limited sample of satellite DNAs, sequence analyses of monomer variants at array junctions confirm the above prediction [57–59]. However, some satellite DNA array ends are abrupt, indicating thus contribution of other mechanisms [60]. It was also shown that mutated variants nascent at array ends can be amplified into a new satellite DNA, as they are probably too divergent to be further homogenized with the set of original repeats [59]. In this way, homogenization mechanisms maintain high sequence similarity of satellite monomers in the array and at the same time provide divergent repeat variants that may be used as a source of new satellite DNAs. Accumulation of mutations and sequence rearrangements in monomers at array ends and their amplification into a new satellite, when they occur, are results of stochastic events, and comparisons of such sequences consequently do not follow species phylogeny.

Monomer sequences of some satellite DNAs are complex higher-order repeat (HOR) units, formed by concurrent amplification and homogenization of 2 or more monomers adjacent in the original satellite DNA. Concerted evolution of satellite DNAs in a higher-order register is characteristic, for instance, of the alpha satellite of primates [55], and of satellite families in bovids [61]. HORs can be built in several steps, first by forming a complex monomer composed of short repeated segments, and then by merging these monomers into a more complex HOR unit. Formation of HORs can be sometimes accompanied by inversions of whole subunits [33, 34]. When a new unit of sequence homogenization is established, constituent subunits are excluded from the process of concerted evolution and start to accumulate sequence divergence. In the same time, sequence variability among HOR units remains low, as typical for satellite DNAs [55]. Evolutionary trends towards increasing monomer length and complexity can be predicted by theoretical models based on assumed ratios between crossover and mutation rates [62]. However, it is not clear what triggers formation of HORs, since this is not the feature of all satellite DNAs. It was suggested that complex repeating units are favored by mechanisms homogenizing the alpha satellite in proximal centromeric locations, while arrays of the same satellite are composed of monomeric repeats when located distally [48]. Although an effect of stochastic events in sequence dynamics cannot be excluded, the preferential formation of HORs in proximal pericentromeric and centromeric chromatin of human chromosomes strongly indicates some structural and functional preferences related to centromere environment.

Genome-Wide Homogeneity of Satellite Arrays

While the pattern of chromosome-specific subfamilies is typical for many satellite DNAs, including the human alpha satellite, some satellites do not show any signs of intragenomic diversification. Instead, they seem to be quite uniformly homogenized across the entire chromosomal set. A detailed study of this organizational pattern was done on satellite DNAs of *Tenebrio molitor* and several other tenebrionid beetles. A specificity of satellite DNAs explored in these species is localization in pericentromeric heterochromatin of all chromosomes and lack of subgroups of randomly cloned satellite DNA monomers that would indicate subfamilies. Analysis of the distribution of *T. molitor* satellite DNA monomer variants suggested random dispersal of monomers across the genome [63]. In addition, sequence variability of adjacent monomers in satellite arrays of *T. molitor* satellite DNA is equivalent to the variability of randomly cloned monomers [Plohl M., unpublished data]. Alterations in copy numbers do not seem to affect significantly, if at all, the localization pattern of beetle satellites. In this regard, studies of expansions and contractions of DNA sequences in the library of beetle satellites revealed chromosomal localization unaffected even when genomic abundance of the satellite DNA was significantly changed, from <0.5% to >30% [7, 64].

Distribution of satellite DNAs and their monomer variants on beetle chromosomes indicates enhanced efficiency of sequence homogenization regardless of the location within an array, between homologous or among heterologous chromosomes. Although this effect can be explained as a result of specificities in homogenization mechanism(s), similar efficiency of intra- and interchromosomal spread of satellite

DNA variants can be obtained in a stage of meiotic bouquets. Chromosomal bouquets are regularly observed in the first meiotic division of all examined beetle species and probably represent the characteristic trait of the family [34]. At this stage, all chromosomes of the set align with their massive pericentromeric regions, with chromosomal arms sticking out from the cluster. Uniform distribution of satellite DNAs in (peri)centromeric heterochromatin may represent the cause and the consequence of the bouquet association, due to a synergic effect between the bouquet stage and satellite DNA sequence dynamics. According to the proposed model [34], a positive feedback loop may be established in which sequence similarity facilitates the alignment of heterologous chromosomes, while at the same time the alignment is required for efficient satellite DNA sequence homogenization. In this process, an initial spread of a satellite DNA to heterologous chromosomes may be mediated by some other mechanism, such as eccDNA.

In parallel to monomer variants of a single satellite DNA, non-homologous pericentromeric satellite DNAs coexisting in the beetle genome are also uniformly distributed on all chromosomes. The result is an irregular organizational pattern built of relatively short arrays of each satellite DNA, mutually interspersed in pericentromeric heterochromatin. The longest uninterrupted arrays of the highly abundant satellite DNA (30% of the genome) of the beetle *Tribolium madens* were estimated not to exceed 70 kb [65]. On the contrary, long homogeneous domains of a single satellite are usually several hundreds of kb long, such as in humans [66], *Drosophila* [67], and *Arabidopsis* [68]. Contrasting the chromosomal specificity of these arrays, arrays of highly abundant satellites alternate with less represented repeats in pericentromeric heterochromatin of all *T. madens* and *T. audax* chromosomes [34, 65]. Satellite DNAs in these species share sufficiently high sequence similarity to allow homologous recombination between satellite arrays, resulting in this 'mosaic' organizational pattern [34, 59]. However, hybridization experiments indicate that similar patterns were also formed when unrelated satellites are involved [7, 64]. Detailed analysis of a similar interspersion pattern of 2 satellites on *Drosophila* minichromosomes suggests its origin through illegitimate recombination [60]. All these results indicate high rates of different recombinational processes in heterochromatin, opposing the traditional view on heterochromatin as a recombinationally inert genome compartment.

Evolutionarily Conserved Satellite DNA Sequences

Appearance, spread and accumulation of mutations in the course of molecular drive lead to divergent evolution of satellite repeats in reproductively separated groups of organisms. As predicted by the model of concerted evolution, some satellites accumulate mutations rapidly, with a rate enabling detection of a sequence divergence at the species level [38, 69] or below the species level, in populations or in ecotypes [70]. However, high rates of sequence change are not characteristic of all satellite DNAs and in these cases sequence variability among satellite monomers can differentiate organisms of higher taxonomic ranks, such as families [71]. As a consequence of low rates of evolution, many satellite DNAs can be found in groups of species, indicating their persistence in genomes during long evolutionary periods [72, 73]. Among the most extreme cases are, for example, the simple dodeca satellite, detected in *D. melanogaster, Arabidopsis thaliana* and *Homo sapiens* [74] and the family of BIV160 satellite DNAs in bivalve mollusks, estimated to persist for over 500 Myr [35].

Search for sequences homologous to highly abundant satellite DNAs by PCR revealed almost perfectly conserved low-copy counterparts in a group of related species [7, 17, 64] (see also section 'Satellite DNA Library Concept'). Sequence variability among monomer variants within and between species is found to be equivalent in these satellites, and their sequence is apparently long-time conserved or frozen in evolution. For example, monomers of the major satellite DNA from the beetle *Palorus ratzeburgii* could not be resolved from monomers detected as low-copy tandem repeats in the pericentromeric heterochromatin of *Pimelia elevata*, although both species shared a common ancestor about 60 Myr ago [64]. Similarly, satellite DNAs in the *Drosophila virilis* species group remained conserved for a period of about 20 Myr [75]. In sturgeons, satellite DNA remained practically unchanged over 100 Myr [76, 77]. Evolutionary pathways supporting extreme stability of satellite DNA variability profiles are hard to understand, in particular if we envisage homogenization and fixation of mutations as a random process, dependent entirely on stochastic nature of molecular mechanisms and population dynamics, as proposed in the original model of molecular drive [37]. One explanation of the phenomenon can be given under the hypothesis that some monomer sequences might be evolutionarily preferred in comparison with others [41]. Some monomers may be preferred because of their functional potential and/or simply because particular combinations of nucleotides and structural features of the DNA molecule are favored by homogenization mechanisms (see also [2]). According to an alternative view, heterochromatin environment and different aspects of species biology can slow down mutation and homogenization/fixation rates of DNA sequences and cause extreme conservation of satellite DNAs [40, 77]. Nevertheless, the ability to preserve DNA sequence over long time spans can represent a unique feature which may be accomplished only when DNA sequences are repeated in tandem.

In addition to the enormously large span in evolutionary rates of diverse satellite sequences, differences in sequence dynamics can be observed for a single satellite. For example, variability of a satellite DNA differs in closely related *Vicia* species [78], and even arrays of the same satellite at different genomic locations have their own rates of concerted evolution [79]. In this regard, different scenarios of satellite DNA evolution were observed in the satellite of the *Drosophila buzzatii* species cluster. A pool of ancestral monomers is maintained without homogenizing novel mutations, concerted evolution concomitantly favors existing sets of monomer variants, and mutated monomer variants are preferentially homogenized in other subfamilies [80]. It can be concluded that different rates are determined by a complex network of interactions depending on a satellite DNA sequence and its putative functional interactions, genome location, rates of homogenization mechanisms in a genome and population factors [81]. However, details concerning determinants involved in selection of evolutionary pathways and switches between the types of sequence management remain unclear for now.

Copy Number Changes

Specific features of satellite DNA evolution are extreme variations in copy number and, consequently, a high polymorphism in length of satellite arrays. For instance, tremendous variability in length, from 10- to 20-fold, is marked in alpha satellite DNA array size in human chromosomes [82]. Up to 3-fold alterations in length of alpha satellite arrays specific for the X chromosome were detected among human individuals [83]. There is also substantial difference in rice centromeres, where CentO satellite DNA arrays on different chromosomes vary in size from only 60 kb to 1.9 Mb. The (peri)centromeric region of *O. sativa* chromosome 6 contains 4 times more copies of CentO satellite DNA in the *japonica* subspecies than in he *indica* subspecies [18]. Given that these subspecies diverged from each other less than 1 Myr ago, it is clear that rapid amplification/contraction of satellite DNA arrays can appear within short evolutionary time. Dramatic amplifications and deletions of satellite DNAs have been documented on the species level, such as in satellites of rye [73], cattle [84], and species of the rodent genus *Ctenomys* [85].

It is considered that copy number changes are a predominant consequence of unequal crossover occurring between highly homologous arrays [46]. In addition, it is highly probable that large-scale changes, such as segmental duplications and mechanisms based on rolling circle replication, contribute significantly to the array length polymorphism. High copy number polymorphism in a common set of satellite DNAs can be a basis for rapid evolutionary alterations among species, as explained in the following section.

Satellite DNA Library Concept

Differences in composition of dominant satellite DNA sequences among closely related species are traditionally interpreted as a consequence of rapid gradual sequence evolution in separate lineages [36, 37, 41]. In addition to the large potential of the sequence change, satellite DNAs are permanently altered in copy number by expanding and contracting arrays of satellite monomers (see also section 'Copy Number Changes'). The library concept of satellite DNA evolution explains the occurrence of a species-specific profile of satellite DNAs as a result of differential amplifications and/or contractions within a pool of sequences shared by related genomes [7, 17, 86]. The library of satellite DNAs represents a permanent source of sequences that can be independently amplified in each species into a dominant, high copy number satellite DNA. Because usually more than 1 satellite DNA exists in a genome, fluctuations in their copy numbers can change very efficiently and rapidly any profile of genomic satellite DNA. This observation points out that a high rate of sequence changes in the course of concerted evolution is not the only possible cause that may explain diversity among dominant satellite DNAs in related species.

The library hypothesis of satellite DNA evolution had originally been suggested by Fry and Salser [86] analyzing a satellite DNA from the kangaroo rat, but the first experimental evidence of this concept was provided by studying satellite DNAs in insects of the genus *Palorus* [7] (fig. 1b). Study of the 4 unrelated species-specific dominant satellite DNAs revealed presence of low-copy counterparts of each of them in every examined species. Comparisons of high-copy and low-copy monomer variants of these satellites showed high interspecific sequence conservation and the complete lack of any species-diagnostic mutations. Not only different satellite DNAs but also monomer variants of a single satellite can be distributed in related species in variant copy numbers, forming a library of variants [87]. In this regard, the most extreme example is a widely distributed library of satellite DNA variants found in 3 main bivalve clades [35]. Until now, satellite DNA libraries were detected in various plant and animal taxa, probably representing the most common mode of satellite DNA evolution (e.g. [88–90]). Satellite DNA evolution through rapid changes in copy numbers can trigger rapid evolution of the genome as a whole. For example, in the genus *Ctenomys*, recurrent amplification and deletion of RPCP pericentromeric satellite DNA is associated with extensive chromosomal rearrangements [85]. Another important example linking satellite DNA library dynamics and genome evolution is shown by the analysis of 3 satellite DNAs in the group of marsupial species [91]. The authors showed that every change in karyotypic evolution of the genus was accompanied by a change in the predominant centromeric sequence component from the library. It appears that contractions in centromeric satellite DNAs are often associated with chromosomal fusions, while expansions of these satellite DNAs lead to translocation events. In contrast, differential amplifications in the library of unrelated satellite DNAs that occupy (peri)centromeric regions of 4 examined *Palorus* species do not affect the karyotype stability [7]. Although the most of so far analyzed satellite DNA sequences that evolve according to the library model were located mainly in pericentromeric chromosome regions, subtelomerically located satellite DNAs compose a library in the allopolyploid plant genus *Nicotiana* [90], indicating applicability of the concept regardless of the genomic location. In the library of *Nicotiana* allopolyploids, parental satellite DNAs are first redistributed within the genome, while contraction of high copy parental satellite DNAs is associated with the rise of new satellite sequences at the same chromosomal loci.

Study of the library of 3 related satellite DNAs differently amplified in taxa of root-knot nematodes from the genus *Meloidogyne* indicates selection as a limiting factor responsible for formation and persistence of satellite DNAs in the library [17]. Analysis of the distribution of sequence variability among these related satellite DNAs revealed highly structured monomers, composed of alternating lowly variable, moderately variable and highly variable domains. Interestingly, comparison of satellite DNA sequences cloned from each species revealed that the entire monomer sequence is uniformly conserved, even in domains characterized as highly variable, although species are separated for about 45 Myr. These data strongly suggest 2 distinct phases in satellite DNA evolution [17]. The first phase is library constitution; formation of a set of new satellite DNAs in a common ancestor, driven by the process of concerted evolution and modulated by functional constraints. During the second phase, satellite DNA sequences remain preserved for long time periods (see also section 'Evolutionary Conserved Satellite DNA Sequences') and become subject to differential amplification in related genomes. Analyses of mutations suggest that in the second phase changes accumulate but do not spread among monomers, indicating that homogenization mechanisms act on satellite monomers as a whole and favor persistence of ancestral monomer sequences. Conserved distribution of variability of monomers in the library might indicate a complex pattern of functional interactions which limit the range and phasing of allowed changes. Besides conserved sequence domains, other features of satellite DNA monomers as monomer length, tertiary or secondary structure can be critical for entry and survival in the library [2].

Additional study of 7 satellite DNAs in congeneric *Meloidogyne* species revealed their long term conservation in the library and distribution of satellite DNAs in related species in the course of lineage diversification [92]. In this case, instead of sequence divergence, the distribution of satellite DNAs in the library, in terms of their presence/absence in related genomes, is shown to be an informative character in phylogenetic studies. It has been shown that phylogenetic information based on satellite DNA distribution can supplement the data inferred from classical phylogenetic markers [92].

To conclude, the satellite DNA library contributes to genomic stability through its pool of functionally adapted and conserved satellite DNA sequences. At the same time, it also offers significant variability through the process of differential amplification which can completely change satellite DNA profiles in closely related species.

Satellite DNA Evolution and Centromere

The centromere is a complex, specialized locus crucial for proper chromosomal segregation in cell divisions. Since functional centromeres are usually positioned within large domains of satellite DNAs, our knowledge about structural features, organization and evolution of centromeric satellite sequences is essential for a comprehensive understanding of centromere function. In general, satellite arrays in centromeric regions are much longer than necessary for centromere function. While centromeric regions typically encompass megabases of satellite DNA sequences, functional centromere domains in *Drosophila* comprise only 15–40 kb, corresponding to the minimum length of 30–70 kb of alpha satellite DNA in a functional centromere of the artificial human chromosomes [93]. Dramatic alteration in copy number of satellite DNAs is a general characteristic of all known (peri)centromeric regions, among homologous and heterologous chromosomes, on the individual, subspecies and species level (see also section 'Copy Number Changes'). The same satellite DNA sequence very often constitutes 2 structurally and functionally different chromatin forms: centrochromatin and pericentromeric heterochromatin, distinct by the presence of specialized histone H3-like proteins and nucleosomal epigenetic modifications (see also section 'Structure, Organization and Function of Human Alpha Satellite DNA').

The highly conserved function of centromeres opposes rapid evolution of underlying satellite DNA sequences, the phenomenon known as 'centromeric paradox' [13]. Therefore, the central issue in research of many satellite DNAs is the conflict between importance of centromeric DNA and the fact that it evolves so rapidly. In the wild rice, for example, functional centromeric CentO satellite DNA is completely replaced with unrelated CentF during the period of 7–9 Myr [94]. An interesting feature of many centromeric satellite DNAs is similarity in monomer length, which in turn corresponds to the nucleosomal unit length, thus supporting a structural role of these sequences in the centromeric region [13]. In addition, as already discussed above (see section 'Satellite DNA Sequence Features'), uneven distribution of variability along monomer sequences, which results in conserved domains, was observed in centromeric satellite DNAs from various organisms. Different conserved motifs can be involved in various complex interactions in centromeric heterochromatin, participating in that way in centromere function.

DNA sequences in (peri)centromeric chromatin show considerable diversity, not only in terms of different satellite DNAs and their organizational forms (see section 'Evolution of Alpha Satellite Repeats'), but also in contribution of other sequences. Distal pericentromeric regions of human alpha satellite are enriched in mobile elements, while central domains, including the domain of the functional centromere, are composed of homogeneous alpha satellite arrays organized in the HOR form [48]. Five single intact transposons are directly inserted at different locations in the AATAT satellite arrays in *Drosophila* centromeres [67]. In grasses, maize and rice, beside species-specific satellite DNAs, centromeres contain substantial portions of species-specific retrotransposons [18, 95]. Retroelements are extensively intermingled with satellite DNAs and both sequence types mark functional parts of the plant centromeres [96]. Moreover, the contribution of satellite DNAs can be only sporadic in some centromeric regions. An illustrative example is chromosome 8 in rice with only 68.4 kb of satellite DNA arrays [97]. The peculiarity of these regions is the presence of active genes which require heterochromatin environment for their activity. Recent assembly and mapping of non-satellite DNA components in *Drosophila* centromeres revealed the presence of more than 200 coding genes [67]. Transcriptionally active genes have been also found in the centromere of chromosome 8 in rice [97]. On the contrary, analyses of *Arabidopsis* and human centromeres have not evidenced the presence of unique sequences or gene candidates [48, 98].

Phenomena that challenge the role of satellite DNAs in centromere function are de novo formed centromeres or neocentromeres. Neocentromeres arise in satellite-devoid regions as stable and functional centromeres, formed in the region of non-repetitive (euchromatic) genomic DNA which may also include transcriptionally active genes. Since the majority of natural centromeres contain satellite DNAs, it seems that neocentromeres become fixed in a population only after incorporation of repetitive sequences. Marshall et al. [99] hypothesized that satellite DNAs help to increase loading of histone H3-like protein CENP-A at the centromere (observed level of CENP-A is lower in neocentromeres than in satellite-rich centromeres) or promote establishing chromatin environment favorable for sister chromatid cohesion. It was also proposed that incorporation of satellite DNAs increases accuracy of chromosome segregation and aids to the stability of chromosomes during mitosis and meiosis [99].

The impact of rapidly evolving satellite DNAs can be viewed in the context of species radiation. A tandemly repeated form of sequences can be evolutionarily favored in (peri)centromeric regions because long arrays exhibit dual features: maintenance of a sequence homogeneity crucial for centromere stability, while in the same time it can be a source of extremely rapid changes. New dominant (peri)centromeric satellite DNAs can be established by gradual evolution of a nucleotide sequence in existing satellites, by expansion from the library or by recruitment of a new satellite DNA sequence (fig. 1). Fixation of new dominant satellite DNAs which correspond to centromeric function in a population offers a significant change in the DNA structure of the (peri)centromeric regions and can lead to reproductive isolation and species radiation.

Evolution of Alpha Satellite Repeats

Structure, Organization and Function of Human Alpha Satellite DNA

One of the most extensively studied repetitive DNA families is certainly the primate-specific alpha satellite DNA. Initially discovered and described in the African green monkey genome [100, 101], it has since been detected in a number of primate species studied to date, including humans, great apes, lesser apes, Old World monkeys and New World monkeys. Such comprehensive analyses establish alpha satellite as a paradigm, not only for understanding structure and organization, but also for interpreting evolutionary behavior of other satellite sequences in complex genomes.

In humans, alpha satellite is defined by 171-bp-long, AT-rich monomers categorized into 2 basic types according to their genomic organization and sequence properties: monomeric and HOR arrays [10]. The monomeric portion of alpha satellite comprises individual repeats that are 50–100% identical to one another, with an average pairwise similarity of 72%. These heterogeneous arrays lack any ordered periodicity or hierarchy, their monomers occasionally change orientation, and they are also

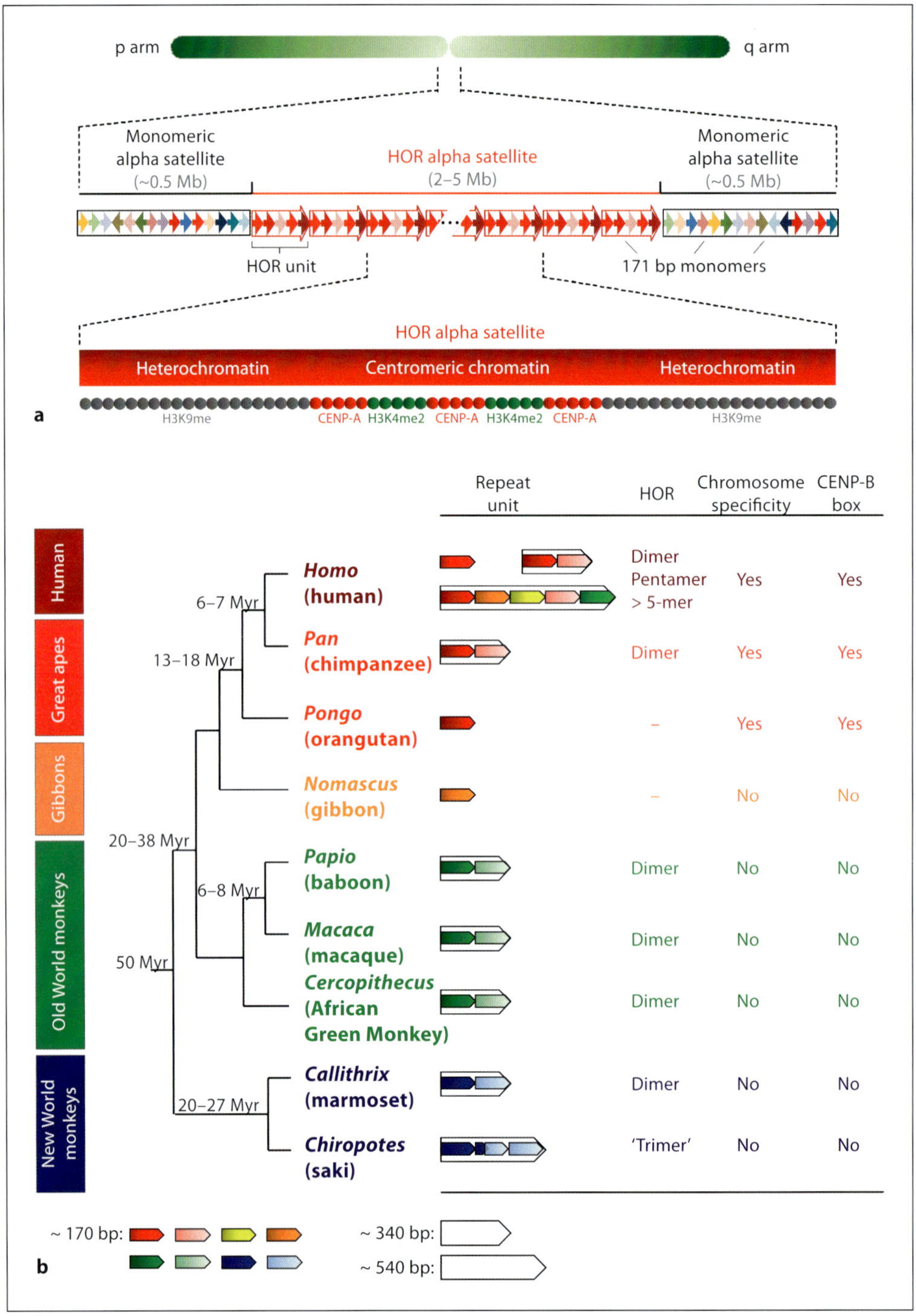

	Repeat unit	HOR	Chromosome specificity	CENP-B box
Homo (human)		Dimer Pentamer > 5-mer	Yes	Yes
Pan (chimpanzee)		Dimer	Yes	Yes
Pongo (orangutan)		–	Yes	Yes
Nomascus (gibbon)		–	No	No
Papio (baboon)		Dimer	No	No
Macaca (macaque)		Dimer	No	No
Cercopithecus (African Green Monkey)		Dimer	No	No
Callithrix (marmoset)		Dimer	No	No
Chiropotes (saki)		'Trimer'	No	No

Fig. 3. **a** Structural organization of human (peri)centromeric regions. A typical human chromosome is schematically delineated, emphasizing (peri)centromeric regions. Small arrows in different colors represent single monomers of alpha satellite DNA, while higher-order repeat (HOR) units are indicated by large red arrows. A fraction of HOR alpha satellite forms centromeric chromatin, built from subdomains of nucleosomes containing centromeric histone CENP-A (red circles) interspersed with

frequently intruded by other satellites and interspersed repetitive elements such as LINEs, SINEs, and LTRs (fig. 3a). Stretches of monomeric alpha satellite have been identified within the pericentromeric regions of human chromosomes where they flank the higher-order blocks of alpha satellite. On the other hand, higher-order alpha satellite is based on multimeric repeat units made up of 2 to over 20 diverged successive monomers. These multimeric HORs are tandemly organized into extremely homogeneous arrays, typically spanning up to several megabases across the centromere locus (fig. 3a). Although fundamental 171-bp monomers within a multimeric HOR show an average pairwise sequence similarity of ~70%, multimeric HOR units are 97–100% identical. HOR units of individual chromosomes differ in their monomer composition and length, what makes them chromosome-specific. At some chromosomes, there is more than 1 HOR array; for instance, the centromeric region of human chromosome 17 contains 2 HOR alpha blocks, D17Z1 and D17Z1-B, composed of HOR units of 16 and 14 monomers, respectively [102]. Albeit in most cases HOR alpha satellite arrays are chromosome-specific and therefore have been used for diagnostic purposes such as aneuploidy screening for many years, the length of HOR arrays between homologous chromosomes and among individuals varies up to several times [83, 103].

Monomeric and higher-order alpha satellites differ not only in sequence organization and localization, but also in their functionality. Namely, the HOR fraction of human alpha satellite DNA is associated with centromere function, while there is no evidence for the direct involvement of monomeric arrays in centromeric activity. Monomers within HOR units contain the CENP-B box, a 17-bp sequence motif responsible for binding foundation kinetochore protein CENP-B, whereas monomeric arrays are deprived of CENP-B boxes [104]. Further, active centromeres are marked by nucleosomes assembled with CENP-A, a centromere-specific histone H3 variant, which associates normally with HOR alpha satellite [105]. CENP-A nucleosomes alternate with nucleosomes containing histone H3 dimethylated at lysine 4, distinguishing centromeric chromatin from flanking heterochromatin that is defined by H3 lysine 9 methylation [106] (see also fig. 3a). Interestingly, the centromeric chromatin domain occupies just 35–50% of the HOR array [107], while the rest of HOR alpha satellite contributes to the assembly of the surrounding heterochromatin (fig. 3a). It is remarkable that alpha satellite as an underlying DNA component is able to form different chromatin domains via interactions with variously modified histones. In other words, different histone modifications

histone H3 dimethylated at lysine 4 (H3K4me2) (green circles). The remainder of alpha satellite DNA is assembled into heterochromatin enriched for nucleosomes containing histone H3 methylated at lysine 9 (H3K9me) (grey circles). Image is based on Schueler and Sullivan [106]. **b** Structural properties of alpha satellite DNA in primates. Schematic illustration of repeating units. The form of HOR units, chromosome specificity of satellite suprachromosomal families as well as the presence of CENP-B box within the sequence is indicated. Phylogenetic relationships and approximate divergence dates are derived from the tree of living primates [110].

as different epigenetic signatures on the same DNA sequence prove the capacity of satellite sequences to adopt different functional roles, enabling thus the flexibility of centromere positioning.

In the human genome, higher-order alpha satellite is a prevailing type of satellite DNA. Monomeric arrays nevertheless are predominantly represented in the current genome assembly, reflecting difficulties in accurate sequencing and assembling higher-order sequences of the functional centromere. In fact, there is no complete map for any of the 24 human centromeres. Pericentromeric regions that link centromeric HOR arrays and euchromatic arms on both sides have been mapped only for chromosomes 8, 17 and X [102, 108, 109]. The genome assembly data on chromosome 8 and X show the same orientation of HOR units on both the p and q arm sides, suggesting a continuous directionality across the array [108, 109]. This indication strongly implies the homogeneity of the centromeric DNA area, yet the exact alpha satellite maps are still waiting to be revealed.

Evolution of Alpha Satellite DNA in Primates

Alpha satellite DNA as a widespread repetitive family within the primate lineage clearly illustrates concerted manner of evolution, showing greater sequence similarity within species than between species [55]. In all characterized species, the common leitmotif is a ~170-bp monomer; however, during the course of primate evolution this fundamental seeding unit has experienced a number of sequence and structural variations (fig. 3b).

Comparison of alpha satellite profiles in a variety of primate species suggests a scenario according to which at the time of the very first amplification steps of a primordial monomer several divergent variants emerged [111]. By combining different monomer variants, more complex repeating units ascended, among which the simplest are dimeric (~342 bp) HOR units found in the genomes of Old World monkeys and New World monkeys (fig. 3b). It has been assumed that the dimeric structure of New World monkeys is more ancient, as it has been revealed that marmoset dimeric HORs show much lower similarity between first and second monomers (40–50%) than it was reported for macaques and humans (75–89%) [112, 113]. Some of the New World primates, such as saki monkeys, possess even more complex ~540-bp HOR alphoid units, composed of 340-bp dimers united with 33-bp and 168-bp-long duplications of the first and second monomers, respectively [114]. Although amplification in a higher-order register is by far the most influential mechanism of alpha satellite shaping, the example of saki monkeys shows that rearrangements also play a significant role, at least in Neotropical primates. Bioinformatics approaches using whole-genome shotgun data disclosed the dimeric HOR as a repeat structure common to all Old World monkey species, including African green monkey [115]. Besides the dimeric form, the additional common feature of alpha satellite DNAs in Old World monkeys is that their repeating units, in spite of higher-order organization, are the same among non-homologous chromosomes. In other words, they are not chromosome-specific, due to efficient interchromosomal homogenization. Chromosome specificity of alpha satellite repeats has also found to be absent in the white-cheeked gibbon [112] (fig. 3b).

It has been postulated that a change from genome-wide to chromosome-specific alpha satellite homogenization happened within the last 25–35 Myr of primate evolution when human and ape genomes adopted chromosome-specific HORs. Analysis of hominoid alpha satellite monomers proposed the classification of repeats into 5 suprachromosomal families, reflecting chromosome affiliation and evolutionary trends [111]. The most advanced alpha satellite setup is the one described in humans, uniting monomeric and the most complex higher-order types of alpha satellite. From the evolutionary point of view, it can be assumed that HORs originated from the 'library' of monomeric arrays. The evidence of short (up to 10 kb) HOR zones embedded within larger arrays of monomeric alpha blocks speaks in favor of this hypothesis [10]. It has been predicted that such 'HOR seeding' zones were generated by local homogenization events, thus initiating transition phase in the early stages of sequence family homogenization. Since the HOR arrays harbor centromere loci, it is intuitive to propose that suitability for being promoted into a centromere-relevant HOR structure is predetermined by functionality potential of the sequence. Notably, the emergence of CENP-B box in primates [20] coincides with the arising of chromosome-specific higher-order structures in the humans-apes linage (fig. 3b), and could represent an advantageous feature that, once acquired, promoted a dominant centromeric sequence. Phylogenetic analysis of alpha satellite monomers from the (peri)centromeric region of the human X chromosome suggests that the centrally located HOR domain, which currently functions as a centromere, is evolutionarily the most recent. This central region is flanked by gradual layers of older alpha satellite arrays that 'move' distally, losing their centromere competence as new, more successful variants arise and intrude into the central position [58, 116]. Once these 'ex-centromeres' are dismissed, they are not subject to intensive homogenization anymore, so they start to accumulate mutations and interspersed repetitive elements; at the same time new, 'centromere-busy' HOR arrays are homogenized very efficiently [111]. Comparison of human and chimpanzee alpha satellite revealed a higher rate of divergence among HORs as compared to monomeric arrays [102]. Interestingly, despite the fact that major human and chimpanzee alpha satellite suprachromosomal families share a common origin, they map to non-orthologous chromosomes [117], reflecting the rapid evolution of alpha satellite DNA in Hominidae.

Finally, alpha satellite DNA, declared to be primate-specific, has not yet been documented in basal primates. Recent characterization of CENP-A-associated satellite DNA in the lemur species *Daubentonia madagascariensis* revealed 2 satellite families related to each other but unrelated in sequence to alpha satellite DNA [118], broadening thus the frame for studying satellite DNA evolution in primate genomes.

References

1 Charlesworth B, Sniegowski P, Stephan W: The evolutionary dynamics of repetitive DNA in eukaryotes. Nature 1994;371:215–220.

2 Plohl M, Luchetti A, Meštrović N, Mantovani B: Satellite DNAs between selfishness and functionality: structure, genomics and evolution of tandem repeats in centromeric (hetero)chromatin. Gene 2008;409:72–82.

3 Ohno S: So much 'junk' DNA in our genome. Brookhaven Symp Biol 1972;23:366–370.

4 Stanley HP, Kasinsky HE, Bols NC: Meiotic chromatin diminution in a vertebrate, the holocephalan fish *Hydrolagus collie* (Chondrichthyes, Holocephali). Tissue Cell 1984;16:203–215.

5 Grewal SI, Elgin SC: Transcription and RNA interference in the formation of heterochromatin. Nature 2007;447:399–406.

6 Ferree PM, Barbash DA: Species-specific heterochromatin prevents mitotic chromosome segregation to cause hybrid lethality in *Drosophila*. PLoS Biol 2009;7:e1000234.

7 Meštrović N, Plohl M, Mravinac B, Ugarković Đ: Evolution of satellite DNAs from the genus *Palorus* – experimental evidence for the 'library' hypothesis. Mol Biol Evol 1998;15:1062–1068.

8 Malik HS, Henikoff S: Adaptive evolution of Cid, a centromere-specific histone in *Drosophila*. Genetics 2001;157:1293–1298.

9 Wang S, Lorenzen MD, Beeman RW, Brown SJ: Analysis of repetitive DNA distribution patterns in the *Tribolium castaneum* genome. Genome Biol 2008;9:R61.

10 Rudd MK, Willard HF: Analysis of the centromeric regions of the human genome assembly. Trends Genet 2004;20:529–533.

11 Hoskins RA, Carlson JW, Kennedy C, Acevedo D, Evans-Holm M, et al: Sequence finishing and mapping of *Drosophila melanogaster* heterochromatin. Science 2007;316:1625–1628.

12 Palomeque T, Lorite P: Satellite DNA in insects: a review. Heredity 2008;100:564–573.

13 Henikoff S, Ahmad K, Malik HS: The centromere paradox: stable inheritance with rapidly evolving DNA. Science 2001;293:1098–1102.

14 Romanova LY, Deriagin GV, Mashkova TD, Tumeneva IG, Mushegian AR, et al: Evidence for selection in evolution of alpha satellite DNA: the central role of CENP-B/pJ alpha binding region. J Mol Biol 1996;261:334–340.

15 Heslop-Harrison JS, Murata M, Ogura Y, Schwarzacher T, Motoyoshi F: Polymorphisms and genomic organization of repetitive DNA from centromeric regions of *Arabidopsis* chromosomes. Plant Cell 1999;11:31–42.

16 Hall SE, Luo S, Hall AE, Preuss D: Differential rates of local and global homogenization in centromere satellites from *Arabidopsis* relatives. Genetics 2005; 170:1913–1927.

17 Meštrović N, Castagnone-Sereno P, Plohl M: Interplay of selective pressure and stochastic events directs evolution of the MEL172 satellite DNA library in root-knot nematodes. Mol Biol Evol 2006; 23:2316–2325.

18 Cheng Z, Dong F, Langdon T, Ouyang S, Buell CR, et al: Functional rice centromeres are marked by a satellite repeat and a centromere-specific retrotransposon. Plant Cell 2002;14:1691–1704.

19 Zhang W, Lee HR, Koo DH, Jiang J: Epigenetic modification of centromeric chromatin: hypomethylation of DNA sequences in the CENH3-associated chromatin in *Arabidopsis thaliana* and maize. Plant Cell 2008;20:25–34.

20 Haaf T, Mater AG, Wienberg J, Ward DC: Presence and abundance of CENP-B box sequences in great ape subsets of primate-specific alpha-satellite DNA. J Mol Evol 1995;41:487–491.

21 Masumoto H, Masukata H, Muro Y, Nozaki N, Okazaki T: A human centromere antigen (CENP-B) interacts with a short specific sequence in alphoid DNA, a human centromeric satellite. J Cell Biol 1989;109:1963–1973.

22 Masumoto H, Nakano M, Ohzeki J: The role of CENP-B and alpha-satellite DNA: de novo assembly and epigenetic maintenance of human centromeres. Chromosome Res 2004;12:543–556.

23 Yoda K, Ando S, Okuda A, Kikuchi A, Okazaki T: In vitro assembly of the CENP-B/alpha-satellite DNA/core histone complex: CENP-B causes nucleosome positioning. Genes Cells 1998;3:533–548.

24 Kipling D, Warburton PE: Centromeres, CENP-B and Tigger too. Trends Genet 1997;13:141–145.

25 Broccoli D, Miller OJ, Miller DA: Relationship of mouse minor satellite DNA to centromere activity. Cytogenet Cell Genet 1990;54:182–186.

26 Canapa A, Barucca M, Cerioni PN, Olmo E: A satellite DNA containing CENP-B box-like motifs is present in the antarctic scallop *Adamussium colbecki*. Gene 2000;247:175–180.

27 Henikoff S, Dalal Y: Centromeric chromatin: what makes it unique? Curr Opin Genet Dev 2005;15:177–184.

28 Martínez-Balbás A, Rodríguez-Campos A, García-Ramírez M, Sainz J, Carrera P, et al: Satellite DNAs contain sequences that induced curvature. Biochemistry 1990;29:2342–2348.

29 Fitzgerald DJ, Dryden GL, Bronson EC, Williams JS, Anderson JN: Conserved patterns of bending in satellite and nucleosome positioning DNA. J Biol Chem 1994;269:21303–21314.
30 Radic MZ, Saghbini M, Elton TS, Reeves R, Hamkalo BA: Hoechst 33258, distamycin A, and high mobility group protein I (HMG-I) compete for binding to mouse satellite DNA. Chromosoma 1992;101:602–608.
31 Plohl M, Meštrović N, Bruvo B, Ugarković Đ: Similarity of structural features and evolution of satellite DNAs from *Palorus subdepressus* (Coleoptera) and related species. J Mol Evol 1998; 46:234–239.
32 Jonstrup AT, Thomsen T, Wang Y, Knudsen BR, Koch J, Andersen AH: Hairpin structures formed by alpha satellite DNA of human centromeres are cleaved by human topoisomerase IIalpha. Nucleic Acids Res 2008;36:6165–6174.
33 Mravinac B, Ugarković Đ, Franjević D, Plohl M: Long inversely oriented subunits form a complex monomer of *Tribolium brevicornis* satellite DNA. J Mol Evol 2005;60:513–525.
34 Mravinac B, Plohl M: Parallelism in evolution of highly repetitive DNAs in sibling species. Mol Biol Evol 2010;27:1857–1867.
35 Plohl M, Petrović V, Luchetti A, Ricci A, Šatović E, et al: Long-term conservation vs high sequence divergence: the case of an extraordinarily old satellite DNA in bivalve mollusks. Heredity 2010;104: 543–551.
36 Dover GA: Molecular drive: a cohesive mode of species evolution. Nature 1982;299:111–117.
37 Dover GA: Molecular drive in multigene families: how biological novelties arise, spread and are assimilated. Trends Genet 1986;2:159–165.
38 Bachmann L, Sperlich D: Gradual evolution of a specific satellite DNA family in *Drosophila ambigua, D. tristis,* and *D. obscura*. Mol Biol Evol 1993;10: 647–659.
39 Mantovani B: Satellite sequence turnover in parthenogenetic systems: the apomictic triploid hybrid *Bacillus lynceorum* (Insecta, Phasmatodea). Mol Biol Evol 1998;15:1288–1297.
40 Luchetti A, Marini M, Mantovani B: Non-concerted evolution of the RET76 satellite DNA family in *Reticulitermes* taxa (Insecta, Isoptera). Genetica 2006;128:123–132.
41 Strachan T, Webb D, Dover GA: Transition stages of molecular drive in multiple-copy DNA families in *Drosophila*. EMBO J 1985;4:1701–1708.
42 Pons J, Petitpierre E, Juan C: Evolutionary dynamics of satellite DNA family PIM357 in species of the genus *Pimelia* (Tenebrionidae, Coleoptera). Mol Biol Evol 2002;19:1329–1340.
43 López-Flores I, de la Herrán R, Garrido-Ramos MA, Boudry P, Ruiz-Rejón C, Ruiz-Rejón M: The molecular phylogeny of oysters based on a satellite DNA related to transposons. Gene 2004;339:181–188.
44 Mravinac B, Plohl M, Ugarković Đ: Preservation and high sequence conservation of satellite DNAs suggest functional constraints. J Mol Evol 2005;61: 542–550.
45 Smith GP: Evolution of repeated DNA sequences by unequal crossover. Science 1976;191:528–535.
46 Stephan W: Recombination and the evolution of satellite DNA. Genet Res 1986;47:167–174.
47 Okumura K, Kiyama R, Oishi M: Sequence analyses of extrachromosomal Sau3A and related family DNA: analysis of recombination in the excision event. Nucleic Acids Res 1987;15:7477–7489.
48 Schueler MG, Higgins AW, Rudd MK, Gustashaw K, Willard HF: Genomic and genetic definition of a functional human centromere. Science 2001;294: 109–115.
49 Miller WJ, Nagel A, Bachmann J, Bachmann L: Evolutionary dynamics of the SGM transposon family in the *Drosophila obscura* species group. Mol Biol Evol 2000;17:1597–1609.
50 Cafasso D, Cozzolino S, De Luca P, Chinali G: An unusual satellite DNA from *Zamia paucijuga* (Cycadales) characterised by two different organisations of the repetitive unit in the plant genome. Gene 2003;311:71–79.
51 Macas J, Koblízková A, Navrátilová A, Neumann P: Hypervariable 3′ UTR region of plant LTR-retrotransposons as a source of novel satellite repeats. Gene 2009;448:198–206.
52 Gaffney PM, Pierce JC, Mackinley AG, Titchen DA, Glenn WK: Pearl, a novel family of putative transposable elements in bivalve mollusks. J Mol Evol 2003;56:308–316.
53 Kejnovsky E, Kubat Z, Macas J, Hobza R, Mracek J, Vyskot B: Retand: a novel family of gypsy-like retrotransposons harboring an amplified tandem repeat. Mol Genet Genomics 2006;276:254–263.
54 Cohen S, Segal D: Extrachromosomal circular DNA in eukaryotes: possible involvement in the plasticity of tandem repeats. Cytogenet Genome Res 2009;124: 327–338.
55 Willard HF, Waye JS: Hierarchical order in chromosome-specific human alpha satellite DNA. Trends Genet 1987;3:192–198.
56 Stephan W: Tandem-repetitive noncoding DNA: forms and forces. Mol Biol Evol 1989;6:198–212.
57 McAllister BF, Werren JH: Evolution of tandemly repeated sequences: What happens at the end of an array? J Mol Evol 1999;48:469–481.

58 Schueler MG, Dunn JM, Bird CP, Ross MT, Viggiano L, et al: Progressive proximal expansion of the primate X chromosome centromere. Proc Natl Acad Sci USA 2005;102:10563–10568.

59 Mravinac B, Plohl M: Satellite DNA junctions identify the potential origin of new repetitive elements in the beetle *Tribolium madens*. Gene 2007;394:45–52.

60 Kuhn GC, Teo CH, Schwarzacher T, Heslop-Harrison JS: Evolutionary dynamics and sites of illegitimate recombination revealed in the interspersion and sequence junctions of two nonhomologous satellite DNAs in cactophilic *Drosophila* species. Heredity 2009;102:453–464.

61 Modi WS, Ivanov S, Gallagher DS: Concerted evolution and higher-order repeat structure of the 1.709 (satellite IV) family in bovids. J Mol Evol 2004; 58:460–465.

62 Stephan W, Cho S: Possible role of natural selection in the formation of tandem-repetitive noncoding DNA. Genetics 1994;136:333–341.

63 Plohl M, Borštnik B, Lucijanić-Justić V, Ugarković Đ: Evidence for random distribution of sequence variants in *Tenebrio molitor* satellite DNA. Genet Res 1992;60:7–13.

64 Mravinac B, Plohl M, Meštrović N, Ugarković Đ: Sequence of PRAT satellite DNA 'frozen' in some Coleopteran species. J Mol Evol 2002;54:774–783.

65 Žinić SD, Ugarković Đ, Cornudella L, Plohl M: A novel interspersed type of organization of satellite DNAs in *Tribolium madens* heterochromatin. Chromosome Res 2000;8:201–212.

66 Shiels C, Coutelle C, Huxley C: Contiguous arrays of satellites 1, 3, and beta form a 1.5-Mb domain on chromosome 22p. Genomics 1997;44:35–44.

67 Sun X, Le HD, Wahlstrom JM, Karpen GH: Sequence analysis of a functional *Drosophila* centromere. Genome Res 2003;13:182–194.

68 Heslop-Harrison JS, Brandes A, Schwarzacher T: Tandemly repeated DNA sequences and centromeric chromosomal regions of *Arabidopsis* species. Chromosome Res 2003;11:241–253.

69 Garrido-Ramos MA, de la Herrán R, Jamilena M, Lozano R, Ruiz Rejón C, Ruiz Rejón M: Evolution of centromeric satellite DNA and its use in phylogenetic studies of the Sparidae family (Pisces, Perciformes). Mol Phylogenet Evol 1999;12:200–204.

70 Hall SE, Kettler G, Preuss D: Centromere satellites from *Arabidopsis* populations: maintenance of conserved and variable domains. Genome Res 2003; 13:195–205.

71 Arnason U, Grétarsdóttir S, Widegren B: Mysticete (baleen whale) relationships based upon the sequence of the common cetacean DNA satellite. Mol Biol Evol 1992;9:1018–1028.

72 King K, Jobst J, Hemleben V: Differential homogenization and amplification of two satellite DNAs in the genus *Cucurbita* (Cucurbitaceae). J Mol Evol 1995;41:996–1005.

73 Vershinin AV, Alkhimova EG, Heslop-Harrison JS: Molecular diversification of tandemly organized DNA sequences and heterochromatic chromosome regions in some Triticeae species. Chromosome Res 1996;4:517–525.

74 Abad JP, Carmena M, Baars S, Saunders RD, Glover DM, et al: Dodeca satellite: a conserved G+C-rich satellite from the centromeric heterochromatin of *Drosophila melanogaster*. Proc Natl Acad Sci USA 1992;89:4663–4667.

75 Heikkinen E, Launonen V, Müller E, Bachmann L: The pvB370 *Bam*HI satellite DNA family of the *Drosophila virilis* group and its evolutionary relation to mobile dispersed genetic pDv elements. J Mol Evol 1995;41:604–614.

76 de La Herrán R, Fontana F, Lanfredi M, Congiu L, Leis M, et al: Slow rates of evolution and sequence homogenization in an ancient satellite DNA family of sturgeons. Mol Biol Evol 2001;18:432–436.

77 Robles F, de la Herrán R, Ludwig A, Ruiz Rejón C, Ruiz Rejón M, Garrido-Ramos MA: Evolution of ancient satellite DNAs in sturgeon genomes. Gene 2004;338:133–142.

78 Macas J, Navrátilová A, Koblízková A: Sequence homogenization and chromosomal localization of VicTR-B satellites differ between closely related *Vicia* species. Chromosoma 2006;115:437–447.

79 Kuhn GC, Küttler H, Moreira-Filho O, Heslop-Harrison JS: The 1.688 repetitive DNA of *Drosophila*: concerted evolution at different genomic scales and association with genes. Mol Biol Evol 2012;29:7–11.

80 Kuhn GC, Sene FM, Moreira-Filho O, Schwarzacher T, Heslop-Harrison JS: Sequence analysis, chromosomal distribution and long-range organization show that rapid turnover of new and old pBuM satellite DNA repeats leads to different patterns of variation in seven species of the *Drosophila buzzatii* cluster. Chromosome Res 2008;16:307–324.

81 Navajas-Pérez R, Quesada del Bosque ME, Garrido-Ramos MA: Effect of location, organization, and repeat-copy number in satellite-DNA evolution. Mol Genet Genomics 2009;282:395–406.

82 Choo KH: The Centromere. Oxford, Oxford University Press, 1997.

83 Mahtani MM, Willard HF: Pulsed-field gel analysis of alpha-satellite DNA at the human X chromosome centromere: high-frequency polymorphisms and array size estimate. Genomics 1990;7:607–613.

84 Nijman IJ, Lenstra JA: Mutation and recombination in cattle satellite DNA: a feedback model for the evolution of satellite DNA repeats. J Mol Evol 2001; 52:361–371.

85 Slamovits CH, Cook JA, Lessa EP, Rossi MS: Recurrent amplifications and deletions of satellite DNA accompanied chromosomal diversification in South American tuco-tucos (genus *Ctenomys*, Rodentia: Octodontidae): a phylogenetic approach. Mol Biol Evol 2001;18:1708–1719.

86 Fry K, Salser W: Nucleotide sequences of HS-alpha satellite DNA from kangaroo rat *Dipodomys ordii* and characterization of similar sequences in other rodents. Cell 1977;12:1069–1084.

87 Cesari M, Luchetti A, Passamonti M, Scali V, Mantovani B: Polymerase chain reaction amplification of the Bag320 satellite family reveals the ancestral library and past gene conversion events in *Bacillus rossius* (Insecta Phasmatodea). Gene 2003; 312:289–295.

88 Lin CC, Li YC: Chromosomal distribution and organization of three cervid satellite DNAs in Chinese water deer (*Hydropotes inermis*). Cytogenet Genome Res 2006;114:147–154.

89 del Bosque ME, Navajas-Pérez R, Panero JL, Fernández-González A, Garrido-Ramos MA: A satellite DNA evolutionary analysis in the North American endemic dioecious plant *Rumex hastatulus* (Polygonaceae). Genome 2011;54:253–260.

90 Koukalova B, Moraes AP, Renny-Byfield S, Matyasek R, Leitch AR, Kovarik A: Fall and rise of satellite repeats in allopolyploids of *Nicotiana* over c. 5 million years. New Phytol 2010;186:148–160.

91 Bulazel KV, Ferreri GC, Eldridge MD, O'Neill RJ: Species-specific shifts in centromere sequence composition are coincident with breakpoint reuse in karyotypically divergent lineages. Genome Biol 2007;8:R170.

92 Meštrović N, Plohl M, Castagnone-Sereno P: Relevance of satellite DNA genomic distribution in phylogenetic analysis: a case study with root-knot nematodes of the genus *Meloidogyne*. Mol Phylogenet Evol 2009;50:204–208.

93 Okamoto Y, Nakano M, Ohzeki J, Larionov V, Masumoto H: A minimal CENP-A core is required for nucleation and maintenance of a functional human centromere. EMBO J 2007;26:1279–1291.

94 Lee HR, Zhang W, Langdon T, Jin W, Yan H, et al: Chromatin immunoprecipitation cloning reveals rapid evolutionary patterns of centromeric DNA in *Oryza* species. Proc Natl Acad Sci USA 2005;102: 11793–11798.

95 Zhong CX, Marshall JB, Topp C, Mroczek R, Kato A, et al: Centromeric retroelements and satellites interact with maize kinetochore protein CENH3. Plant Cell 2002;14:2825–2836.

96 Ma J, Wing RA, Bennetzen JL, Jackson SA: Plant centromere organization: a dynamic structure with conserved functions. Trends Genet 2007;23:134–139.

97 Wu J, Yamagata H, Hayashi-Tsugane M, Hijishita S, Fujisawa M, et al: Composition and structure of the centromeric region of rice chromosome 8. Plant Cell 2004;16:967–976.

98 Copenhaver GP, Nickel K, Kuromori T, Benito MI, Kaul S, et al: Genetic definition and sequence analysis of *Arabidopsis* centromeres. Science 1999;286: 2468–2474.

99 Marshall OJ, Chueh AC, Wong LH, Choo KH: Neocentromeres: new insights into centromere structure, disease development, and karyotype evolution. Am J Hum Genet 2008;82:261–282.

100 Maio JJ: DNA strand reassociation and polyribonucleotide binding in the African green monkey, *Cercopithecus aethiops*. J Mol Biol 1971;56:579–595.

101 Rosenberg H, Singer M, Rosenberg M: Highly reiterated sequences of SIMIANSIMIANSIMIANSI MIANSIMIAN. Science 1978;200:394–402.

102 Rudd MK, Wray GA, Willard HF: The evolutionary dynamics of alpha-satellite. Genome Res 2006;16: 88–96.

103 Wevrick R, Willard HF: Long-range organization of tandem arrays of alpha satellite DNA at the centromeres of human chromosomes: high-frequency array-length polymorphism and meiotic stability. Proc Natl Acad Sci USA 1989;86:9394–9398.

104 Ikeno M, Masumoto H, Okazaki T: Distribution of CENP-B boxes reflected in CREST centromere antigenic sites on long-range alpha-satellite DNA arrays of human chromosome 21. Hum Mol Genet 1994;3:1245–1257.

105 Lam AL, Boivin CD, Bonney CF, Rudd MK, Sullivan BA: Human centromeric chromatin is a dynamic chromosomal domain that can spread over noncentromeric DNA. Proc Natl Acad Sci USA 2006; 103:4186–4191.

106 Schueler MG, Sullivan BA: Structural and functional dynamics of human centromeric chromatin. Annu Rev Genomics Hum Genet 2006;7:301–313.

107 Sullivan LL, Boivin CD, Mravinac B, Song IY, Sullivan BA: Genomic size of CENP-A domain is proportional to total alpha satellite array size at human centromeres and expands in cancer cells. Chromosome Res 2011;19:457–470.
108 Nusbaum C, Mikkelsen TS, Zody MC, Asakawa S, Taudien S, et al: DNA sequence and analysis of human chromosome 8. Nature 2006;439:331–335.
109 Ross MT, Grafham DV, Coffey AJ, Scherer S, McLay K, et al: The DNA sequence of the human X chromosome. Nature 2005;434:325–337.
110 Perelman P, Johnson WE, Roos C, Seuánez HN, Horvath JE, et al: A molecular phylogeny of living primates. PLoS Genet 2011;7:e1001342.
111 Alexandrov I, Kazakov A, Tumeneva I, Shepelev V, Yurov Y: Alpha-satellite DNA of primates: old and new families. Chromosoma 2001;110:253–266.
112 Cellamare A, Catacchio CR, Alkan C, Giannuzzi G, Antonacci F, et al: New insights into centromere organization and evolution from the white-cheeked gibbon and marmoset. Mol Biol Evol 2009;26:1889–1900.
113 Pike LM, Carlisle A, Newell C, Hong SB, Musich PR: Sequence and evolution of rhesus monkey alphoid DNA. J Mol Evol 1986;23:127–137.
114 Alves G, Seuánez HN, Fanning T: A clade of New World primates with distinctive alphoid satellite DNAs. Mol Phylogenet Evol 1998;9:220–224.
115 Alkan C, Ventura M, Archidiacono N, Rocchi M, Sahinalp SC, Eichler EE: Organization and evolution of primate centromeric DNA from whole-genome shotgun sequence data. PLoS Comput Biol 2007;3:1807–1818.
116 Shepelev VA, Alexandrov AA, Yurov YB, Alexandrov IA: The evolutionary origin of man can be traced in the layers of defunct ancestral alpha satellites flanking the active centromeres of human chromosomes. PLoS Genet 2009;5:e1000641.
117 Warburton PE, Haaf T, Gosden J, Lawson D, Willard HF: Characterization of a chromosome-specific chimpanzee alpha satellite subset: evolutionary relationship to subsets on human chromosomes. Genomics 1996;33:220–228.
118 Lee HR, Hayden KE, Willard HF: Organization and molecular evolution of CENP-A-associated satellite DNA families in a basal primate genome. Genome Biol Evol 2011;3:1136–1149.

Miroslav Plohl
Department of Molecular Biology
Bijenička 54
HR–10002 Zagreb (Croatia)
Tel. +385 1 4561 083, E-Mail plohl@irb.hr

Garrido-Ramos MA (ed): Repetitive DNA.
Genome Dyn. Basel, Karger, 2012, vol 7, pp 153–169

Satellite DNA-Mediated Effects on Genome Regulation

Ž. Pezer[a] · J. Brajković[a] · I. Feliciello[a,b] · Đ. Ugarković[a]

[a]Department of Molecular Biology, Ruđer Bošković Institute, Zagreb, Croatia; [b]Dipartimento di Medicina Clinica e Sperimentale, Università degli Studi di Napoli Federico II, Napoli, Italy

Abstract

Being the major heterochromatin constituents, satellite DNAs serve important roles in heterochromatin establishment and regulation. Their transcripts act as epigenetic signals required for organization of pericentromeric heterochromatin during embryogenesis and are necessary for developmental progression. In addition, satellite DNAs and their transcripts potentially play an active role in modulating gene expression and epigenetic states of a genome. Due to the presence of promoter elements and transcription factor binding sites within a sequence, satellite DNAs can interfere with the expression of nearby genes. Gene activity can be directly controlled by the number of repeats in a section of satellite DNA. In the case of stress, transcriptional activation of pericentromeric satellite DNAs seems to be part of a general stress response program activated by environmental stimuli. Such diverse forms of genome regulation modulated by satellite DNAs may be controlled by selective pressures and could influence the adaptability of the organism.

Copyright © 2012 S. Karger AG, Basel

Even though some time ago satellite DNAs were considered useless evolutionary remnants, the functional significance of their sequences is continuously becoming clearer. The existence of conserved motifs and structural properties, as well as emerging evidence on their widespread transcriptional activity, prompted us to re-examine these highly abundant eukaryotic sequences. Ugarković [1] first hypothesized that satellite DNAs could influence nearby gene expression and therefore designated them as regulatory elements. The explosion of epigenetic research over the last years gave clear evidence that satellite DNAs are targets of silencing mechanisms, including RNA interference (RNAi) and involving epigenetic modifications [2, 3]. Tandem arrangement of satellite DNAs predisposed them to elicit RNAi-based silencing mechanisms and nucleate formation of heterochromatin with consequences for the regulation of genes in the vicinity. On the other hand, presence of functional promoters and transcription factor binding sites within satellite sequences could also influence nearby genes, probably by the mechanism of transcriptional interference. Based on these observations and experiments performed, we can hypothesize now that the influence of satellite DNAs on nearby genes, as well as on heterochromatin, is epigenetic in nature and could be modulated by changes in environment [4]. In this chapter we give a comprehensive view on the role of satellite DNAs and their transcripts in heterochromatin formation and regulation as well as in modulation of gene expression.

Functional Constraints on Satellite DNA Sequence

In tandemly repeated satellite DNAs, mutations that change the sequence of one repeat are less common than recombination-induced replacement of one repeat by another, and therefore, the repeats resemble each other much more than they would if they had been evolving independently. This phenomenon, known as concerted evolution, is usually random and every version of repeat has an equal probability of being the one that replaces the others [5, 6]. However, comparison of monomer sequences within different satellite DNAs reveals that some monomer regions are more conserved while others show higher mutation rates [7–9]. Such a non-uniform rate of evolution along the sequence indicates the functional constraint on a part of satellite sequences, probably induced by interaction with satellite DNA-bound proteins. In addition to relatively conserved regions found in diverse centromeric and pericentromeric satellites, other more variable regions also exist. Variable regions might be functionally important owing to their interaction with rapidly evolving proteins. Such an example is the centromere-specific histone, CenH3, which replaces histone H3 in centromeric nucleosomes and is required for proper chromosome distribution during cell division [10]. Unlike the highly conserved histone H3, CenH3 is divergent and subject to the influence of positive selection [11]. According to the 'centromere drive' model [12], satellite DNA changes to enhance binding and subvert meiosis in its own favor while centromeric proteins adapt, either by increasing or reducing binding, to suppress the deleterious effects of a 'selfish centromere'. In addition to CenH3, other kinetochore proteins exhibit rapid sequence evolution in the fly *Drosophila melanogaster* as well as in the worm *Caenorhabditis elegans*, while in mammals, plants and fungi the rate of evolution is much lower [13].

Some satellite DNAs exhibit sequence conservation of the whole monomer sequence for long evolutionary periods. Extreme sequence conservation of 2 satellite DNAs that represent major pericentromeric repeats in the coleopteran insects *Palorus ratzeburgii* and *P. subdepressus* has been reported [14, 15]. These satellites are present in many coleopteran species at a low copy number and their sequences have remained unchanged for 60 million years. This remarkable antiquity and sequence conservation is also characteristic of human alpha satellite DNA, which has been detected as a rare, highly conserved repeat in evolutionarily distant species such as chicken and zebrafish [16]. This complete sequence conservation and the wide evolutionary distribution of some satellite sequences has led to the assumption that in addition to participating in centromere formation, they could perform some other role, possibly acting as *cis*-regulatory elements of gene expression [1]. Ancient satellite DNAs, up to 80 million years old, have also been reported in some species of fish [17] and whales [18] and their remarkable preservation is thought to be related to a low mutation rate generally observed in aquatic environments [17]. However, functional constraints have been implicated in the preservation of salamander satellite 2 sequence – its promoter activity and the self-cleavage ability of its transcripts have remained conserved for 200 million years [19].

Due to possible functional constraints on satellite DNAs, it is not surprising that some characteristics of satellite DNAs are shared between many eukaryotic organisms [1]. Probably the most common feature of satellite DNA is its intrinsic curvature. Satellite repeats are generally AT-rich and the periodical distribution of AT tracts causes DNA bending into a super-helical tertiary structure [20]. This sequence-dependent property is thought to be responsible for the tight packing of DNA and proteins in heterochromatin [20, 21]. Conserved CENP-B box-like motifs have been identified within satellite DNA of mammals and some invertebrates (see for example [9, 22, 23]). The CENP-B box is a 17-bp motif in human alpha satellite DNA and a binding site for centromere protein B (CENP-B) [24], homologs of which have been found in many eukaryotes. Not every repeat of alpha satellite contains a functional CENP-B box, but they appear at regular intervals in human centromeres and seem to be essential for centromeric chromatin assembly [25].

Palindromic sequences, which could potentially lead to the formation of dyad structures, are common elements of centromeric and pericentromeric satellite DNAs in budding yeast, insects and human [26–28]. It is not clear if they perform some function, but it can be hypothesized that some palindromic sequences could be recognized by DNA-binding proteins such as transcription factors. Some homeodomain proteins like Pax3, which is known to play an important role during neurogenesis, bind short palindromes present within major mouse satellite DNA (pers. commun. T. Jenuwein). A recent investigation has revealed that topoisomerase II recognizes and cleaves a specific hairpin structure formed by alpha satellite DNA [29]. It has been suggested that a subpopulation of the cellular topoisomerase II located at centromeres plays a role for sister chromatid cohesion in the centromeric region. The hairpin cleavage therefore could be connected to a cohesion role of topoisomerase II at centromeres.

Transcription of Satellite DNAs

Given their relatively simple sequence and the lack of any significant open reading frame, previously reported transcription of satellite DNA has been ascribed to read-through from upstream genes and transposable elements [30–32]. However, promoter elements and transcription start sites, as well as binding motifs for transcription factors, have been mapped within some satellites. Putative internal promoters have been reported in the wasp *Diadromus pulchellus* [33] where motifs cognate to RNA polymerase (Pol) II and III are present within the satellite monomer sequence. In schistosome satellite DNA, which encodes an active ribozyme, a functional RNA Pol III promoter is present [34]. The sequence of the highly conserved satellite 2 found in distant families of salamanders shares structural and functional properties with the typical vertebrate small nuclear RNA promoter [35]. Promoters for RNA Pol II are characteristic for the centromeric and pericentromeric satellite DNAs PRAT and PSUB from the beetle species *P. ratzeburgii* and *P. subdepressus*, respectively [36, 37]. In addition, motifs similar to A and B boxes, associated with RNA Pol III transcription, are also located within PSUB and PRAT satellite sequences [37].

The *Drosophila* GAGA transcription factor, that binds GA/CT-rich elements in promoters of many *Drosophila* genes and activates transcription by opening chromatin structure, was found associated with heterochromatin throughout the cell cycle. It is proposed that the GAGA factor directly interacts with a GC/AT-rich subset of satellite DNA repeats and modifies heterochromatin structure [38]. Human satellite III has a binding motif for the heat-shock transcription factor 1, which drives its RNA Pol II-dependent transcription in stress conditions [39]. Gamma satellite DNA, the abundant pericentromeric sequence of all murine chromosomes, contains conserved binding sites for the ubiquitous transcription factor Yin Yang 1 (YY1) [40]. YY1 belongs to the Polycomb group of proteins involved in gene regulation during development. It has been found associated with gamma satellite DNA in proliferating cells, whereas the association strongly diminished during transition to the quiescent state (G0). It has been proposed that the interaction of YY1 with gamma satellite DNA could lead to the targeting of proteins required for heterochromatinization or to the silencing of euchromatic genes by bringing them in close proximity of pericentromeric heterochromatin.

Transcripts of satellite DNAs have been reported in many organisms including vertebrates, invertebrates and plants. Transcripts are usually heterogeneous in size and are in some cases strand-specific, while in others transcription proceeds from both DNA strands. Some transcripts are present as polyadenylated RNA in the cytoplasm, while some others are found exclusively in the nucleus [4, 41]. Polyadenylated transcripts of the GC-rich satellite DNA of the Bermuda land crab are present in the cytoplasm of different tissues [42]. Satellite 2, an abundant tandemly repeated sequence distributed in clusters throughout the genome of the newt *Notophthalmus viridescens*, is transcribed on lampbrush chromosomes and stable, strand-specific transcripts are present in the cytoplasm in a variety of different tissues [43]. Human satellite III DNA is transcribed in response to stress, generating heterogeneously-sized RNAs that contain a polyA tail but remain in the nucleus [44]. On the other hand, abundant satellites of *Palorus* beetles, PRAT and PSUB, are continuously expressed during larval, pupal and imago stages. The transcripts are of variable size, originate from both strands of satellite DNA, but differ in expression between the 2 strands. Most of the transcripts are detected in the nucleus and are not polyadenylated [36, 37].

Transcription of many satellite DNAs is gender- or stage-specific and is often associated with differentiation and development. In mammals, accumulation of centromeric and pericentromeric transcripts occurs at the transcriptional level in the course of proliferation and cell cycle [45], differentiation of myoblasts [46], in heat-shocked cells [47, 48] and in cancer cells [49], and is mediated by RNA Pol II. The most abundant mouse gamma satellite DNA is differentially expressed in cells of the developing central nervous system as well as in adult liver and testis [50]. It has been recently shown that transcription of pericentromeric gamma satellite DNA is required for organization of pericentromeric chromatin into chromocenters in early mouse embryos and is necessary for developmental progression [51]. In chicken and zebrafish, transcription of alphoid repeat sequences also displays a specific temporal and spatial expression pattern during embryogenesis [16].

Transcription of satellite DNAs is also regulated by the cell cycle. In mouse, gamma satellite DNA transcription occurs with the highest rates in early S phase and in mitosis, while it is downregulated at the metaphase-anaphase transition. The transcription proceeds in the form of small, ~200 nucleotides (nt) long RNA during mitosis, while abundant heterogeneously-sized transcripts, ~500–10 000 nt long, are induced in G1 phase [46]. Besides being cell-cycle regulated, transcription of mouse gamma satellite DNA is also linked to cellular proliferation. Transcription of pericentromeric heterochromatin is found to be cell cycle-regulated in fission yeast *Schizosaccharomyces pombe*, exhibiting the highest level during S phase and the lowest in G2, while in mitosis transcripts were not observed [52, 53].

The hammerhead ribozyme structures associated with transcribed satellite DNA sequences have been found in salamanders [54], schistostomes [34] and *Dolichopoda* cave crickets [55]. All hammerhead ribozymes detected in animal satellite DNAs so far have been shown to self-cleave in *cis* long multimeric satellite transcripts into monomers, but the physiological role of ribozymes is not known.

Satellite RNAs as Epigenetic Regulators of Heterochromatin Establishment

Heterochromatin plays an essential role in preservation of epigenetic information, transcriptional repression of repetitive DNA and proper chromosome segregation. A surprising role of satellite DNA transcripts in heterochromatin establishment was revealed by Volpe et al. [2]. They showed that transcripts derived from tandemly repeated centromeric DNA of the fission yeast *S. pombe* exist in the form of small, 20–25 bp long RNAs and are involved in specific chromatin modifications through RNAi. In *S. pombe*, analysis of small interfering RNAs (siRNAs) involved in heterochromatin formation showed that they derive preferentially from the most conserved regions of repeats [56]. This indicates that conservation of parts of satellite repeats is rather due

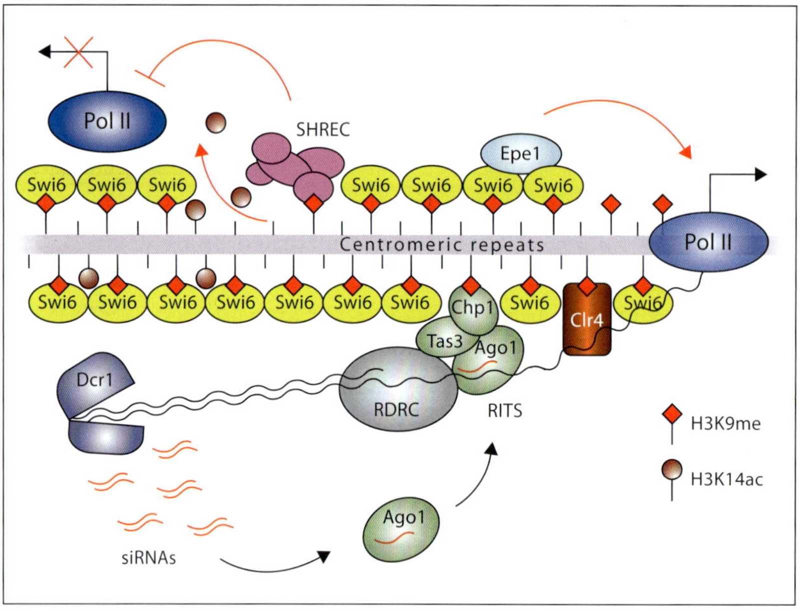

Fig. 1. Mechanism of heterochromatin formation in fission yeast *S. pombe*. Centromeric repeat (*dg* and *dh*) transcripts produced by Pol II are processed by the RNAi machinery, including the complexes RITS and RDRC (which interact with each other and localize across heterochromatic regions). The slicer activity of Ago1 (a component of RITS) and the RNA-directed RNA polymerase activity of Rdp1 (a component of RDRC) are required for processing the repeat transcripts into siRNAs. The siRNA-guided cleavage of nascent transcripts by Ago1 might make these transcripts preferential substrates for Rdp1 to generate double-stranded RNA, which in turn is processed into siRNAs by Dcr1. The targeting of histone-modifying effectors, including the Clr4-containing complex, is thought to be mediated by siRNAs. This process most probably involves the base-pairing of siRNAs with nascent transcripts, but the precise mechanism remains undefined. siRNAs produced by heterochromatin-bound RNAi 'factories' might also prime the assembly of RISC-like complexes capable of mounting a classic RNAi response. Methylation of H3K9 by histone methyltransferase Clr4 is necessary for the stable association of RITS with heterochromatic loci, apparently through binding to the chromodomain of Chp1. This methylation event also recruits Swi6, which, together with other factors, mediates the spreading of various effectors, such as SHREC. SHREC might facilitate the proper positioning of nucleosomes to organize the higher-order chromatin structure that is essential for the diverse functions of heterochromatin, including transcriptional gene silencing. Swi6 also recruits an anti-silencing protein, Epe1, which modulates heterochromatin to facilitate the transcription of repeat elements, in addition to other functions. A dynamic balance between silencing and anti-silencing activities determines the expression state of a locus within a heterochromatic domain [69].

to functional constraints than to frequent events of homologous recombination causing sequence identity. Therefore, conserved regions found in different satellite DNAs could be functional in the sense that they represent a preferential source of siRNAs that recruit protein complexes responsible for heterochromatin formation.

The chromatin silencing mechanism is best described in fission yeast *S. pombe* (fig. 1). It is initiated by long double-stranded RNA that arises from bidirectional transcription of repeated centromeric DNA and is further processed by the RNAse III-like ribonuclease Dicer into siRNAs. siRNAs are then loaded into the RNA-induced transcriptional silencing complex (RITS) through their association with the Argonaute protein. RITS also interacts with the RNA-directed RNA polymerase complex (RDRC) which is required for the production of secondary double-stranded RNA and amplification of the silencing signal [57]. Both RITS and RDRC associate with the nascent non-coding centromeric RNA transcript, and binding to RITS is probably achieved through the base-pairing of siRNA molecules with nascent RNA and by direct contact with the RNA Pol II elongation complex. In addition to siRNAs, the association of RITS with chromatin also requires a histone methyltransferase. Histone H3 methylation at lysine 9 is essential for the recruitment of heterochromatin protein 1 (HP1) or Swi6, an *S. pombe* counterpart of HP1. This represents an initial step in the formation of heterochromatin. HP1 has several functions at the centromere, such as silencing gene expression and recombination, promotion of kinetochore assembly and prevention of erroneous microtubule attachment to the kinetochores [58].

Mutations in components of the RNAi pathway lead to the loss of pericentromeric heterochromatin in fission yeast, resulting in missegregation of chromosomes [2, 59]. *S. pombe* cells deficient in pericentromeric heterochromatin are unable to recruit the chromosome cohesin to centromeres and fail to maintain centromere cohesion [60]. It was also revealed that heterochromatic proteins and RNAi machinery promote CENP-A deposition and kinetochore assembly over the central domain of the fission yeast centromere [61]. However, absence of these factors does not affect CENP-A deposition on endogenous centromeres or on minichromosome centromeres which have incorporated CENP-A in previous generations. In general, pericentromeric heterochromatin appears to be an absolute requirement for the establishment of centromeres in fission yeast, together with the central DNA region which binds CENP-A (*cnt* region) and the *otr* region which contains *dg-dh* repeats [61]. In addition to fission yeast, pericentromeric heterochromatin seems to be required for the accurate segregation of chromosomes during mitosis in many eukaryotes, including *Drosophila* and mammals [62, 63].

The RNAi machinery has been shown to be evolutionarily conserved and is proposed to be responsible for pericentromeric heterochromatin formation in different animal species. Analysis of *D. melanogaster* heterochromatin revealed its prominent pericentromeric localization and prevalent DNA composition based on satellite DNAs and transposable elements (TE). As in fission yeast *S. pombe*, *D. melanogaster* heterochromatin is associated with histone H3 methylation on lysine 9 (H3K9) by the histone methylase Su(var)3-9 that enables recruitment of HP1, necessary to maintain and spread the heterochromatic state [64]. It has been speculated for long time if endogenous siRNA pathways, similar to those in *S. pombe*, are involved in formation of heterochromatin in *Drosophila*. Small RNA molecules related to several types of repetitive DNA have been isolated from *D. melanogaster* [65]. These repeat-associated RNAs, ranging from 23–26 nt in size, are most abundant in testes and early embryos, which may be related to regulation of transposon activity and dramatic changes in heterochromatin structure that occur in these stages.

Examination and analysis of small RNA libraries obtained from different developmental stages of fly revealed presence of TE-derived small RNAs in all stages: in early embryos most of them correspond to 25 nt long piRNAs. They are formed in gonads from long transcripts of TEs and induce silencing of TEs through a feedback regulatory mechanism involving the Piwi subfamily of Argonaute proteins [66]. In other developmental phases 25-nt piRNAs are partially replaced by a population of 21 nt long RNAs that also derive from long TE transcripts. Due to the limitation of methods of high-throughput deep sequencing, that is restricted to non-tandemly repeated DNA, small RNAs that derive from satellite DNAs were not systematically examined. However, siRNAs deriving from 1.688 satellite have a size range between 19 and 28 nt and were detected in early embryos as well as in larvae [65]. It has been shown that a nuclear pool of TE-derived 21 nt long siRNAs is involved in heterochromatin formation in somatic cells of *Drosophila* and that components of the RNAi pathway participate in heterochromatin maintenance [67]. This implicates similarity between mechanisms of heterochromatin formation in *S. pombe* and *Drosophila* and points to the role of pericentromeric transcripts, either satellite DNA or transposon-derived, in heterochromatin formation. The possible mechanism by which repeat-derived siRNAs could promote heterochromatin formation in *Drosophila* is by tethering complementary nascent transcripts of satellite DNAs and transposons, and guiding chromatin modifiers such as histone methylase Su(var)3-9 that induces H3K9 methylation. Identification of proteins that tether siRNAs to chromatin in *Drosophila* and other animals needs, however, to be elucidated. There are also experimental indications that in *D. melanogaster* RNAi is involved in establishment of heterochromatin in the early embryo, but once set, heterochromatin can be maintained in the absence of RNAi in somatic tissues [68].

In addition to *S. pombe* and *D. melanogaster*, siRNAs cognate to satellite DNAs seem to be involved in the epigenetic process of chromatin modification in plants such as *Arabidopsis* and rice, as well as in nematodes such as *C. elegans* [3, 69]. Different from *D. melanogaster* and many other insects, plants often contain a high portion of methylated repetitive DNA. In plants, siRNAs were found to promote heterochromatin formation not only by directing histone methylation, but also by directing DNA methylation at the loci they were derived from [70]. Regulation of primary transcripts by RNAi and corresponding siRNAs of 21–24 nt, deriving from *Arabidopsis*, rice and sugar beet satellite DNAs, respectively, are readily experimentally proved [70–72]. Small RNAs cognate to the abundant satellite TCAST [27, 73] have been detected in the red flour beetle *Tribolium castaneum* (unpublished results). Small RNAs are more abundant in embryos than in later developmental stages, ranging in size between 21–26 nt, with a predominant size of 24 nt. Some components of the RNAi machinery have been identified in the sequenced genome of *T. castaneum*, such as Dicer and Argonaute protein families, but not the RNA-dependent RNA polymerase (RdRP) gene [74]. RdRP transcribes single-stranded RNA from an RNA template and is important for the production of siRNA as well as amplification of the RNAi effect in fungi, protists, nematodes and plants. However, it seems to be lacking in insects and vertebrates.

In mammals, however, siRNAs seem not to elicit chromatin modification, although an unidentified RNA component appears to be required for maintaining pericentromeric heterochromatin [75, 76]. In mouse pericentromeric heterochromatin, gamma satellite DNA as its major constituent is transcribed as small, ~200 nt long RNA during mitosis, while during G1 and S phase transcription occurs in the form of long, heterogeneous RNAs [45]. However, no evidence for siRNA-sized molecules at any time during the cell cycle exists, indicating that there could be a difference in heterochromatin expression and establishment between mammals on one side and fission yeast, plants and insects on the other side.

Satellite RNA as Structural Component of Centromeres

In addition to the role of satellite DNA transcripts in heterochromatin formation, many examples illustrate the involvement and possible importance of longer RNAs for centromere/kinetochore formation and function. Long, single-stranded alpha satellite DNA transcripts encompassing a few satellite monomers are shown to be functional components of the human kinetochore [77]. Centromere alpha satellite RNA is required for the assembly of CENP-C1, INCENP (inner centromere protein) and survivin (an INCENP-interacting protein) at the metaphase centromere. It also directly facilitates the accumulation and assembly of centromere-specific nucleoprotein components at the interphase nucleolus. The nucleolus sequesters centromeric components such as alpha satellite RNA and centromere proteins for timely delivery to the chromosomes for kinetochore assembly at mitosis. CENP-C has been shown to be an RNA-associating protein that binds alpha satellite RNA, as revealed by in vitro binding assay. The same protein also binds alpha satellite DNA in vivo and obviously has dual RNA- and DNA-binding function [78]. In mammals, CENP-C is evolving rapidly and, different from CENP-A (vertebrate CenH3), shows evidence of positive selection [79]. It is possible that a pool of CENP-C has a centromere DNA-binding role that persists throughout the cell cycle. The other pool of CENP-C is involved in relocation of alpha satellite RNA and centromere proteins from the nucleolus onto the mitotic centromere. CENP-B and CENP-C recognize the same subfamilies of alpha satellite DNA, but it is not clear whether CENP-C preferentially recognizes a specific sequence within satellite DNA or RNA. In vitro experiments indicate that CENP-C does not bind a specific DNA sequence, similar to CENP-A which also seems to be a sequence non-specific binding protein [78]. However, the existence of binding sites for different proteins in alpha satellite DNA could explain the non-random distribution of mutations within a sequence and can give strong support for the influence of selection on the evolution of this satellite DNA sequence.

RNA encoded by centromeric satellite DNA and retrotransposons, ranging in size between 40 and 200 nt, has been shown to be an integral component of the kinetochore in maize, tightly bound to centromeric histone H3 [80]. At the centromere of the marsupial tammar wallaby, satellite DNA and retroviral transcripts are accumulated. The transcripts are double-stranded, bound by centromere proteins and are processed into a small RNA of 34–42 nt [81]. Interestingly, transcripts of a similar size, i.e. 40 nt, were produced from rice centromeric satellite DNA repeats together with 21–24 nt long siRNAs that might derive from the pericentromeric portion of the same satellite [72]. Murine minor satellite DNA associated with the centromeric region is transcribed from both strands and transcripts are processed into 120-nt RNA which localizes to the centromere [82]. The overexpression of satellite transcripts is impaired by mislocalization of centromere-associated proteins essential for the formation of centromeric heterochromatin. In addition, forced accumulation of transcripts leads to defects in chromosome segregation and impaired centromere function resulting in aneuploidy. The absence of siRNAs homologous to murine minor satellite indicates that the longer non-coding RNA plays a role in heterochromatin formation and centromere establishment in the murine system. Long, stable transcripts of centromeric satellite DNAs are also characteristic for some beetle species [36, 37]. Based on studies in mammalian and insect systems, it appears that aberrant transcription of non-coding centromeric satellite DNA affects heterochromatin maintenance and fidelity of mitosis [83, 84]. This indicates that centromeric RNA is an important functional component of the centromere/kinetochore complex, probably tightly bound to proteins, and subtle changes in centromeric RNA/kinetochore protein ratio affect chromosome stability and segregation. Stoichiometric expression of all kinetochore components, including proteins and non-coding centromeric RNA, seems to be important for normal kinetochore assembly and function.

Mitotic and chromosome segregation defects have been reported for fission yeast mutants defective in RNA metabolism [85]. RNase activity of Dis3, a core component of the exosome that is required for processing of different RNAs, is shown to be necessary for heterochromatin silencing within the centromere as well as for proper kinetochore formation and establishment of kinetochore-microtubule interactions [86, 87]. Thus, RNAi-independent degradation of centromeric transcripts also contributes to heterochromatin formation and proper centromere function. All these examples demonstrate the importance of cellular RNA metabolism for proper chromosome segregation during mitosis. In addition to the relatively well-understood RNAi mechanism which moderates heterochromatin establishment in different eukaryotic systems, other mechanisms involving longer RNAs also operate in centromeric chromatin assembly and kinetochore formation. Although these mechanisms are poorly understood, it seems that centromere-encoded longer RNAs could serve as a scaffold for chromatin-remodeling complexes at centromeres as well as a structural component of kinetochores. It can be proposed that specific secondary and tertiary structures of centromeric RNAs are important for assembly of such complexes.

Overexpression of non-coding satellite DNAs is characteristic of some tumors. Analysis of transcription of human satellite 2 and alpha satellite, which are located in pericentromeric and centromeric heterochromatin, respectively, revealed an elevated level of their expression in ovarian epithelial carcinomas and Wilms tumors [88]. Aberrant overexpression of pericentromeric satellite DNAs was observed in epithelial cancers in mouse and human [49, 89]. Tumor-associated derepression of satellites was highly correlated with increased expression of LINE-1 retrotransposons, along with a subset of cellular genes in close proximity of LINE-1. It can be hypothesized that increased accumulation of non-coding RNA deriving from pericentromeric and centromeric satellite DNAs interferes with heterochromatin formation, centromere integrity and kinetochore establishment, affecting in this way mitotic segregation and genomic stability. Abnormal chromosome segregation is a common characteristic of human tumors and many of the molecular origins of chromosome missegregation derive from defective centromere and kinetochore function [83].

Satellite DNAs as *cis*-Regulatory Elements of Gene Expression

Satellite DNA repeats are preferentially accumulated in the regions characterized by a low recombination rate, such as pericentromeric and subtelomeric heterochromatic portions of chromosomes. However, there are several exceptions involving minor amounts of satellite sequences present in euchromatin. Such examples of limited localization of satellite sequences in euchromatin involve simple and complex satellite repeats. In the yeast *Saccharomyces cerevisiae*, as many as 25% of gene promoters contain tandem repeat sequences and variation in repeat length results in changes in gene expression [90]. Gene activity on a yeast chromosome is therefore directly controlled by the number of repeats in a section of non-coding DNA. On the other hand, great propensity of tandem repeats to change the copy number facilitates evolutionary tuning of gene expression in yeast. In *D. melanogaster*, 8 tandem repeats of simple AATAC satellite are found in front of the s38 chorion gene on the X chromosome [91] and could potentially affect gene expression. The 359 bp repeats of the 1.688 satellite, located predominantly in pericentromeric heterochromatin of the X chromosome, are also found in other positions of the same chromosome [92]. In the beetle *T. castaneum*, 360-bp repeats of abundant centromeric and pericentromeric satellite TCAST are found dispersed in the vicinity of genes on all chromosomes (unpublished results). The discovery of short satellite segments interspersed among the genes in the euchromatic portion of genomes suggests a possible regulatory role of these sequences, since they are often source of regulatory elements such as promoters and/or transcription factor binding sites (fig. 2) [1]. Recently, a regulatory role of 32-bp satellite repeats, located in the intron of the major histocompatibility complex gene (*MHIIβ*) of the

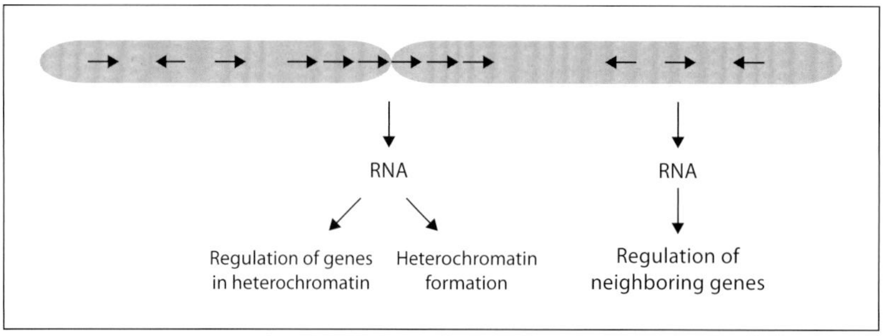

Fig. 2. Regulatory role of satellite DNAs and their transcripts. Transcripts of tandemly repeated satellite repeats, located in (peri)centromeric regions, play a role in heterochromatin formation as well as in the regulation of the genes located in heterochromatin. Transcripts of satellite DNA repeats dispersed within the euchromatin could play a role in the regulation of the neighboring genes. Transcription of satellite repeats is temperature-sensitive, and the role of transcripts in environmental stress response is proposed.

fish *Salvelinus fontinalis*, on *MHIIβ* gene expression was demonstrated [93]. The level of gene expression depends on temperature, being higher at lower temperatures, as well as on the length of the satellite repeat, where a longer satellite array induces reduced expression. Although the mechanism of *cis*-acting satellite gene regulation is not clear, there is evidence that temperature-sensitive satellite DNA could play an important role in the gene regulation of the adaptive immune response. Because the number of satellite repeats changes more frequently than other stretches of DNA do, this setup allows the organism to evolve more quickly. In such a way, non-coding satellite DNA can help organisms to adapt to changing environments.

Heterochromatin and Satellite DNA Response to Environmental Stimuli

Heterochromatin structure and level of expression of heterochromatic DNA seem to be highly sensitive to environmental conditions, in particular to heat-shock. In *Arabidopsis* plants subjected to prolonged heat stress, heterochromatin-associated silencing is released and transcription of satellite DNAs is significantly increased [94]. The increase of transcription is transient and after few days of recovery, transcripts returned to their previous level. Activation of transcription of repetitive elements in heterochromatin of *Arabidopsis* occurs without loss of epigenetic marks such as DNA methylation or histone modifications (H3K9), but is accompanied by heterochromatin decondensation and loss of nucleosomes [95]. Expression of human pericentromeric satellite III located on chromosome 9 is not detected under standard conditions, but is also transiently activated by heat stress [47]. Changes in expression of satellite DNAs, induced by environmental stress, influence heterochromatin structure and this can further reflect on centromere function as well as on function of genes located in heterochromatin. It is known that important developmental genes are located in heterochromatin as revealed for *D. melanogaster* [96] and that the proximity of heterochromatin is an important regulatory requirement for their function [97]. It is unknown whether the reorganization of heterochromatin domains is part of a physiological gene expression program or whether it is an undesirable product in pathological situations. In this light, it is noteworthy that the 'euchromatinization' of specific blocks of pericentromeric heterochromatin elicited by heat shock and other stress treatments can be part of a general stress response program activated in cells to cope with harmful conditions [44].

Heterochromatin is involved in gene silencing and this process is developmentally programmed in *Drosophila* and mammals [98]. Heterochromatin formation in *D. melanogaste*r is influenced by transcripts of satellite DNA elements and transposons present in the heterochromatin. On the other hand, insect as well as plant development is very sensitive to changes in environment, particularly temperature. With a lowering of the temperature, the length of the developmental period is prolonged, and at a critical temperature development ceases altogether. Recently it has been shown that in the beetle *T. castaneum* expression of satellite DNA is temperature-sensitive (unpublished results). Temperature-sensitive expression of heterochromatic satellite DNAs indicates also involvement of their transcripts in the signaling mechanisms responsible for insect development, differentiation and stress response (fig. 2).

Dynamic changes of heterochromatin structure and its 'euchromatization' in response to environmental cues may also trigger amplification of TEs or recombination within tandem arrays of satellite DNAs. This could provoke structural reshuffling of the genome and could lead to establishment of new structural domains and regulatory circuits.

Acknowledgements

This work was supported by EU FP6 Marie Curie Transfer of Knowledge Grant MTKD-CT-2006-042248 and grant 00982604 from the Croatian Ministry of Science. I.F. is Marie Curie Fellow at Ruder Boskovic Institute.

References

1 Ugarković Đ: Functional elements residing within satellite DNAs. EMBO Rep 2005;6:1035–1039.

2 Volpe TA, Kidner C, Hall IM, Teng G, Grewal SIS, Martienssen RA: Regulation of heterochromatic silencing and histone H3 lysine-9 methylation by RNAi. Science 2002;297:1833–1837.

3 Bernstein E, Allis CD: RNA meets heterochromatin. Genes Dev 2005;19:1635–1655.

4 Pezer Z, Brajković J, Feliciello I, Ugarković Đ: Transcription of satellite DNAs in insects. Prog Mol Subcell Biol 2011;51:161–178.

5 Malik HS: The centromere-drive hypothesis: a simple basis for centromere complexity. Prog Mol Subcell Biol 2009;48:33–52.

6 Ugarkovic Đ: Centromere-competent DNA: structure and evolution. Prog Mol Subcell Biol 2009;48: 53–76.

7 Borstnik B, Pumpernik D, Lukman D, Ugarković D, Plohl M: Tandemly repeated pentanucleotides in DNA sequences of eucaryotes. Nucleic Acids Res 1994;22:3412–3417.

8 Romanova LY, Deriagin GV, Mashkova TD, Tumeneva IG, Mushegian AR, et al: Evidence for selection in evolution of alpha satellite DNA: the central role of CENP-B/pJ alpha binding region. J Mol Biol 1996;261:334–340.

9 Mravinac B, Plohl M, Ugarković Đ: Conserved patterns in the evolution of *Tribolium* satellite DNAs. Gene 2004;332:169–177.

10 Henikoff S, Dalal Y: Centromeric heterochromatin: what makes it unique? Curr Opin Genet Dev 2005;15:177–184.

11 Cooper JL, Henikoff S: Adaptive evolution of the histone fold domain in centromeric histones. Mol Biol Evol 2004;21:1712–1718.

12 Henikoff S, Malik HS: Centromeres: selfish drivers. Nature 2002;417:227.

13 Meraldi P, McAinsh AD, Rheinbay E, Sorger PK: Phylogenetic and structural analysis of centromeric DNA and kinetochore proteins. Genome Biol 2006; 7:R23.

14 Mravinac B, Plohl M, Meštrović N, Ugarković Đ: Sequence of PRAT satellite DNA 'frozen' in some coleopteran species. J Mol Evol 2002;54:774–783.

15 Mravinac B, Plohl M, Ugarković Đ: Preservation and high sequence conservation of satellite DNAs suggest functional constraints. J Mol Evol 2005; 61:542–550.

16 Li YX, Kirby ML: Coordinated and conserved expression of alphoid repeat and alphoid repeat-tagged coding sequences. Dev Dyn 2003;228:72–81.

17 De la Herrán R, Fontana F, Lanfredi M, Congiu L, Leis M, et al: Slow rates of evolution and sequence homogenization in an ancient satellite DNA family of sturgeons. Mol Biol Evol 2001;18:432–436.

18 Arnason U, Höglund M, Widegren B: Conservation of highly repetitive DNA in cetaceans. Chromosoma 1984;89:238–242.

19 Green B, Pabon-Pena LM, Graham TA, Peach SE, Coats SR, Epstein LM: Conserved sequence and functional domains in satellite 2 from three families of salamanders. Mol Biol Evol 1993;10:732–750.

20 Fitzgerald DJ, Dryden GL, Bronson EC, Williams JS, Anderson JN: Conserved pattern of bending in satellite and nucleosome positioning DNA. J Biol Chem 1994;269:21303–21314.

21 Ugarković D, Plohl M, Lucijanić-Justić V, Borštnik B: Detection of satellite DNA in *Palorus ratzeburgii*: analysis of curvature profiles and comparison with *Tenebrio molitor* satellite DNA. Biochimie 1992;74: 1075–1082.

22 Kipling D, Warburton PE: Centromeres, CENP-B and *Tigger* too. Trends Genet 1997;13:141–145.

23 Lorite P, Carrillo JA, Tinaut A, Palomeque T: Comparative study of satellite DNA in ants of the *Messor* genus. Gene 2002;297:113–122.

24 Masumoto H, Masukata H, Muro Y, Nozaki N, Okazaki T: A human centromere antigen (CENP-B) interacts with a short specific sequence in alphoid DNA, a human centromeric satellite. J Cell Biol 1989;109:1963–1973.

25 Ohzeki J, Nakano M, Okada T, Masumoto H: CENP-B box is required for de novo centromere chromatin assembly on human alphoid DNA. J Cell Biol 2002;159:765–775.

26 Tal M, Shimron F, Yagil G: Unwound regions in yeast centromere IV DNA. J Mol Biol 1994;243: 179–189.

27 Ugarković D, Podnar M, Plohl M: Satellite DNA of the red flour beetle *Tribolium castaneum* – comparative study of satellites from the genus *Tribolium*. Mol Biol Evol 1996;13:1059–1066.

28 Zhu L, Chou SH, Reid BR: A single G-to-C change causes human centromere TGGAA repeats to fold back into hairpins. Proc Natl Acad Sci USA 1996;93: 12159–12164.

29 Jonstrup AT, Thomsen T, Wang Y, Knudsen BR, Koch J, Andersen AH: Hairpin structures formed by alpha satellite DNA of human centromeres are cleaved by human topoisomerase IIα. Nucleic Acids Res 2008;36:6165–6175.

30 Diaz MO, Barsacchi-Pilone G, Mahon KA, Gall JG: Transcripts from both DNA strands of a satellite DNA occur on lampbrush chromosome loops of the newt *Notophthalmus*. Cell 1981;24:649–659.

31 Wu ZG, Murphy C, Gall JG: A transcribed satellite DNA from the bullfrog *Rana catesbeiana*. Chromosoma 1986;93:291–297.

32 Gaubatz JW, Cutler RG: Mouse satellite DNA is transcribed in senescent cardiac muscle. J Biol Chem 1990;265:17753–17758.

33 Renault S, Rouleux-Bonnin F, Periquet G, Bigot Y: Satellite DNA transcription in *Diadromus pulchellus* (Hymenoptera). Insect Biochem Mol Biol 1999;29: 103–111.

34 Ferbeyre G, Smith JM, Cedergren R: Schistosome satellite DNA encodes active hammerhead-ribozymes. Mol Cell Biol 1998;18:3880–3888.

35 Coats SR, Zhang Y, Epstein LM: Transcription of satellite 2 DNA from the newt is driven by a snRNA type of promoter. Nucleic Acids Res 1994;22:4697–4704.

36 Pezer Z, Ugarkovic D: RNA Pol II promotes transcription of centromeric satellite DNA in beetles. PLoS One 2008;3:e1594.

37 Pezer Z, Ugarkovic D: Transcription of pericentromeric heterochromatin in beetles – satellite DNAs as active regulatory elements. Cytogenet Genome Res 2009;124:268–276.
38 Raff JW, Kellum R, Alberts B: The *Drosophila* GAGA transcription factor is associated with specific regions of heterochromatin throughout the cell cycle. EMBO J 1994;13:5977–5983.
39 Metz A, Soret J, Vourc'h C, Tazi J, Jolly C: A key role for stress-induced satellite III transcripts in the relocalization of splicing factors into nuclear stress granules. J Cell Sci 2004;117:4551–4558.
40 Shestakova EA, Mansuroglu Z, Mokrani H, Ghinea N, Bonnefoy E: Transcription factor YY1 associates with pericentromeric γ-satellite DNA in cycling but not in quiescent (G0) cells. Nucleic Acids Res 2004;32:4390–4399.
41 Vourc'h C, Biamonti G: Transcription of satellite DNAs in mammals. Prog Mol Subcell Biol 2011;51: 95–118.
42 Varadaraj K, Skinner DM: Cytoplasmic localization of transcripts of a complex G+C-rich crab satellite DNA. Chromosoma 1994;103:423–431.
43 Epstein LM, Mahon KA, Gall JG: Transcription of a satellite DNA in the newt. J Cell Biol 1986;103:1137–1144.
44 Valgardsdottir R, Chiodi I, Giordano M, Cobianchi F, Riva S, Biamonti G: Structural and functional characterization of noncoding repetitive RNAs transcribed in stressed human cells. Mol Biol Cell 2005;16:2597–2604.
45 Lu J, Gilbert DM: Proliferation-dependent and cell cycle-regulated transcription of mouse pericentromeric heterochromatin. J Cell Biol 2007;179:411–421.
46 Terranova R, Sauer S, Merkenschlager M, Fisher AG: The reorganisation of constitutive heterochromatin in differentiating muscle requires HDAC activity. Exp Cell Res 2005;310:344–356.
47 Rizzi N, Denegri M, Chiodi I, Corioni M, Valgardsdottir R, et al: Transcriptional activation of a constitutive heterochromatic domain of the human genome in response to heat shock. Mol Biol Cell 2004;15:543–551.
48 Jolly C, Metz A, Govin J, Vigneron M, Turner BM, et al: Stress induced transcription of satellite III repeats. J Cell Biol 2004;164:25–33.
49 Eymery A, Horard B, El Atifi-Borel M, Fourel G, Berger F, et al: A transcriptomic analysis of human centromeric and pericentric sequences in normal and tumor cells. Nucleic Acids Res 2009;37:6340–6354.
50 Rudert F, Bronner S, Garnier J-M, Dollé P: Transcripts from opposite strands of gamma satellite DNA are differentially expressed during mouse development. Mamm Genome 1995;6:76–83.
51 Probst AV, Okamoto I, Casanova M, El Marjou F, Le Baccon P, Almouzni G: A strand-specific burst in transcription of pericentric satellites is required for chromocenter formation and early mouse development. Dev Cell 2010;19:625–638.
52 Chen ES, Zhang K, Nicolas E, Cam HP, Zofall M, Grewal SI: Cell cycle control of centromeric repeat transcription and heterochromatin assembly. Nature 2008;451:734–737.
53 Kloc A, Zaratiegui M, Nora E, Martienssen R: RNA interference guides histone modification during the S phase of chromosomal replication. Curr Biol 2008;18:490–495.
54 Epstein LM, Gall JG: Self-cleaving transcripts of a satellite DNA in a newt. Cell 1987;48:535–543.
55 Rojas AA, Vázquez-Tello A, Ferbeyre G, Venanzetti F, Bachmann L, et al: Hammerhead-mediated processing of satellite pDo500 family transcripts from *Dolichopoda* cave crickets. Nucleic Acids Res 2000; 28:4037–4043.
56 Djupedal I, Kos-Braun IC, Mosher RA, Söderholm N, Simmer F, et al: Analysis of small RNA in fission yeast; centromeric siRNAs are potentially generated through a structured RNA. EMBO J 2009;28:3832–3844.
57 Verdel A, Jia S, Gerber S, Suglyama T, Gygi S, et al: RNAi-mediated targeting of heterochromatin with the RITS complex. Science 2004;303:672–676.
58 Yamagishi Y, Sakuno T, Shimura M, Watanabe Y: Heterochromatin links to centromeric protection by recruiting shugoshin. Nature 2008;455:251–256.
59 Allshire RC, Nimmo ER, Ekwall K, Javerzat JP, Cranston G: Mutations derepressing silent centromeric domains in fission yeast disrupt chromosome segregation. Genes Dev 1995;9:218–233.
60 Bernard P, Maure JF, Partridge JF, Genier S, Javerzat JP, Allshire RC: Requirement of heterochromatin for cohesion at centromeres. Science 2001;21:2539–2542.
61 Folco HD, Pidoux AL, Urano T, Allshire RC: Heterochromatin and RNAi are required to establish CENP-A chromatin at centromeres. Science 2008;319:94–97.
62 Kellum R, Alberts BM: Heterochromatin protein 1 is required for correct chromosome segregation in *Drosophila* embryos. J Cell Sci 1995;108:1419–1431.
63 Peters AH, O'Carroll D, Scherthan H, Mechtler K, Sauer S, et al: Loss of the Suv39h histone methyltransferases impairs mammalian heterochromatin and genome stability. Cell 2001;107:323–337.

64 Ebert A, Lein S, Schotta G, Reuter G: Histone modification and the control of heterochromatin gene silencing in *Drosophila.* Chromosome Res 2006;14:377–392.

65 Aravin AA, Lagos-Quintana M, Yalcin A, Zavolan M, Marks D, et al: The small RNA profile during *Drosophila melanogaster* development. Dev Cell 2003;5:337–350.

66 Brennecke J, Aravin AA, Stark A, Dus M, Kellis M, et al: Discrete small RNA generating loci as master regulators of transposon activity in *Drosophila.* Cell 2007;128:1089–1103.

67 Fagegaltier D, Bougé AL, Berry B, Poisot E, Sismeiro O, et al: The endogenous siRNA pathway is involved in heterochromatin formation in *Drosophila.* Proc Natl Acad Sci USA 2009;106:21258–21263.

68 Huisinga KL, Elgin SCR: Small RNA-directed heterochromatin formation in the context of development: What flies might learn from fission yeast. Biochim Biophys Acta 2008;1789:3–16.

69 Grewal SI, Elgin SC: Transcription and RNA interference in the formation of heterochromatin. Nature 2007;447:399–406.

70 Zakrzewski F, Weisshaar B, Fuchs J, Bannack E, Minoche AE, et al: Epigenetic profiling of heterochromatic satellite DNA. Chromosoma 2011;120: 409–422.

71 May BP, Lippman ZB, Fang Y, Spector DL, Martienssen RA: Differential regulation of strand-specific transcripts from *Arabidopsis* centromeric satellite repeat. PLoS Genet 2005;1:e79.

72 Lee HR, Neumann P, Macas J, Jiang J: Transcription and evolutionary dynamics of the centromeric satellite repeat CentO in rice. Mol Biol Evol 2006;23: 2505–2520.

73 Feliciello I, Chinali G, Ugarković Đ: Structure and population dynamics of the major satellite DNA in the red flour beetle *Tribolium castaneum.* Genetica 2011;139:999–1008.

74 Tomoyasu Y, Miller SC, Tomita S, Schoppmeier M, Grossman D, Bucher G: Exploring systemic RNA interference in insects: a genome-wide survey for RNAi genes in *Tribolium.* Genome Biol 2008;9:R10.

75 Maison C, Bailly D, Peters AH, Quivy JP, Roche D, et al: Higher-order structure in pericentromeric heterochromatin involves a distinct pattern of histone modification and an RNA component. Nat Genet 2002;30:329–334.

76 Wang F, Koyama N, Nishida H, Haraguchi T, Reith W, Tsukamoto T: The assembly and maintenance of heterochromatin initiated by transgene repeats are independent of the RNA interference pathway in mammalian cells. Mol Cell Biol 2006;26:4028–4040.

77 Wong LH, Brettingham-Moore KH, Chan L, Quach JM, Anderson MA, et al: Centromere RNA is a key component for the assembly of nucleoproteins at the nucleolus and centromere. Genome Res 2007; 17:1146–1160.

78 Politi V, Perini G, Trazzi S, Pliss A, Raska I, et al: CENP-C binds the alpha-satellite DNA in vivo at specific centromere domains. J Cell Sci 2002;11: 2317–2327.

79 Talbert PB, Bryson TD, Henikoff S: Adaptive evolution of centromere proteins in plants and animals. J Biol 2004;3:18.

80 Topp CN, Zhong CX, Dawe RK: Centromere-encoded RNAs are integral components of the maize kinetochore. Proc Natl Acad Sci USA 2004; 101:15986–15991.

81 Carone DM, Longo MS, Ferreri GC, Hall L, Harris M, et al: A new class of retroviral and satellite encoded small RNAs emanates from mammalian centromeres. Chromosoma 2009;118:113–125.

82 Bouzinba-Segard H, Guais A, Francastel C: Accumulation of small murine minor satellite transcripts leads to impaired centromeric architecture and function. Proc Natl Acad Sci USA 2006;103: 8709–8714.

83 Pezer Ž, Ugarković Đ: Role of non-coding RNA and heterochromatin in aneuploidy and cancer. Semin Cancer Biol 2008;18:123–130.

84 Frescas D, Guardavaccaro D, Kuchay SM, Kato H, Poleshko A, et al: KDM2A represses transcription of centromeric satellite repeats and maintains the heterochromatic state. Cell Cycle 2008;7:1–9.

85 Win TZ, Stevenson AL, Wang SW: Fission yeast Cid12 has dual functions in chromosome segregation and checkpoint control. Mol Cell Biol 2006; 26:4435–4447.

86 Murakami H, Goto DB, Toda T, Chen ES, Grewal SI, et al: Ribonuclease activity of Dis3 is required for mitotic progression and provides a possible link between heterochromatin and kinetochore function. PLoS One 2007;3:e317.

87 Bühler M, Haas W, Gygi SP, Moazed D: RNAi-dependent and -independent RNA turnover mechanisms contribute to heterochromatic gene silencing. Cell 2007;129:707–721.

88 Alexiadis V, Ballestas ME, Sanchez C, Winokur S, Vedanarayanan V, et al: RNAPol-ChIP analysis of transcription from FSHD-linked tandem repeats and satellite DNA. Biochim Biophys Acta 2007;1796:29–40.

89 Ting DT, Lipson D, Paul S, Brannigan BW, Akhavanfard S, et al: Aberrant overexpression of satellite repeats in pancreatic and other epithelial cancers. Science 2011;331:593–596.

90 Vinces MD, Legendre M, Caldara M, Hagihara M, Verstrepen KJ: Unstable tandem repeats in promoters confer transcriptional evolvability. Science 2009;324:1213–1216.
91 Spradling AC, de Cicco DV, Wakimoto BT, Levine JF, Katfayan LJ, Cooley L: Amplification of the X-linked *Drosophila* chorion gene cluster requires a region upstream from the s38 chorion gene. EMBO J 1987;6:1045–1053.
92 Tartof KD, Hobbs C, Jones M: A structural basis for variegating position effects. Cell 1984;37:869–878.
93 Croisetiere S, Bernatchez L, Belhumeur P: Temperature and length-dependent modulation of the MH class IIβ gene expression in brook charr *(Salvelinus fontinalis)* by a *cis*-acting minisatellite. Mol Immun 2010;47:1817–1829.
94 Tittel-Elmer M, Bucher E, Broger L, Mathieu O, Paszkowski J, Vaillant I: Stress-induced activation of heterochromatic transcription. PLoS Genet 2010;6: e1001175.
95 Pecinka A, Dinh HQ, Baubec T, Rosa M, Lettner N, Mittelsten Scheid O: Epigenetic regulation of repetitive elements is attenuated by prolonged heat stress in *Arabidopsis*. Plant Cell 2010;22:3118–3129.
96 Pimpinelli S, Sullivan W, Prout M, Sandler L: On biological functions mapping to the heterochromatin of *Drosophila melanogaster*. Genetics 1985;109: 701–724.
97 Dimitri P, Caizzi R, Giordano E, Carmela Accardo M, Lattanzi G, Biamonti G: Constitutive heterochromatin: a surprising variety of expressed sequences. Chromosoma 2009;118:419–435.
98 Lu BY, Ma J, Eissenberg JC: Developmental regulation of heterochromatin-mediated silencing in *Drosophila*. Development 1998;125:2223–2234.

Đ. Ugarković
Department of Molecular Biology
Ruđer Bošković Institute
Bijenička 54, HR–10002 Zagreb (Croatia)
Tel. +385 1 4561197, E-Mail ugarkov@irb.hr

Garrido-Ramos MA (ed): Repetitive DNA.
Genome Dyn. Basel, Karger, 2012, vol 7, pp 170–196

The Birth-and-Death Evolution of Multigene Families Revisited

J.M. Eirín-López[a] · L. Rebordinos[b] · A.P. Rooney[d] · J. Rozas[c]

[a]CHROMEVOL-XENOMAR Group, Departamento de Biología Celular y Molecular, Universidade da Coruña, A Coruña, [b]Área de Genética, Facultad de Ciencias del Mar y Ambientales, Universidad de Cádiz, Cádiz, [c]Departament de Genètica y Institut de Recerca de la Biodiversitat (IRBio), Universitat de Barcelona, Barcelona, Spain; [d]Crop Bioprotection Research Unit, National Center for Agricultural Utilization Research, Agricultural Research Service, US Department of Agriculture, Peoria, Ill., USA

Abstract

For quite some time, scientists have wondered how multigene families come into existence. Over the last several decades, a number of genomic and evolutionary mechanisms have been discovered that shape the evolution, structure and organization of multigene families. While gene duplication represents the core process, other phenomena such as pseudogene formation, gene loss, recombination and natural selection have been found to act in varying degrees to shape the evolution of gene families. How these forces influence the fate of gene duplicates has ultimately led molecular evolutionary biologists to ask the question: How and why do some duplicates gain new functions, whereas others deteriorate into pseudogenes or even get deleted from the genome? What ultimately lies at the heart of this question is the desire to understand how multigene families originate and diversify. The birth-and-death model of multigene family evolution provides a framework to answer this question. However, the growing availability of molecular data has revealed a much more complex scenario in which the birth-and-death process interacts with different mechanisms, leading to evolutionary novelty that can be exploited by a species as means for adaptation to various selective challenges. Here we provide an up-to-date review into the role of the birth-and-death model and the relevance of its interaction with forces such as genomic drift, selection and concerted evolution in generating and driving the evolution of different archetypal multigene families. We discuss the scientific evidence supporting the notion of birth-and-death as the major mechanism guiding the long-term evolution of multigene families.

Copyright © 2012 S. Karger AG, Basel

Forty-one years ago Susumu Ohno [1] stated that gene and genome duplications are the major evolutionary mechanisms for generating functional innovation. Since then, we have learned much regarding the evolutionary processes that influence nucleotide and amino acid substitution, both at the intraspecific and interspecific levels [2]. However, our current understanding of gene duplication dynamics is considerably less [3]. Despite the fact that a number of models and hypothesis have been developed to describe the evolutionary dynamics of gene duplications within and between species, the lack of readily available, high quality data limited our ability to test the applicability of most models to real data in past studies of the 'pre-genomic' era. The 2 main sources of problems were (1) the lack of complete genome information for many, if not most, gene families, and (2) the lack of accurate methods for inferring orthologous-paralogous gene relationships [4].

Gene families can be classified according to a number of criteria [3, 5, 6]. Such criteria may include, for example, (1) function, (2) how members are distributed across the genome, and (3) the primary mechanism responsible for generating the families in question. For instance, gene families have been categorized separating those organized into gene clusters from those with members at dispersed locations across the chromosomes. Yet, a classification based on the underlying mechanism for the origin of the family members is, in many cases, much more informative: not only does it explain the chromosomal distribution of family members, but it also provides insights into their evolutionary fate. Gene families essentially arise by 2 basic gene duplication mechanisms: unequal crossing-over and retroposition [7]. The first mechanism usually creates tandem repeats physically linked on the chromosomes and, therefore, in a non-random fashion. The family members in this case may have introns (if the original gene had introns) and non-coding regulatory sequences. In contrast, retroposition results in the insertion of an intronless cDNA with losses of upstream non-coding regions and with poly(A) tracts, more or less at random, at locations dispersed across the genome. The knowledge of the mechanism of origin is critical to understanding the forces that drive the generation of gene clusters; for example, a particular cluster of genes might have arisen simply due to random chance of having been located in a region of the genome more prone to unequal crossing-over than in other regions.

The recent availability of complete genomes from closely related species has provided valuable opportunities to conduct extensive studies of gene family evolution [8]. The analyses of these new data, however, also present a number of difficulties that remain to be solved such as, for example, the inability of current assembling algorithms to handle highly repetitive DNA sequences. Another problem concerns the accurate inference of orthologous-paralogous relationships. Currently, gene gain and loss events can be estimated either from the number of gene family members in the extant species of a phylogeny [9, 10], or via gene tree/species tree reconciliation [11]. The latter methods, however, have important limitations [8], such as their dependence on the correct gene tree and the true species tree, as well as the incomplete lineage sorting problem. Although there have been some improvements to minimize the gene tree uncertainty by taking into account clade support values, branch lengths [12] or synteny information [13], gene tree/species tree reconciliation is not well suited in order to conduct statistical hypothesis testing, and as such, it has limitations in its application.

Models of Multigene Family Evolution

The study of the mechanisms governing the evolution of multigene families has constituted a controversial issue ever since sets of functionally related genes were first discovered. The aforementioned limitations and others, such as the lack of detailed knowledge pertaining to the structure, organization and diversity of family members, their functional meaning as well as the lack of accurate methodologies for determining phylogenetic homology among sequences have fueled this controversy. The first efforts focused on deciphering the evolutionary dynamics of gene families date back to the early 1960s, with studies using hemoglobin and myoglobin as model systems [14]. The finding that the genes encoding these proteins are phylogenetically related and that they acquired new gene functions through their gradual divergence led to the proposal of the first general model of evolution of these multigene families, referred to as 'divergent evolution'.

The validity of the divergent evolution model was quickly challenged by the growing amount of data collected from studies on additional families, especially those displaying tandemly arrayed organizations (i.e. ribosomal DNA (rDNA) and histones). Within this context, the development of DNA sequencing techniques during the 1970s helped researchers to analyze the patterns of variation in coding and noncoding regions, unveiling that nucleotide sequences of different multigene family members are more closely related within species than between species. Such observations (which deviate from the predictions made by the divergent evolution model) were explained by an alternative model of multigene family evolution termed 'concerted evolution'. According to this model, after the split of an ancestral species into 2 descendent ones, the members of a repeated gene family would evolve together as a block, displaying a high degree of homogeneity within a given descendant species as they gradually diverged with respect to repeats from closely related species. Under this model, sequence homogenization results from random unequal crossing-over and gene conversion among gene family members, although some gene variants are expected to occur due to mutation.

The apparent efficiency of the concerted evolution model in explaining the observed patterns of molecular variation quickly overshadowed any alternative explanation throughout the 1970s and 1980s, consolidating the notion that most multigene families evolve following this model. Indeed, it was not until the early 1990s that concerted evolution began to be seriously questioned, especially as a result of the growing availability of molecular data coming from the dawn of the genomic era. Surprisingly, these data revealed that far from being conserved and homogeneous, most multigene families encompassed far too much intraspecific diversity (genetic and functional) to be consistent with a homogenizing mechanism. These conclusions, together with other atypical features observed across multigene family members (most notably the presence of between-species clustering patterns in phylogenies and the presence of pseudogenes), motivated the proposal of a new model termed 'birth-and-death evolution' [15]. In contrast to the concerted evolution model, the birth-and-death model promotes genetic diversification and provides an explanation for the generation of new gene families.

The Birth-and-Death Model of Evolution

Over the last 2 decades, Nei and colleagues conducted a number of key studies that provide the foundation for the theory that underlies the birth-and-death model. Since then, a number of multigene families have been identified that undergo birth-and-death evolution (reviewed in [6]). The basic foundational elements of the model are the differential levels of gene duplication and subsequent loss or maintenance of gene copies within a multigene family. Accordingly, when duplication gives rise to new copies of a gene, and these copies do not evolve in concert as discussed in the previous section, some of the copies may persist in the genome for long periods of time. Eventually, the copies diverge in sequence such that they no longer are identical nor do they possess extensive regions of similarity. On the other hand, some copies of the original 'parent' gene may degenerate into pseudogenes or they may get deleted from the genome through, for example, unequal crossing-over. Consequently, the most common way to determine if birth-and-death evolution characterizes a multigene family is to look for the 2 hallmark features of the model: (1) an interspecific gene clustering pattern and (2) the presence of pseudogenes.

There are cases in which an interspecific phylogenetic clustering pattern and/or pseudogene formation are not detectable. While the latter is dependent mostly on intrinsic genome dynamics and random chance, both are dependent upon proper analytical techniques. Still there are instances in which even thorough, proper analyses can lead to the erroneous conclusion that birth-and-death evolution does not occur, simply because an intraspecific gene clustering pattern was observed in the reconstructed multigene family phylogeny. Such false conclusions can arise from (1) recent gene duplication within a species; (2) strong purifying selection; and (3) rapid gene turnover. With respect to recent gene duplication, enough nucleotide substitutions will accumulate over time such that divergence between gene duplicates eventually becomes detectable (albeit over hundreds of thousands, if not millions, of years). Thus, a pattern of between-species gene clustering will characterize the phylogeny of the multigene family, provided that enough time has elapsed for the divergence of gene duplicates and/or their orthologs present in different genomes [6]. In cases involving strong purifying selection, one must consider the differences in the way in which substitutions accumulate and are distributed between protein-coding and non-protein-coding genes. For protein-coding genes, an analysis of divergence levels at synonymous versus non-synonymous sites will reveal if purifying selection, and not concerted evolution, is the cause for sequence constraint; indeed, under purifying selection synonymous sites will have some divergence levels even when non-synonymous sites show no variation [16]. In the case of genes that do not encode protein, such as ribosomal RNA (rRNA) genes, the analysis is more difficult and often requires study of nucleotide substitution levels in the regions immediately flanking the genes as well as in introns or intergenic spacer regions, if present, followed by comparison to sequence divergence within the coding region of the gene [17]. Differential levels of nucleotide sequence conservation between the coding and non-coding regions may reveal if purifying selection is the determinant for sequence conservation. Finally, when rapid gene turnover occurs within a multigene family, deletion and duplication are so frequent that orthologous gene pairs are quickly lost between species, so a within-species clustering pattern predominates [17–20]. But some amount of nucleotide substitution should still be observable and there may be at least some between-species clustering events, both of which are indicators that birth-and-death evolution has occurred. The aforementioned examples are only a few, and there may be more that can produce potentially misleading results in analyses designed to detect birth-and-death evolution.

It should be noted that, while considerable effort has been given to the study of gene duplication, little attention has been paid to the effects of gene deletion on multigene family evolution. Thus, much like the failure to recognize recent gene duplication, strong purifying selection and rapid gene turnover as causes of within-species gene clustering patterns, the failure to recognize the importance of gene loss may result in phylogenetic patterns that could be misinterpreted as, for instance, lateral transfer events [21]. In this case, however, phylogenetic analyses may not help and the problem is rendered intractable (see [21] for a review of this topic).

There are a number of genomic and evolutionary mechanisms that can shape the structure, organization and evolution of multigene families (see [6]). For the last decades, concerted evolution has prevailed as the 'default' long-term evolutionary model for the evolution of most (if not all) multigene families. We nowadays know that multigene families encompass too much genetic diversity to be generated and maintained by means of such a homogenizing mechanism. Indeed, comprehensive studies conducted during the last 10 years, addressing the evolution of multigene families, usually support the birth-and-death process as the underlying mechanism. However, in spite of the evidence gathered in favor of this latter model, birth-and-death has only shyly replaced concerted evolution as the 'default' model of long-term evolution of multigene families. In the present chapter we provide an up-to-date review into the role of the birth-and-death model and its interaction with forces such as genomic drift, natural selection and concerted evolution in generating and driving the evolution of different archetypal multigene families. Here we show empirical evidence supporting the concept of birth-and-death as the major mechanism underlying the long-term evolution of multigene families.

Rates of Birth-and-Death Evolution: Lessons from Gene Families of the Chemosensory System

Despite the availability of complete genome information for a number of eukaryotic species, we are far from understanding the forces that have driven the lineage-specific expansions (or contractions) of many multigene families, as well as more general features that characterize their evolution. The most important limitations are the following: (1) quite often, so-called complete genomes are not fully completed and are very fragmented. This is a very important problem since repetitive DNA regions are usually the worst assembled and, therefore, often incompletely represented in a genome annotation; this limitation is more critical for tandemly distributed repetitive regions than for those showing a more dispersed distribution across the genome. Hence, we lack detailed information (number of copies, physical location on the chromosomes) for many gene families, but especially those that exist in large clusters of tandemly repeated genes. (2) Many species for which completed genomes are available are separated by vast evolutionary times. Indeed, there are few cases in which we have genome information from relatively closely related species (e.g. within a genus or within a family); the genome sequence of 12 *Drosophila* species is one of such few examples [22]. Since many gene families have relatively high gene turnover rates (birth-and-death rates), information from highly divergent genomes can confound fine and exhaustive lineage-specific analyses (e.g. the accurate determination of the numbers of gene gains and losses might be highly inaccurate depending upon the rate of gene turnover). (3) Current methods for inferring orthologous-paralogous relationships may have low accuracy [4] (e.g. gene tree and species tree problem), when gene conversion is frequent or when large numbers of gene gains and losses have occurred. Regardless, limitation 2 likely will no longer be a problem in the near future, but limitations 1 and 3 may take longer to be resolved.

A comparative genome analysis using the complete set of genes in a phylogenetic framework provides the most conclusive evidence on the gene family's origin and evolutionary fate. In particular, analyses including genomes from closely and distantly related species have been shown to constitute a very successful approach. The genome analyses of the major gene families involved in the chemosensory system of the insects represent a good example to illustrate the state-of-the-art of gene family evolutionary analysis using complete genome DNA sequence data [13, 20, 23]. The most important proteins implicated in the early chemoreception steps in insects are encoded by gene families of moderate size. This process, which occurs inside the aqueous fluid of the chemosensory hair-like structures named sensilla, comprises the first contact of the external chemical signals (the odorant in the olfactory system) with membrane chemoreceptor proteins (the olfactory receptors in the olfactory system).

These multigene families can be classified into 2 main functional groups, the odorant-binding (OBPs) and chemosensory (CSPs) proteins (involved in the transport of the chemical signals through the sensillar lymph), and the chemosensory receptors that recognize the external cues and translate this information into an electrical signal (a dendritic spike) to the central nervous system, which elicit the appropriate behavior. In insects, there are 3 chemosensory receptor gene families: the olfactory (ORs) and gustatory (GRs) receptors, which in turn encompass the chemoreceptor superfamily, and the ionotropic receptors (IRs). Comparative

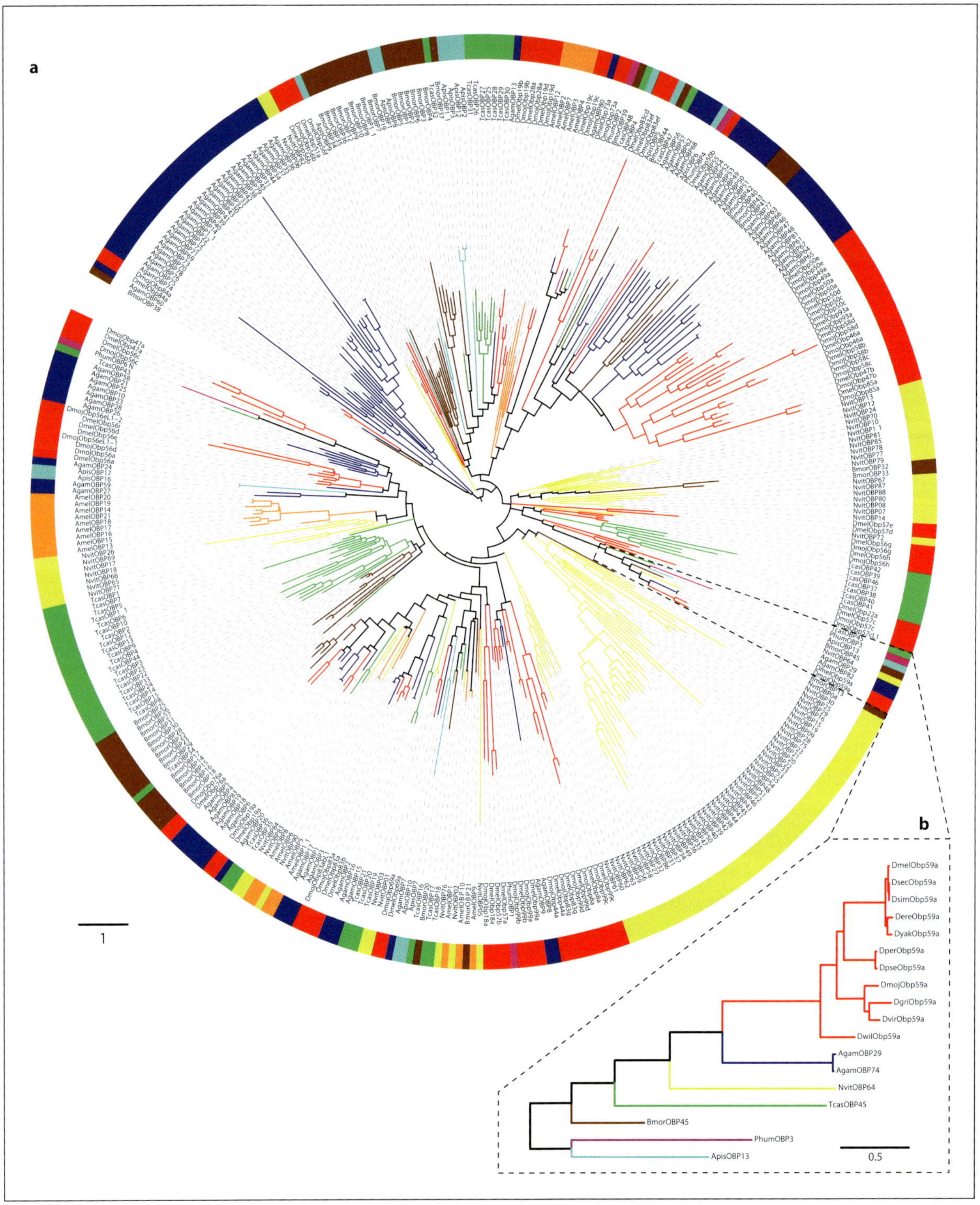

a
1
b
DmelObp59a
DsecObp59a
DsimObp59a
DereObp59a
DyakObp59a
DperObp59a
DpseObp59a
DmojObp59a
DgriObp59a
DvirObp59a
DwilObp59a
AgamOBP29
AgamOBP74
NvitOBP64
TcasOBP45
BmorOBP45
PhumOBP3
ApisOBP13
0.5

genome analyses revealed that the size of these multigene families differs markedly across species [20, 23] (fig. 1). While the number of genes of the OBP family ranges from 21 (in *Apis mellifera*) to 83 (in *Anopheles gambiae*), the CSP numbers range from 3 (in *Drosophila ananassae*) to 22 (in *Bombyx mori*); and whereas the ORs vary from 48 (in *Acyrthosiphon pisum* and in *B. mori*) to 265 (in *Tribolium castaneum*), the GRs range from 10 *(A. mellifera)* to 220 *(T. castaneum)*, and the IRs from 10 *(A. mellifera)* to 95 *(Aedes aegypti)*. Furthermore, these figures do not include information for the body louse *(Pediculus humanus)*, which contains a considerably lower number of genes (5 OBPs, 7 CSPs, 10 ORs, 8 GRs and 12 IRs), the cause for which likely stems from its parasitic lifestyle. This disparate number of genes in different insect species, nevertheless, provides a good opportunity to gain insight into the evolutionary mechanisms shaping gene family sizes and, particularly, into the role of natural selection and adaptation. Furthermore, the fact that these gene families include a moderate number of members allows for a comprehensive analysis that combine both automatic and manual 'gene calling' efforts, and also increases the accuracy of the resulting annotation.

It has been shown that the major gene families of the chemosensory system are usually arranged in chromosome clusters [23]. For instance, nearly 70% of the *Drosophila melanogaster* OBP genes (52 genes) are arranged in 10 clusters of 2–6 genes each. Nevertheless, despite the fact that this kind of arrangement also exists in other insect species and in other gene families, the actual fraction of the genes arranged in clusters is highly variable. Interestingly, physically neighboring members of these families are also phylogenetically related; for instance, evolutionarily new OBP duplicates are usually identified in extant chromosomal clusters, whereas phylogenetically close OBP genes are also located in the same cluster. Such data clearly supports unequal crossing-over as the main mechanism that generates tandem gene duplications of the chemosensory gene families.

Phylogenetic Analyses and the Birth-and-Death Process

Phylogenetic analyses including orthologous and paralogous copies show that the actual number of members is relatively conserved across the *Drosophila* genus, with few examples of species-specific expansions. However, a fine-scale investigation

Fig. 1. Phylogenetic analysis of the insect OBP genes. **a** Amino acid sequences of *A. gambiae* (Agam), *A. mellifera* (Amel), *A. pisum* (Apis), *B. mori* (Bmor), *D. melanogaster* (Dmel), *D. mojavensis* (Dmoj), *Nasonia vitripennis* (Nvit), *P. humanus* (Phum), and *T. castaneum* (Tcas). **b** Phylogenetic relationships of the OBP59a orthologous group in species of panel **a** and the following *Drosophila* species: *D. erecta* (Dere), *D. grimshawi* (Dgri), *D. persimilis* (Dper), *D. pseudoobscura* (Dpse), *D. sechellia* (Dsec), *D. simulans* (Dsim), *D. virilis* (Dvir), *D. willistoni* (Dwil), and *D. yakuba* (Dyak) . The *OBP59a* gene is absent in *A. mellifera*. The phylogenetic branches (and the outer ring) of the different species are depicted in colors: red, *Drosophila* species; blue, *A. gambiae*; brown, *B. mori*; green, *T. castaneum*; orange, *A. mellifera*; yellow, *N. vitripennis*; cyan, *A. pisum;* and pink, *P. humanus*. The scale bar represents 1 (**a**) or 0.5 (**b**) amino acid substitutions per site.

uncovers a large number of gene gains, gene losses and pseudogenization events, although these events have different frequency among gene families. Noticeably, gene losses and pseudogenization events are unequally distributed across the *Drosophila* phylogeny; indeed, the later events are mainly inferred in the terminal branches, suggesting that pseudogenes have a very short half-life. Across this genus, furthermore, it is reasonably easy to observe orthologous groups including all *Drosophila* species and, for a particular orthologous group there usually exists a good reconciliation between gene and species trees (fig. 1b). This data strongly suggests that these genes have diverged independently since their origin. These figures, however, are different from those found when distantly related species are compared (e.g. between insect orders) (fig. 1a). Indeed, there is a dramatic variation in gene family size as well as few examples of genes with orthologous copies across insects and many lineage-specific gene expansions (fig. 1a). Both features, however, are caused by the same basic evolutionary mechanism, the birth-and-death model (see below).

Current analyses of the chemosensory gene families (mostly from the OBP gene family data) within a phylogenetic framework largely support the birth-and-death model of evolution [6], specifically: (1) several gene gain and loss events have occurred in the evolution of the gene family; (2) a number of nonfunctional members (pseudogenes) can be identified across the phylogeny (mostly in terminal phylogenetic branches); (3) the phylogenetic trees inferred from orthologous genes fit well with the accepted species phylogeny; (4) there is no evidence for a major impact of gene conversion in the evolution of paralogous genes (although current methods for detecting gene conversion may be insufficient); (5) the number of orthologous groups including representatives of all surveyed species gradually decreases with increasing divergence time; (6) there is an uneven phylogenetic subfamily distribution across species; and (7) several gene expansions and contractions are identified across large (e.g. within-class or within-order) but not across short (e.g. across a genus) evolutionary times.

Birth-and-Death Rates and the Impact of Natural Selection

Methods and software have been developed to estimate birth-and-death rates (e.g. [13, 24]). The CAFE software [24] implements a stochastic birth-and-death model which allows an estimation of birth-and-death rates using a maximum likelihood approach (λ is the birth-and-death rate per gene and per million years) under the assumption of equal birth-and-death rates. Although this assumption may not always hold (e.g. in the presence of family expansions), it is a useful method for comparing birth-and-death rates across gene families or across species. For example, the birth-and-death rate for the complete set of gene families of *Drosophila* has been estimated as $\lambda = 0.0012$ [25], which indicates that there have been ~17 new gene gains or ~17 losses every million years during the evolution of any one *Drosophila* species' genome. In addition, the birth-and-death rates for the chemosensory gene families are noticeably larger than the estimates for the complete *Drosophila* genomes (OBPs, $\lambda = 0.005$; ORs, $\lambda = 0.006$; GRs, $\lambda = 0.011$; IRs, $\lambda = 0.0023$) [13, 23]; for instance, the value of λ = 0.005 inferred for the OBP gene family (assuming ~50 members) suggests that there has been an OBP gene gain (or a loss) every 4 million years. Such features, therefore, indicate that these gene families have a highly dynamic mode of evolution through which new members are continuously counterbalancing gene losses or non-functionalizations and pseudogenizations.

These high gene turnover rates exhibited by the chemosensory gene families additionally are shaped by natural selection. Indeed, natural selection can modify the rate of fixation in the population of newly duplicated copies, and it also can contribute to the functional diversification associated with sequence divergence. The levels of functional constraint and functional divergence can be analyzed through the comparative analysis of the ratio of non-synonymous (d_N) to synonymous (d_S) divergence ($\omega = d_N/d_S$), in which the ω value serves as a proxy for gauging levels of functional constraint. This method allows for the quantification of the impact of purifying (negative) and adaptive (positive) selection as well as for the testing of contrasting alternative evolutionary hypotheses. In the absence of selection the expected value of ω is 1, whereas statistically significant values lower (or higher) than 1 might be indicative of purifying (or positive) selection. The ω estimates for the OBPs, ORs and GRs of *Drosophila* clearly point to purifying selection as the main evolutionary force (OBPs, ω = 0.15; ORs, ω = 0.14; GRs, ω = 0.22). These ω values, furthermore, differ significantly among genes within a particular gene family. For instance, the ω values among the OBP orthologous groups range from 0.003 to 0.11. Among the ORs, the *Obp83b* gene has the smallest ω ratio, which is consistent with its critical function and its strong conservation across the insects. There are also strong differences among GR members; for instance, the sweet taste and the carbon dioxide receptors display low ratios. The functional constraint levels can also vary across positions of the coding region. Indeed, the specific molecular fingerprint of positive selection could even be detected in amino acids located in the putative odorant binding pocket of some OBPs. Since these changes likely affect the sensitivity or specificity in detecting odorants, these regions may be more likely to evolve by positive selection.

Birth-and-Death Evolution and Genomic Drift: Evolving Evolutionary Novelty in the Fatty Acid Reductase Multigene Family

During the evolutionary history of a multigene family that evolves under a birth-and-death model, the random occurrence of gene duplication and loss can lead to a change in the number of gene copies (i.e. dosage repetition) or paralogous family members (i.e. variant repetition) present within a genome. Thus, if one tallies the number of gene copies or family members present in a species' genome and compares it to a different species' genome, the numbers may be different. Nei [26] termed this 'genomic drift' and likened it to the random change of allele frequencies at a single gene produced by genetic drift.

For the most part, one expects the number of genes that are present in a genome to be determined solely through random chance. Dosage repetition, however, is one instance in which selection may play a role in determining gene copy number. For example, it is generally accepted that a large number of rRNA gene copies facilitates mRNA transcription, and therefore there exists a lower limit on the number of copies that a genome will tolerate. Consequently, the bobbed mutant phenotype of *D. melanogaster* appears when there is a loss of 50% or more of wild-type rRNA genes; in cases in which less than 15% of the wild-type rRNA genes remain, the mutation is lethal [27]. Likewise, adaptation to a novel environment or set of ecological circumstances can also drive changes in gene copy number [28]. For example, the evolution of tetrapods from a fish ancestor was accompanied by a concomitant increase in the number of paralogous olfactory genes present in the ancestral tetrapod genome, presumably in response to the increased number of odorants found on land versus in the aquatic environment [28]. Accordingly, once these new gene duplicates began to diverge from their parental gene, novel functions were acquired and, presumably, the number of odorants that could be detected subsequently increased.

The extent to which genomic drift influences a multigene family can be studied through the inference of the number of gene duplication and loss events that have occurred during the evolutionary history of the family [28, 29]. This is accomplished through the 'reconciliation' of the gene tree (i.e. the multigene family phylogeny) with the species tree [30–33]. In short, this procedure involves inferring the lowest number of duplication and loss events required to produce the observed gene tree given the assumed species tree. The procedure is too laborious to carry out by hand even when there are relatively small numbers of paralogous gene copies; thus, the use of computer software to conduct these analyses is highly recommended (e.g. NOTUNG [30]). To demonstrate how such an analysis is conducted, below we present a case study of the fatty acyl-coenzyme A reductase, or fatty acid reductase (FAR), multigene family using sequences extracted from the complete genomes of representative species of eukaryotes (fig. 2).

Genomic Drift Between Multigene Families

FAR enzymes catalyze the reduction of fatty acids to fatty alcohols in a reaction that is dependent upon NADPH as a cofactor. The number of FAR genes per genome can vary greatly between organisms. In vertebrates, there are 2 reductase genes present in the genome, whereas there are more than a dozen present in the silkworm. The evolutionary origins of this gene family are not well understood, but we found that acyl-CoA synthetase, acyltransferase and oxidoreductase gene families are close relatives of this family on the basis of protein sequence similarity (data not shown) and thus form a superfamily. If we examine the phylogenetic relationships of representatives of this superfamily (fig. 2a), we see evidence of birth-and-death evolution as shown through a pattern of between-species gene clustering. There are a couple of instances in which large single-species gene clusters were found (e.g. slime mold genes). This, however, is not unexpected since there is a lack of a closely related species to include in the comparison in this case. There is no evidence for concerted evolution of these genes, as the branch lengths found within these clusters are all relatively long, indicating that at least a moderate amount of divergence has occurred.

We can examine the question of gene turnover dynamics in more detail through an analysis of gene gain and loss. The analysis shown in figure 2b reveals that a varying amount of activity has taken place over the evolution of this superfamily. At the root of the phylogeny, which represents the last common ancestor shared between the 'lower' (i.e. slime mold and amoeba) and 'higher' eukaryote representatives studied, the ancestral genome was inferred to have possessed 11 genes constituting this superfamily. A considerable amount of gene turnover can be inferred to have occurred as shown through the different numbers of genes present in the various ancestral genomes (internal nodes) in the phylogeny (fig. 2b). Of particular interest is the observation that insects gained substantially higher amounts of genes than the other lineages, whereas vertebrates and nematodes (as represented by *Caenorhabditis elegans*) lost substantially more. As the FAR gene family dominates this superfamily in terms of total numbers of genes (fig. 2a), we can assume that most of this activity involves that family. To test this hypothesis and possibly determine the cause for the pattern, we conducted a separate analysis of the FAR multigene family.

Genomic Drift Within a Multigene Family

The results of our analysis of the FAR multigene family are presented in figures 2c and 2d. Expectedly, the FAR gene family undergoes birth-and-death evolution (fig. 2c) in accordance with the pattern inferred in figure 2a for the superfamily as a whole. However, the pattern of gene gain and loss is substantially different (fig. 2d). Virtually no gene loss was found to have occurred since the divergence of the slime mold from 'higher' eukaryotes (fig. 2d). In fact, the only lineage in which gene loss was found to happen was the nematode lineage (as represented by *C. elegans*), which involved the loss of only a single gene. In contrast, the pattern of gene gain is more dynamic. Two bursts are notable: (1) plants apparently gained 5 genes since they diverged from their last common ancestor shared with animals (fig. 2d), and (2) insects gained a substantial number of genes since they diverged from other animals: 6 genes were gained after they diverged from their last common ancestor shared with vertebrates, and another 5 genes were gained after the divergence of the honeybee from the silkworm and the fruit fly and mosquito. However, the gain of 6 genes along the lineage leading to insects from their last common ancestor with vertebrates must be interpreted with some caution, because there are a substantial number of other insect orders that are not represented in this phylogeny as well as other invertebrate and vertebrate lineages. Consequently, this number could be the result of 'summing' across other internal branches not found within this phylogeny due to missing taxa. This caveat may also hold for the number of gains along the branch leading to plants. In contrast, the gain of 5 genes in the common ancestor of

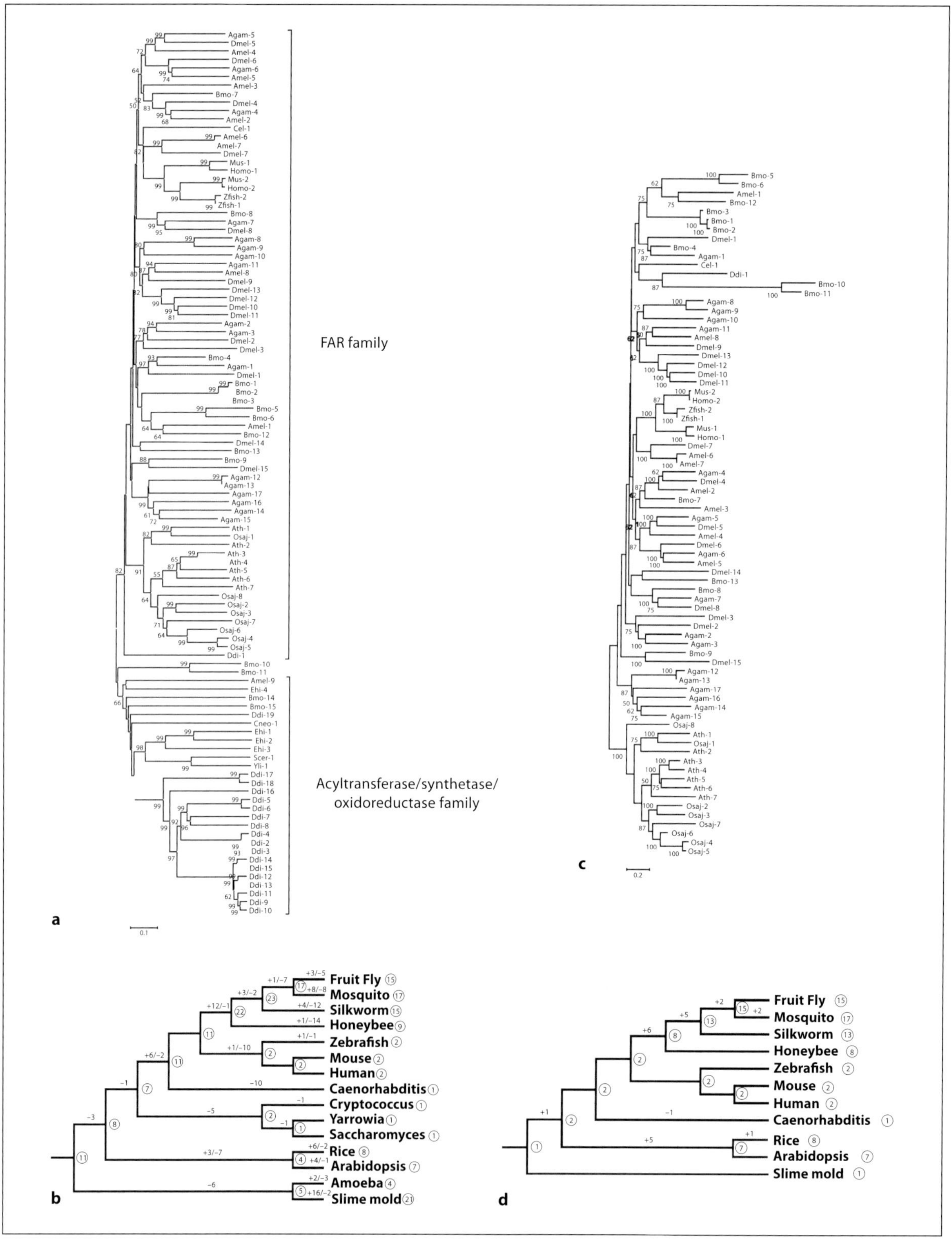

FAR family
Acyltransferase/synthetase/oxidoreductase family
a
b
c
d
Fruit Fly
Mosquito
Silkworm
Honeybee
Zebrafish
Mouse
Human
Caenorhabditis
Cryptococcus
Yarrowia
Saccharomyces
Rice
Arabidopsis
Amoeba
Slime mold

the silkworm, fruit fly and mosquito subsequent to their divergence from the honeybee is likely more reliable, since there are fewer missing taxa relative to the taxonomic rank (order) represented by the species in the study and, therefore, unlikely to alter the number much.

Regardless, simple calculation of the number of FAR genes present in the genomes of the species studied clearly indicates that plants and insects have undergone large expansions relative to the other taxa examined, which is consistent with the genomic drift hypothesis of Nei [26] for multigene families undergoing birth-and-death evolution. The possibility that these expansions facilitated the adaptive evolution of a variety of specialized functions that involve precursors upon which FAR genes act is especially interesting. For example, FAR genes have been shown to function in pheromone biosynthesis in moth species directly through the production an alcohol that confers species-specificity or indirectly through the biosynthesis of precursor compounds [34]. If we can assume that the silkworm is truly representative of moth species, the large number of FAR genes present in moth genomes (13 in the silkworm; fig. 2d) and the variety in substrate specificity that these genes have been shown to display [34, 35] suggest a number of different specialized functions have evolved. Similarly, plant FAR genes also have been shown to have evolved a number of specialized functions, such as the biosynthesis of wax esters used for storage in developing seeds [36], the biosynthesis of the lipid component used in the outer pollen wall, and the biosynthesis of cuticular wax lipids [37]. In contrast, the other species studied have very few FAR genes or even no genes. For example, *C. elegans* and the slime mold were found to have only 1 gene, and vertebrates only have 2, whereas the 3 fungi *(Cryptococcus neoformans, Yarrowia lipolytica*, and *Saccharomyces cerevisiae)* and the amoeba *(Entamoeba histolytica)* did not have any FAR homologues. It is possible that these species rely less on FAR genes to synthesize the fatty alcohol-containing compounds that these species require and other genes have evolved to take over these functions, or perhaps these species

Fig. 2. Birth-and-death evolution of FAR genes. **a**, **b** Phylogenetic analysis of the FAR/acyltransferase/oxidoreductase superfamily (**a**) and the associated gene tree reconciliation analysis for the superfamily (**b**). **c**, **d** Phylogenetic analysis of the FAR multigene family (**c**) and the associated gene tree reconciliation analysis for this family (**d**). **a**, **b** The computer program MEGA 4 [65] was used to reconstruct trees from Poisson amino acid distances using the neighbor-joining method. Numbers along branches represent bootstrap percentage values generated from 1,000 pseudoreplicates; only numbers greater than 50% are shown. **c**, **d** The computer program NOTUNG 2.6 [30] was used to conduct gene tree reconciliation analyses. The phylogenies shown are species trees based on [66]. Numbers along branches denote gene gains (+) or losses (–). Numbers shown in circles are the total number of genes present in the extant species or ancestral species (represented as nodes within the phylogeny) genome. Species abbreviations: Agam, *Anopheles gambiae* (mosquito); Amel, *Apis mellifera* (honeybee); Ath, *Arabidopsis thaliana*; Bmo, *Bombyx mori* (silkworm); Cel, *Caenorhabditis elegans*; Cneo, *Cryptococcus neoformans*; Ddi, *Dictyostelium discoideum* (slime mold); Dmel, *Drosophila melanogaster* (fruit fly); Ehi, *Entamoeba histolytica* (amoeba); Homo, *Homo sapiens* (human); Mus, *Mus musculus* (mouse); Osaj, *Oryza sativa* var. *japonica* (rice); Scer, *Saccharomyces cerevisiae*; Yli, *Yarrowia lipolytica*; Zfish: *Danio rerio* (zebrafish).

simply do not need a large and diverse number of fatty-acid containing compounds (in contrast to insects and plants), so only 1 or 2 genes are sufficient to synthesize all that is necessary.

It is difficult to say which of these possibilities is true without further knowledge of the FAR gene complement and associated functionalities from more species. Regardless, it is reasonable to assume that the genomic drift that produced the expansion of FAR genes in plants and insects is the underlying cause for their ability to synthesize and utilize a wide variety of fatty alcohol-based or derived compounds for a number of highly specialized functions.

Birth-and-Death Evolution and Selective Constraints: Histone Variant Diversification in the Germinal Cell Line

Multigene families often consist of structurally and functionally related genes that are usually clustered around specific genomic regions. The traditional view that a gene family producing a large amount of products needs to maintain homogeneity among its members [38] reinforced the notion that most multigene families were subject to concerted evolution, a process in which a mutation occurring in a repeat spreads all through the gene family members by recurrent unequal crossing-over or gene conversion. However, the increase in genomic molecular data during the last decade has revealed that most gene families encompass far too much genetic and functional diversity to be maintained by means of a homogenizing mechanism. Consequently, different alternative hypotheses have been put forward in order to account for the high diversity and functional differentiation exhibited by the members of different eukaryotic gene families. Among them, the birth-and-death model of evolution (which promotes genetic diversity) has often constituted the alternative hypothesis to concerted evolution [15].

Birth-and-Death Long-Term Evolution of Histone Multigene Families

In eukaryotes and some archaebacteria the members of the histone multigene families encode small basic proteins that are associated with the hereditary material in a nucleoprotein complex called chromatin, which allows for a high level of compaction of genomic DNA within the limited space of the nucleus and also provides the scaffolding upon which most DNA metabolic functions (i.e. replication, transcription and repair) take place. However, the different histone families display a high degree of heterogeneity among their members, depending on their structural and functional role in the nucleosome (the chromatin subunit) as well as depending upon whether the chromatin structure is in a somatic or a germinal setup. In addition, post-translational histone modifications also influence changes in chromatin structure both directly and indirectly by targeting or activating chromatin-remodeling complexes. Histone modifications intersect with cell signaling pathways to control gene expression and can act combinatorially to enforce or reverse epigenetic marks in chromatin [39, 40].

Histones have been used (together with rDNA) to showcase archetypal examples of multigene families subject to concerted evolution during the last 4 decades. However, the notion of this mechanism representing the major long-term evolutionary mode of these proteins has been abandoned given the high diversity and functional differentiation exhibited by the members of the different histone families. On the contrary, it has now been clearly demonstrated that the long-term evolution of the histones can be better described by a birth-and-death model of evolution based on recurrent gene duplication events and strong purifying selection acting at the protein level (e.g. [16]). This mode of evolution eventually leads to the functional differentiation of new gene copies through a process of neofunctionalization or subfunctionalization [40].

Selective Constraints and Histone Diversification in Different Chromatin Setups
Eukaryotic DNA is packed into different chromatin configurations in somatic and germinal cells. Somatic chromatin is formed by the repetition of nucleosomes [41], each consisting of an octamer of core histones (2 of each H2A, H2B, H3 and H4) around which 2 left-handed super-helical turns of DNA (approximately 146 bp) are wrapped. The nucleosomes are joined together in the chromatin fiber by short stretches of linker DNA that interact with linker H1 histones, resulting in an additional folding of the chromatin fiber. Germinal chromatin displays a high degree of heterogeneity depending on sex (male or female) and taxonomic group. Thus, while a nucleosome-based chromatin organization is prevalent in the case of the female germinal cell line (i.e. oocytes), the extreme reduction in the size of the sperm nucleus has led to a drastic reorganization in the male-specific chromatin in which nucleosomes have been replaced by nucleoprotein structures able to produce a tighter packaging of DNA [40].

Sperm chromatin is unique in that most, if not all, is tightly heterochromatinized within the highly compacted sperm nuclei thanks to its association with sperm nuclear basic proteins (SNBPs) [42]. In contrast to the proteins of somatic chromatin (histones), SNBPs exhibit a greater compositional heterogeneity and can be grouped into 3 major types based on structural and compositional considerations. The first is the histone type (H-type) SNBPs, which are very similar to histones from somatic tissues and, therefore, produce a chromatin organization identical to that observed in somatic cell nuclei. The second type consists of protamines (P-type SNBPs), which constitute a group of heterogeneous, small, arginine-rich proteins that result in a tighter packaging of DNA within the sperm nucleus. The third type of SNBPs form a group known as the protamine-like proteins (PL-type), which are related to histone H1 and represent a structurally and functionally intermediate group between the H- and P-types [42]. The chromatin fibers resulting from the association of the different SNBP types with DNA all exhibit a fairly constant diameter in the range of 300–500 Å, independent of the extent of protein folding of the SNBP type involved, which decreases from the H- to PL-type and from the PL- to the P-type [39].

Somatic chromatin is characterized by a nucleosome-based organization in which histones associate with each other and with DNA through different protein-protein interactions including those of an electrostatic nature. Histone proteins are thus subject to strong selective constraints in order to preserve their structure along with the nucleoprotein complex they form with DNA. However, the transition from somatic to germinal chromatin setups during spermiogenesis involves the replacement of histones by specialized SNBPs, leading to the progressive loss of a nucleosome-based chromatin configuration [39]. In this scenario, the functional constraints operating on histones in the germinal cell line are expected to be relaxed, allowing for a higher degree of variation within the different histone types (fig. 3).

Increased Birth-and-Death Histone Diversification in the Male Germinal Cell Line
Nucleosomes modulate accessibility of regulatory proteins to DNA and thus influence eukaryotic gene regulation. The evolution of chromatin remodeling mechanisms governing nucleosome organization at promoters, regulatory elements, and other functional regions in the genome unveil an interplay of sequence-based nucleosome preferences and non-nucleosomal factors in determining nucleosome organization within mammalian cells. The genetic diversity observed among histone family members bears critical implications for the structure and function of the nucleosome in different chromatin settings [43], involving the formation of H2A-H2B and H3-H4 dimers through different protein-protein interactions, including those of an electrostatic nature. When looking at the diversity within core histone families (fig. 3a), it seems that although one of each interacting partners is allowed to have a higher extent of variation (H2A and H3), the other maintains a conserved structure (H2B and H4). Molecular evolutionary studies carried out during the last 10 years have revealed that the long-term evolution of the histone H1 family, as well as of H2A, H3, and H4 core histone families, is governed by birth-and-death under a strong purifying selection acting at the protein level, in order to preserve a functional quaternary structure of the nucleosome core particle [40], able to efficiently bind and package the DNA, as well as to mediate different dynamic processes in chromatin metabolism [43].

However, information about the diversity and the evolution of H2B was lacking until very recently. The H2B family stands out among histones because of the low extent of diversification of its members (compared with H1, H2A, and H3 families) and the lack of specialized variants in the somatic cell lineage. Nevertheless, the H2B family is peculiar by displaying variants exclusively restricted to the male germinal cell lineage. For instance, 2 testis-specific variants have been described in humans so far, including TH2B (also referred to as hTSH2B) [44] and H2BFW (also known as H2BFWT) [45], both involved in the reorganization of chromatin during spermatogenesis. Furthermore, additional minor H2B variants with a lower extent of similarity

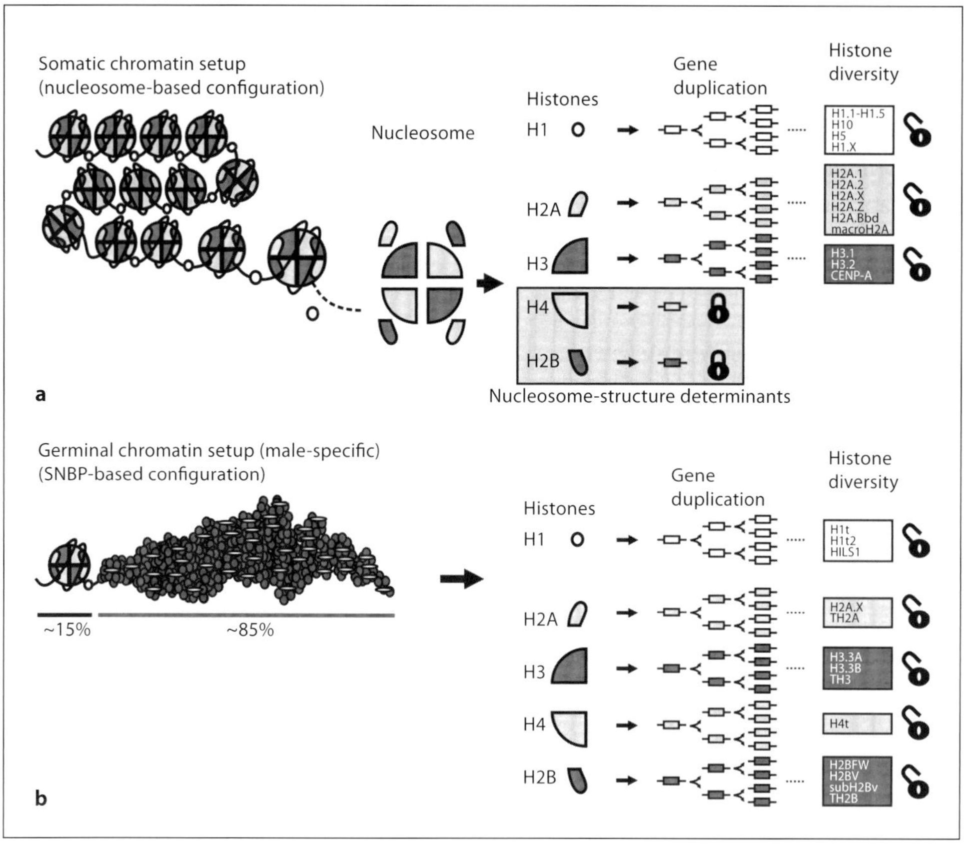

Fig. 3. Chromatin organization and histone diversification in the somatic and male-specific germinal cell line. **a** Histone H2B and H4 variant diversification is locked within a somatic chromatin setup, probably as a consequence of their essential role in maintaining the fundamental structural H2A-H2B and H3-H4 domains of the nucleosome core particle. In contrast, the variation presented by the H2A and H3 counterparts is responsible for imparting different functional and structural specificities to these domains, allowing for the specialization of local chromatin segments genome-wide. **b** The structural reorganization of chromatin during spermiogenesis leads to the loss of a nucleosome-based configuration in the male germinal cell line, lightening the evolutionary constraints operating on histone H2B and H4 evolution. Consequently the process of diversification within these histone families is unlocked allowing for the functional differentiation of germinal variants.

with canonical H2Bs have also been described in the male germinal line including subH2Bv, a sperm-specific histone identified in the bull *Bos taurus*; gH2B, a divergent H2B protein identified in *Lilium longiflorum*, involved in the packaging of chromatin in pollen; and H2BV, a variant first identified in *Trypanosoma brucei* that specifically dimerizes with H2A.Z. In addition, 2 novel H2B variants involved in pericentric heterochromatin reprogramming during mouse spermiogenesis, referred to as H2BL1 and H2BL2, have been recently identified [46], showing resemblance to subH2Bv and H2BFW, respectively.

The constraints driving the long-term evolution of the H2B family in the somatic cell line have been recently investigated, corroborating the presence of birth-and-death evolution under strong purifying selection, maintaining high levels of certain biased amino acids (lysine and alanine) which are important for the establishment of the correct interactions involved in the formation of the nucleosome [47]. On the other hand, and in contrast with other histones, H2B members are also subject to a very rapid process of diversification in the male germinal cell lineage (fig. 3b) involving the functional specialization of different histone variants, probably as a consequence of neofunctionalization and subfunctionalization events after gene duplication [47]. This is specifically evident in the case of the H2BFW variant that evolves almost at the same rate as the quickly evolving histone H2A.Bbd which is also involved in mammalian spermiogenesis [48].

The lack of diversity within the H2B and H4 families has been regarded to be the result of their essential role in the maintenance of the fundamental structural H2A-H2B and H3-H4 domains of the nucleosome. By contrast, the variation presented by the H2A and H3 counterparts would be responsible for imparting different functional and structural specificities to these domains [43]. Such a hypothesis would be consistent with the increase in H2B diversity observed in the male germinal cell line where a dramatic change in chromatin conformation takes place during spermiogenesis. Two conclusions can be drawn from this. First, H2B variation implicitly suggests the possibility of H4 variation. Indeed, the few H4 variants described to date are mostly circumscribed to the testis [49]. Second, the diversification of H2B and H4 histones would be absent from the female germinal cell line (i.e. in oocytes) due to the prevalence of a nucleosome chromatin organization, which would only be compatible with H1 variants such as H1oo and H1M/B4. It thus seems that the reorganization of chromatin structure during spermiogenesis might have affected the evolutionary constraints driving histone H2B evolution, leading to an increase in diversity. However, with the exception of a few structural studies [50], little is known about the specific role performed by the testis-specific H2B variants. Further studies will be needed in order to clearly decipher the connection between the relaxation of the evolutionary constraints described here and the drastic structural chromatin transitions involved in spermiogenesis.

Mixed Effects of Birth-and-Death and Concerted Evolution: the 5S rDNA Gene Family in Fishes and Molluscs

In eukaryotes, rDNA is generally arranged in 2 different gene clusters (multigene families), each composed of hundreds to thousands of gene copies. While the major cluster (45S rDNA) comprises the 18S, 5.8S, and 28S rRNA genes, the minor cluster (5S rDNA) comprises only 5S rRNA genes. The 5S rRNA gene consists of a transcriptional unit of ~120 bp, which is separated from the next unit by a non-transcribed spacer (NTS). Although the 5S rRNA gene is highly conserved, the NTSs are variable both in length and in sequence [51]. Given the apparent homogeneity observed among the different copies, 5S genes have been used to showcase the archetypal example of a gene family subject to concerted evolution. However, the theoretical expectations made by this model are challenged by 3 major molecular evolutionary features displayed by the 5S rDNA family. First, several 5S gene variants have been found, constituting a dual system. Second, 5S rDNA divergent pseudogenes have been found in unrelated taxa. Third, the existence of different types of repeat units has been also corroborated based on the study of spacers. Consequently, different authors have proposed that the variation observed among 5S rDNA members best fits to a birth-and-death model of long-term evolution promoting genetic diversity [6].

Concerted Evolution of 5S rRNA Genes

Concerted evolution has been recently discarded (in favor of a birth-and-death mechanism) as the major model guiding the long-term evolution of several multigene families [6]. However, the case of rDNA seems to be otherwise more complex. Among animals, molluscs and fishes stand out for being the most widely studied groups of organisms with respect to 5S rRNA genes, displaying intense genetic dynamics. Studies on the 5S rDNA from oyster (genus *Crassostrea*) have revealed the existence of (1) two different genes (instead of one, as in the case of the major genes) encoding the minor 5S subunit, and (2) the localization of 5S rRNA genes in 2 pairs of chromosomes different from the chromosome pair (pair 10) where the major genes are located [52]. However, only 1 type of 5S rDNA tandem repeat was found in *Crassostrea* representatives. These results, together with the identification of a microsatellite at the 3′ end of 5S genes (potentially involved in the maintenance of tandem arrays), support the concerted evolution of 5S rRNA genes in these organisms.

Evidence supporting the concerted evolution of 5S rRNA genes has been also found in different fish representatives. For instance, decreased levels of intra- and interspecies nucleotide variation have been recently revealed in the 5S coding regions from fish species belonging to the family Moronidae [53]. Similarly to the case of oyster, the presence of microsatelllite sequences has been also identified at NTS regions. Different authors have suggested that the presence of short microsatellite sequences favors the maintenance of tandem arrays in multigene families. These sequences would act as 'hot spots' for recombination, facilitating gene conversion or unequal crossing-over and therefore, concerted evolution [54, 55].

Mixed Effects of Birth-and-Death and Concerted Evolution

Within molluscs, mussels also attract special interest due to the heterogeneity they display in 5S rDNA organization, including different types of repeat units with divergent spacers such as those identified in *Mytilus* species [56]. Recent studies on *Mytilus* species provided evidence for an apparent absence of interspecies differentiation across 5S coding regions, a notion reinforced by: (1) the lack of fixed differences between species, and (2) the low levels of nucleotide variation found within 5S coding regions in comparisons between different types of units, suggesting the presence of independent evolutionary pathways leading to their differentiation [57]. Although these results do not fit the predictions made by the concerted evolution model, they can be still reconciled with a critical role for this evolutionary model in 5S rDNA evolution. Different studies have put forward a hypothesis in which the homogenization of rDNA units would occur locally within arrays, implying that selective mechanisms operate in the coding region, eliminating mutations without affecting spacer regions [58]. It is thus possible that a first stage of 5S rDNA evolution would had involved the generation of genetic diversity through recurrent gene duplications (birth-and-death), followed by the transposition of several units to different chromosomal locations, leading to the their subsequent independent concerted evolution. However, even though the observed patterns of 5S rDNA evolution could also result from a process of gene duplication and selection without invoking homogenization, a substantial effect of concerted evolution cannot be ruled out until the presence of heterogeneous selective constraints acting on different 5S types is demonstrated [57].

Many studies focused on the molecular organization and evolution of 5S rRNA genes have described the presence of 2 types of 5S rDNA units, especially in the case of fishes [59, 60]. The main difference between these sequences is essentially circumscribed to length polymorphisms in the NTS region, although variation in coding regions is sometimes observed, suggesting that the two 5S rDNA loci evolve independently. However, some reports suggest that both 5S rDNA types are not located in independent clusters, since different 5S variants have been found on the same PCR product displaying a tandem organization [61]. It thus appears that the existence of 2 types of 5S rDNA units constitutes a common trend in fish species [53, 55, 60]. This organization has been commonly referred to as 'dual expression system', where one type is expressed in both the somatic and the germinal (oocyte) cell line, while the other type is specific to oocyte cells. The presence of 5S rDNA units containing divergent types of NTSs was identified in the flatfish *Solea senegalensis*. Furthermore, a repeat unit containing the 5S rRNA gene linked simultaneously to 3 different small nuclear RNA genes (U1, U2, and U5) was described for the first time in this species (U2 snRNA appeared also in the NTS of the oyster *Crassostrea* [54]), probably representing pseudogenes [62]. Sequence divergence among tandemly arranged 5S rRNA and NTS sequences indicates that the rate of concerted evolution is insufficient to homogenize the entire array. Similar results have been described in stingrays [60], a coregonid fish, for which a significant amount of variation was reported in the 5S rRNA coding region and NTS sequences [63], as well as in species belonging to the genus *Brycon*, displaying high levels of divergence in the NTS region [5].

Birth-and-Death Evolution in Dual 5S rDNA Gene Systems

Several species of the family Batrachoididae have traditionally been used as model organisms within teleost fishes. For our studies, we have chosen 4 Venezuelan species *(Amphichthys cryptocentrus, Batrachoides manglae, Porichthys plectrodon, Thalassophryne maculosa)* and the only European species within this family, the toad-fish *Halobatrachus didactylus.* Two types of 5S rDNA units were found in *H. didactylus* and, given the lack of similarity between their NTS sequences, they probably do not share a common ancestral sequence. Although both types seem to represent functional genes, it cannot be concluded that a dual system of 5S rDNA is generally established in the Batrachoididae family since species displaying only one 5S rDNA type have also been found [55]. Given that the sequences of both coding regions and the NTSs are quite conserved in *H. didactylus*, concerted evolution seems to represent the more feasible model for this multigene family.

Although concerted evolution has been traditionally proposed to guide the long-term evolution of 5S rRNA genes, the birth-and-death model of evolution has been recently invoked in order to explain several cases in which homogenization is not observed [60, 64]. Under a birth-and-death model of evolution, 5S rDNA genes would be expected to display divergent variants in the genome, between-species clustering pattern in the phylogenies as well as the presence of pseudogenes. Genome rearrangements (e.g. gene duplications, deletions, insertions) are likely to have been involved in the evolution of 5S rRNA genes in the family Batrachoididae. The results of our analysis suggest that the 5S rRNA genes of the 4 species studied (and also of the European one) are derived from a dual 5S rDNA gene system which was already present in the genome of their common ancestor. However, while *A. cryptocentrus* and *B. manglae* have retained both types of 5S rDNA units, we have found only 1 type in *P. plectrodon* and *T. maculosa*. In these last 2 species, as well as in *B. manglae*, homogenizing mechanisms like those proposed by the concerted evolution model appear to have occurred. While *P. plectrodon* seems to have suffered a recent deletion event (and concerted evolution has not had enough time to act), one of the 5S rDNA types from *A. cryptocentrus* has undergone a higher degree of diversification. Therefore, the emergence of new 5S rDNA variants in *A. cryptocentrus* could be explained by birth-and-death evolution, and these variants could be maintained by purifying selection. Notwithstanding, we cannot exclude the possibility of some homogenization mechanisms reducing sequence divergence within each 5S rDNA unit in this species [61].

The birth-and-death evolution of 5S rDNA in fish species is also supported by the presence of pseudogenes, although the emergence of duplicated pseudogenes can also be explained by unequal crossing-over, one of the main mechanisms acting in concerted evolution [5]. In addition, NTS regions of *A. cryptocentrus* and *B. manglae* display a variable number of $(TG)_n$ or $(AG)_n$ microsatellites which could represent 'hot spots' playing an important role in homogenizing tandem arrays [54]. Furthermore, homogenization resulting from unequal crossing-over or gene conversion during concerted evolution would occur most frequently in regions of chromosomes closer to the telomeres [6]. In this regard, FISH studies using 5S rDNA probes have shown that minor ribosomal genes of *A. cryptocentrus* are located in a subcentromeric position [55], which could hinder the action of the mechanisms that govern concerted evolution.

Birth-and-death has been proposed as a very important mechanism in guiding the long-term evolution of the 5S rDNA family in different organisms. Our results suggest that in many groups of molluscs and fishes the long-term evolution of 5S rRNA genes is most likely mediated by a mixed mechanism in which the generation of genetic diversity is achieved through birth-and-death (recurrent gene duplication), followed by the local homogenization of the different units through concerted evolution (probably after their physical transposition to independent chromosomal locations). In addition, it is important to bear in mind that to completely discern between the relative contributions of concerted evolution and birth-and-death evolution to the overall long-term evolution of 5S rRNA genes, it would be necessary to gather information on the complete set of 5S rRNA genes in different genomes. Although this has not yet been achieved for most 'higher' eukaryotes (including molluscs), it is not the case for certain groups of 'lower' eukaryotes. For example, in a complete genome study of 4 species of fungi, it was shown that the birth-and-death model without contribution of concerted evolution best characterizes the long-term evolution of 5S genes in those organisms [18]. In this case, the apparent homogenization among copies results from a combination of (1) recent gene duplication due to a gene duplication and insertion process similar to retroposon amplification and (2) rapid gene turnover derived from a high frequency of duplication/amplification events. Without a precise knowledge of the complete genome complement of these taxa and the subsequent comparison among closely related species, it is easy to misinterpret that homogenizing forces might also have an important role in the 5S gene evolution of those particular organisms.

Concluding Remarks

Over the long term, the birth-and-death process might result in a large variation in the number of genes or in the number of orthologous copies that would be visualized as gene family expansions (or contractions). The family size, therefore, would result from a trade-off between the stochastic birth-and-death process and the maintenance of genes required for proper function, as depicted by the case of the chemoreceptor system. Hence, the dynamic birth-and-death process has important evolutionary and adaptive implications: both gene gains and losses constitute a significant source of variation for evolutionary change. Indeed, DNA changes (in a particular duplicate) affecting the sensitivity or specificity in the detection of pheromones or related substances as food may be advantageous and might be fostered by shifts in ecological interactions. In so far, as the relevance of gene gains and losses to overall multigene family evolution is concerned, genomic drift plays a clear role in driving the divergence of entire multigene families. As we have shown in the case of the chemoreceptor families, genomic drift can alter the composition of genes within a genome as well as between different species' genomes, encompassing an adaptive value behind these changes. Similarly, drift may have played a part in the case of the FAR gene family in facilitating the ecological adaptation of plants and insects to their environments through the ability to generate a range of fatty alcohols utilized for a variety of physiological purposes. Thus, genomic drift can be viewed as a driving force for evolving evolutionary novelty that can be exploited by a species as means for adaptation to various selective challenges.

Once selection starts operating over a multigene family, changes or shifts in selective constraints will affect the functional dynamics of the birth-and-death process. This mechanism is best exemplified by histone multigene families, where the relaxation of the selective constraints results in higher rates of functional diversification across family members which otherwise must be conserved in order to preserve the nucleosome-based structure of somatic chromatin. However, given the evolutionary patterns observed across 5S rDNA gene family members, an important effect of concerted evolution cannot be ruled out until the presence of heterogeneous selective constraints acting on different 5S types is demonstrated.

Over the last 2 decades many multigene families have been identified that undergo birth-and-death evolution, including former archetypal examples of concerted evolution, such as histones and rRNA genes. Far from the old controversies on the mechanisms driving the evolution of multigene families, the continuous stream of genomic molecular data keeps on creating an increasingly complex canvas of gene families filled with countless evolutionary nuances. In such a complex scenario, the birth-and-death model of evolution provides a framework to understand how multigene families originate and diversify, representing the principal mechanism guiding the long-term evolution of multigene families.

Acknowledgements

This work was supported by grants from the Xunta de Galicia (10-PXIB-103-077-PR to J.M.E.-L.), from the Ministerio de Ciencia e Innovación of Spain-MICINN (CGL2011-24812 to J.M.E.-L., and BFU2010-15484 to J.R.), and from the Junta de Andalucía and CeiA3 (Campus de Excelencia Internacional Agroalimentario to L.R., group BIO-219). J.M.E.-L. was supported by a contract within the Ramon y Cajal Subprogramme (Ministerio de Ciencia e Innovación of Spain-MICINN), and J.R. was partially supported by ICREA Academia (Generalitat de Catalunya).

References

1 Ohno S: Evolution by Gene Duplication. Berlin, Springer-Verlag, 1970.
2 Yang Z: Computational Molecular Evolution. Oxford, Oxford University Press, 2006.
3 Demuth JP, Hahn MW: The life and death of gene families. Bioessays 2009;31:29–39.
4 Gabaldon T: Large-scale assignment of orthology: back to phylogenetics? Genome Biol 2008;9:235.

5 Martins C, Wasko AP: Organization and evolution of 5S ribosomal DNA in the fish genome; in Williams CL (ed): Focus on Genome Research. Hauppauge, Nova Science Publishers, 2004, pp 335–363.

6 Nei M, Rooney AP: Concerted and birth-and-death evolution in multigene families. Annu Rev Genet 2005;39:121–152.

7 Lynch M: The Origins of Genome Architecture. Sunderland, MA, Sinauer Associates, 2007.

8 Hahn MW: Bias in phylogenetic tree reconciliation methods: implications for vertebrate genome evolution. Genome Biol 2007;8:R141.

9 Csuros M: Count: evolutionary analysis of phylogenetic profiles with parsimony and likelihood. Bioinformatics 2010;26:1910–1912.

10 Iwasaki W, Takagi T: Reconstruction of highly heterogeneous gene-content evolution across the three domains of life. Bioinformatics 2007;23:i230–239.

11 Vernot B, Stolzer M, Goldman A, Durand D: Reconciliation with non-binary species trees. Comput Syst Bioinformatics Conf 2007;6:441–452.

12 Dufayard JF, Duret L, Penel S, Gouy M, Rechenmann F, et al: Tree pattern matching in phylogenetic trees: automatic search for orthologs or paralogs in homologous gene sequence databases. Bioinformatics 2005;21:2596–2603.

13 Vieira FG, Sanchez-Gracia A, Rozas J: Comparative genomic analysis of the odorant-binding protein family in 12 *Drosophila* genomes: purifying selection and birth-and-death evolution. Genome Biol 2007;8:R235.

14 Ingram VM: Gene evolution and the haemoglobins. Nature 1961;189:704–708.

15 Nei M, Hughes AL: Balanced polymorphism and evolution by the birth-and-death process in the MHC loci; in Tsuji K, Aizawa M, Sasazuki T (eds): 11th Histocompatibility Workshop and Conference. Oxford, Oxford University Press, 1992, pp 27–38.

16 Rooney AP, Piontkivska H, Nei M: Molecular evolution of the nontandemly repeated genes of the histone 3 multigene family. Mol Biol Evol 2002;19: 68–75.

17 Rooney AP: Mechanisms underlying the evolution and maintenance of functionally heterogeneous 18S rRNA genes in apicomplexans. Mol Biol Evol 2004; 21:1704–1711.

18 Rooney AP, Ward TJ: Evolution of a large ribosomal RNA multigene family in filamentous fungi: birth and death of a concerted evolution paradigm. Proc Natl Acad Sci USA 2005;102:5084–5089.

19 Zhang J, Dyer KD, Rosenberg HF: Evolution of the rodent eosinophil-associated RNase gene family by rapid gene sorting and positive selection. Proc Natl Acad Sci USA 2000;97:4701–4706.

20 Vieira FG, Rozas J: Comparative genomics of the odorant-binding and chemosensory protein gene families across the Arthropoda: origin and evolutionary history of the chemosensory system. Genome Biol Evol 2011;3:476–490.

21 Rooney AP, Ward TJ: Birth-and-death evolution of the internalin multigene family in *Listeria*. Gene 2008;427:124–128.

22 Clark AG, Eisen MB, Smith DR, Bergman CM, Oliver B, et al: Evolution of genes and genomes on the *Drosophila* phylogeny. Nature 2007;450:203–218.

23 Sanchez-Gracia A, Vieira FG, Rozas J: Molecular evolution of the major chemosensory gene families in insects. Heredity 2009;103:208–216.

24 De Bie T, Cristianini N, Demuth JP, Hahn MW: CAFE: a computational tool for the study of gene family evolution. Bioinformatics 2006;22:1269–1271.

25 Hahn MW, Han MV, Han SG: Gene family evolution across 12 *Drosophila* genomes. PLoS Genet 2007;3:e197.

26 Nei M: The new mutation theory of phenotypic evolution. Proc Natl Acad Sci USA 2007;104:12235–12242.

27 Long EO, Dawid IB: Repeated genes in eukaryotes. Annu Rev Biochem 1980;49:727–764.

28 Nei M, Niimura Y, Nozawa M: The evolution of animal chemosensory receptor gene repertoires: roles of chance and necessity. Nat Rev Genet 2008;9:951–963.

29 Nam J, Nei M: Evolutionary change of the numbers of homeobox genes in bilateral animals. Mol Biol Evol 2005;22:2386–2394.

30 Durnad D, Halldórsson BV, Vernot B: A hybrid micro-macroevolutionary approach to gene tree reconstruction. J Comput Biol 2006;13:320–335.

31 Goodman M, Czelusniak J, Moore GW, Romero-Herrera AE, Matsuda G: Fitting the gene lineage into its species lineage, a parsimony strategy illustrated by cladograms constructed from globin sequences. Syst Zool 1979;28:132–163.

32 Page R: Maps between trees and cladistic analysis of historical associations among genes, organisms and areas. Syst Zool 1994;43:58–77.

33 Page R, Charleston M: From gene to organismal phylogeny: Reconciled trees and the gene tree/species tree problem. Mol Phylogenet Evol 1997;7:231–240.

34 Antony B, Fuji T, Moto K, Matsumoto S, Fukuzawa M, et al: Pheromone-gland-specific fatty-acyl reductase in the adzuki bean borer, *Ostrinia scapulalis* (Lepidoptera: Crambidae). Insect Biochem Mol Biol 2009;39:90–95.

35 Moto K, Yoshiga T, Yamamoto M, Takahashi S, Okano K, et al: Pheromone gland-specific fatty-acyl reductase of the silkmoth, *Bombyx mori*. Proc Natl Acad Sci USA 2003;100:9156–9161.
36 Miwa T: Jojoba oil wax esters and derived fatty acids and alcohols: gas chromatographic analyses. J Am Oil Chem Soc 1971;48:259–264.
37 Rowland O, Zheng H, Hepworth SR, Lam P, Jetter R, et al: *CER4* encodes an alcohol-forming fatty acyl-coenzyme A reductase involved in cuticular wax production in *Arabidopsis*. Plant Physiol 2006; 142:866–877.
38 Thatcher TH, Gorovsky MA: Phylogenetic analysis of the core histones H2A, H2B, H3, and H4. Nucleic Acids Res 1994;22:174–179.
39 Eirín-López JM, Ausió J: Origin and evolution of chromosomal sperm proteins. Bioessays 2009;31: 1062–1070.
40 Eirín-López JM, González-Romero R, Dryhurst D, Méndez J, Ausió J: Long-term evolution of histone families: old notions and new insights into their diversification mechanisms across eukaryotes; in Pontarotti P (ed): Evolutionary Biology: Concept, Modeling, and Application. Berlin, Springer-Verlag, 2009, pp 139–162.
41 Zlatanova J, Bishop TC, Victor JM, Jackson V, van Holde K: The nucleosome family: dynamic and growing. Structure 2009;17:160–171.
42 Eirín-López JM, Frehlick LJ, Ausió J: Protamines, in the footsteps of linker histone evolution. J Biol Chem 2006;281:1–4.
43 Talbert PB, Henikoff S: Histone variants – ancient wrap artists of the epigenome. Nat Rev Mol Cell Biol 2010;11:264–275.
44 Zalensky AO, Siino JS, Gineitis AA, Zalenskaya IA, Tomilin NV, et al: Human testis/sperm-specific histone H2B (hTSH2B). Molecular cloning and characterization. J Biol Chem 2002;277:43474–43480.
45 Churikov D, Siino J, Svetlova M, Zhang K, Gineitis A, et al: Novel human testis-specific histone H2B encoded by the interrupted gene on the X chromosome. Genomics 2004;84:745–756.
46 Govin J, Escoffier E, Rousseaux S, Kuhn L, Ferro M, et al: Pericentric heterochromatin reprogramming by new histone variants during mouse spermiogenesis. J Cell Biol 2007;176:283–294.
47 González-Romero R, Rivera-Casas C, Ausió J, Méndez J, Eirín-López JM: Birth-and-death long-term evolution promotes histone H2B variant diversification in the male germinal cell line. Mol Biol Evol 2010;27:1802–1812.
48 Eirín-López JM, Ishibashi T, Ausió J: H2A.Bbd: a quickly evolving hypervariable mammalian histone that destabilizes nucleosomes in an acetylation-independent way. FASEB J 2008;22:316–326.
49 Wolfe SA, Grimes SR: Protein-DNA interactions within the rat histone H4t promoter. J Biol Chem 1991;266:6637–6643.
50 Li A, Maffey AH, Abbott WD, Conde e Silva N, Prunell A, et al: Characterization of nucleosomes consisting of the human testis/sperm-specific histone H2B variant (hTSH2B). Biochemistry 2005;44: 2529–2535.
51 Campo D, Machado-Schiaffino G, Horreo JL, Garcia-Vazquez E: Molecular organization and evolution of 5S rDNA in the genus *Merluccius* and their phylogenetic implications. J Mol Evol 2009;68:208–216.
52 Cross I, Vega L, Rebordinos L: Nucleolar organizing regions in *Crassostrea angulata*: chromosomal location and polymorphism. Genetica 2003;119:65–74.
53 Merlo MA, Cross I, Chairi H, Manchado M, Rebordinos L: Analysis of three multigene families as useful tools in species characterization of two closely-related species, *Dicentrarchus labrax*, *Dicentrarchus punctatus* and their hybrids. Genes Genet Syst 2010;85:341–349.
54 Cross I, Rebordinos L: 5S rDNA and U2 snRNA are linked in the genome of *Crassostrea angulata* and *Crassostrea gigas* oysters: Does the (CT)n(GA)n microsatellite stabilize this novel linkage of large tandem arrays? Genome 2005;48:1116–1119.
55 Ubeda-Manzanaro M, Merlo MA, Palazon JL, Sarasquete C, Rebordinos L: Sequence characterization and phylogenetic analysis of the 5S ribosomal DNA in species of the family Batrachoididae. Genome 2010;53:723–730.
56 Insua A, Freire R, Ríos J, Méndez J: The 5S rDNA of mussels *Mytilus galloprovincialis* and *M. edulis*: sequence variation and chromosomal location. Chromosome Res 2001;9:495–505.
57 Freire R, Arias A, Insua A, Méndez J, Eirin-Lopez JM: Evolutionary dynamics of the 5S rDNA gene family in the mussel *Mytilus*: mixed effects of birth-and-death and concerted evolution. J Mol Evol 2010;70:413–426.
58 Kellogg EA, Appels R: Intraspecific and interspecific variation in 5S RNA genes are decoupled in diploid wheat relatives. Genetics 1995;140:325–343.
59 Pinhal D, Araki CS, Gadig OB, Martins C: Molecular organization of 5S rDNA in sharks of the genus *Rhizoprionodon*: insights into the evolutionary dynamics of 5S rDNA in vertebrate genomes. Genet Res (Camb) 2009;91:61–72.
60 Pinhal D, Yoshimura TS, Araki CS, Martins C: The 5S rDNA family evolves through concerted and birth-and-death evolution in fish genomes: an example from freshwater stingrays. BMC Evol Biol 2011;11:151.

61 Robles F, de la Herran R, Ludwig A, Rejon CR, Rejon MR, et al: Genomic organization and evolution of the 5S ribosomal DNA in the ancient fish sturgeon. Genome 2005;48:18–28.
62 Manchado M, Zuasti E, Cross I, Merlo A, Infante C, et al: Molecular characterization and chromosomal mapping of the 5S rRNA gene in *Solea senegalensis*: A new linkage to the U1, U2, and U5 small nuclear RNA genes. Genome 2006;49:79–86.
63 Sajdak SL, Reed KM, Phillips RB: Intraindividual and interspecies variation in the 5S rDNA of coregonid fish. J Mol Evol 1998;46:680–688.
64 Lopez-Piñon MJ, Freire R, Insua A, Mendez J: Sequence characterization and phylogenetic analysis of the 5S ribosomal DNA in some scallops (Bivalvia: Pectinidae). Hereditas 2008;145:9–19.
65 Tamura K, Dudley J, Nei M, Kumar S: MEGA4: Molecular Evolutionary Genetics Analysis (MEGA) software version 4.0. Mol Biol Evol 2007;24:1596–1599.
66 Hedges SB, Kumar S: The Time Tree of Life. New York, Oxford University Press, 2009.

José M. Eirín-López
Departamento de Biología Celular y Molecular
Universidade da Coruña, Facultade de Ciencias
Campus de A Zapateira s/n, ES–15071 A Coruña (Spain)
Tel. +34 981 167 000 (2257), E-Mail jeirin@udc.es, http://chromevol.udc.es

Garrido-Ramos MA (ed): Repetitive DNA.
Genome Dyn. Basel, Karger, 2012, vol 7, pp 197–221

Chromosomal Distribution and Evolution of Repetitive DNAs in Fish

M.B. Cioffi · L.A.C. Bertollo

Universidade Federal de São Carlos, Departamento de Genética e Evolução, São Carlos, SP, Brazil

Abstract

Fish exhibit the greatest diversity of all vertebrates, making this group extremely attractive for the study of a number of evolutionary questions. Fish genomes have intrinsic characteristics that may be responsible for the amazing diversity of fish species observed, but little is known about their structure and organization. A large amount of data from mapping of repetitive DNA sequences of several species has been generated, providing an important source of information for better understanding the involvement of repetitive DNA sequences in chromosomal organization. Almost all classes of repeated DNAs have been mapped in fishes, and all fish genomes analyzed contain at least one, mostly all types of repetitive DNAs. DNA sequence data combined with the chromosomal mapping of these repeated elements by means of cytogenetic techniques can provide a clearer picture of the genome, which is not yet clearly defined, even if already sequenced. In this chapter, we do not aim to analyze all available data on the chromosomal distribution of repetitive DNAs in fish species, but instead wish to draw attention to the impact of repetitive DNA sequences on fish karyotyping and genome evolution, with a particular focus on B chromosome origin and maintenance and on the differentiation of sex chromosomes. We also discuss the integration of chromosome analysis and genomic data, which represents a promising tool for fish cytogenomics.

Copyright © 2012 S. Karger AG, Basel

The Repetitive Fraction of the Genome

The presence of repetitive DNA sequences in eukaryotic genomes is a common feature; these sequences are characterized by a wide heterogeneity and diversity of repeated families [1]. In many species, repeated sequences comprise a large portion of the genome, causing an important question to arise: why are repeated sequences so prevalent in the genome? During the course of eukaryotic evolution, genes and genome segments seem to have been duplicated, leading to an increase in the DNA content of the cell nucleus. The variation in genome size between different eukaryotes is often reported as differences in the amount of repetitive DNA sequences [2, 3].

What functions these repeated sequences serve, if any, are mostly unknown. For a long time, no function was attributed to some of the repeated sequences, and these elements were regarded as 'junk DNA'. However, in recent years this concept has changed, mostly due to the discovery of transcribed regions within repeated elements and their involvement in genomic functions, demonstrating that such sequences are extremely important for the structural and functional organization of the genome [4].

Repetitive DNAs can be classified as interspersed elements or tandem arrays. The interspersed elements are represented by transposable elements (TEs) and are widely distributed throughout the genome, while the tandemly arrayed sequences include the multigene families, such as ribosomal RNAs (rRNA) and histone genes, and the satellites, micro- and minisatellites. These sequences constitute the nuclear genome architecture, together with the less repeated sequences which include the low copy number sequences and the low repeated DNAs.

In completely sequenced genomes, the repeated elements remain as gaps due to the difficulty of correctly identifying their position and array within the genome. Even the chromosomes that have been reportedly 'sequenced to completion' have multiple gaps in their centromeric regions related to the presence of duplicated and repeated segments [5]. Additional studies of these repeated segments are required to better understand the genome structure and function. Therefore, integrating DNA sequence data with the chromosomal mapping of these repeated elements by means of cytogenetic techniques (cytogenetic mapping) can provide a more comprehensive picture of the genome, which is not yet clearly defined, even in completely sequenced genomes.

Fish Genomes and Their Impact on Evolutionary Studies

Fishes are the most diversified vertebrate group, which makes them extremely attractive for the study of a number of evolutionary questions. The term 'fish' most precisely describes any non-tetrapodal craniate that has gills throughout life and whose limbs, if present, are in the shape of fins. However, fishes do not constitute a monophyletic group but are instead a paraphyletic collection of taxa including hagfish, lampreys, sharks, rays and the lobe-finned and ray-finned bony fishes. The latter is by far the most diverse group and is well represented in freshwaters, while the others are predominantly marine groups. It has been suggested that fishes represent 34,500 out of the approximately 55,000 recognized living vertebrate species [6] and that this diversity might be related to the ability of fish genomes to undergo genetic changes more rapidly than other vertebrate groups [7].

Fish genomes have intrinsic characteristics that might be involved in the formation of the amazing diversity of fish species. For example, the DNA content of haploid fish cells varies from 0.39 pg to 248 pg among species, and the chromosome number can also vary from 2n = 12 to 2n = 446 [8]. Several polymorphisms at the chromosomal level, including the presence of supernumerary chromosomes, polyploidy and structural variations, are also frequent in the group. There is substantial evidence that an ancient tetraploidization event has provided the evolutionary framework for the diversification of gene function and for speciation in ray-finned fishes. The completion and comparison of the genomic sequence of different fish species, as well as subsequent functional genomics approaches, will allow for a better understanding of the maintenance of hundreds of paralogs over hundreds of millions of years of evolution in this group. Such analyses have also provided new information concerning the evolution and evolutionary impact of repetitive DNA in fish genomes.

Although fish have traditionally been the subject of comparative evolutionary studies, they have now drawn attention as models in genomic and molecular genetic research. Many genome sequencing projects have used fish, including the catfish *Ictalurus punctatus*, the rainbow trout *Oncorhynchus mykiss*, the Atlantic salmon *Salmo salar*, the three-spined stickleback *Gasterosteus aculeatus*, the Nile tilapia *Oreochromis niloticus*, the 2 pufferfish species *Takifugu rubripes* and *Tetraodon nigroviridis*, the platyfish *Xiphophorus maculatus*, the medaka *Oryzias latipes*, the spined loach *Cobitis taenia* and the zebrafish *Danio rerio*, which commonly serves as a model organism in studies of vertebrate development and gene function [9]. The pufferfish, which have one of the smallest genomes among vertebrates, offer an interesting model for understanding the evolutionary forces that lead to a reduction in genome size. Among these fishes, the small amount of repetitive elements in comparison to other species is clearly one of the factors that contribute to their compact genome size [7].

Because fishes occupy an ancestral position in vertebrate phylogeny, studies conducted in this group can help to clarify several issues related to genome organization and the evolution of vertebrates. Comprehensive examination of different fish species can provide useful data for comparative genomics within the vertebrate lineage and also contribute to a better understanding of the mechanisms underlying genome evolution as a whole. Furthermore, comparisons of different fish genomes are important for unraveling the molecular mechanisms driving biodiversity within this group, as fishes represent approximately half of the extant vertebrate species and display an enormous diversity in morphology, ecology and behavior [6]. However, this tremendous biodiversity has yet to be widely exploited. In this regard, cytogenetic studies have brought important contributions.

In general, little is known about the structure and organization of fish genomes, as most available information is related to the structure and evolution of chromosomes. Molecular studies focusing on the genes and DNA sequences of this animal group are mainly restricted to repetitive sequences such as satellites, TEs and ribosomal DNA (rDNA). In fact, a large amount of data has been generated by chromosomal mapping of repetitive DNA sequences in several fish species, providing an important source of information for the role of such sequences in the structural and functional organization of the genome [4]. In this chapter, we do not aim to completely analyze the available data on the chromosomal distribution of repetitive DNA in fish species, but instead wish to draw attention to the impact of repetitive DNA sequences on fish karyotyping as well as genome evolution, with a particular focus on B chromosome origin and maintenance and the differentiation of sex chromosomes.

Repeated DNAs in Fish Genomes: A Chromosomal Perspective

Major advances in cytogenetic studies were obtained based on the detection of DNA sequences on chromosomes, whole genomic DNA or even parts of chromosomes by fluorescence in situ hybridization (FISH). FISH has been extensively used to map DNA sequences on chromosomes. It is now possible to map the locations of DNA sequences across related species and genera to show their probable conservation and/or diversification over time. The advent of FISH started the 'molecular cytogenetic era' and then, more recently, when combined with genomic data, the 'phylogenomic era' which allows the integration of the molecular information of DNA sequences with their physical location along chromosomes and within whole genomes [10]. However, although repetitive DNAs have been well studied in invertebrates and mammals, the available data are still limited for fish genomes, considering the large number of existing species. Despite the lack of repetitive DNA data in fishes, almost all classes of repeated DNAs have already been mapped in fish genomes, contributing to the knowledge of the complex organization of the DNA molecule in the cell nucleus (fig. 1).

All analyzed fish genomes contain at least one, but usually every type of repetitive DNA. These repetitions are indeed natural components of the heterochromatin, and each species' genome has a specific library of families of repetitive elements that are preferentially located within the heterochromatin. In fact, heterochromatin represents a conspicuous fraction of the total eukaryotic genome, as it is composed primarily of a variety of accumulated repetitive sequences [1, 11]. One of the major distinctions between heterochromatin and euchromatin is the density of repeated sequences. The great majority of the assembled heterochromatic sequences are repeats which are predominantly organized as scrambled clusters of TEs, whereas only a small percentage of the euchromatin is classified as repetitive [12]. Under the selfish DNA hypothesis, inserted repetitive elements would accumulate in heterochromatin because there are fewer genes. Inserted elements are therefore less likely to be deleterious and more likely to be reproduced. However, the dynamics of such accumulation, the specificities in targeting and location of different sequences, and the possible roles repetitive elements might play within heterochromatin contravene the view of elements being abundant in this region because of the damage they cause in euchromatin [11]. Rather than representing the mere addition of 'junk DNA' to the genomic 'ghost town', the accumulation of repetitive DNAs in heterochromatin might turn out to be an important evolutionary interaction between these 2 ubiquitous and fluid components of the genome [12].

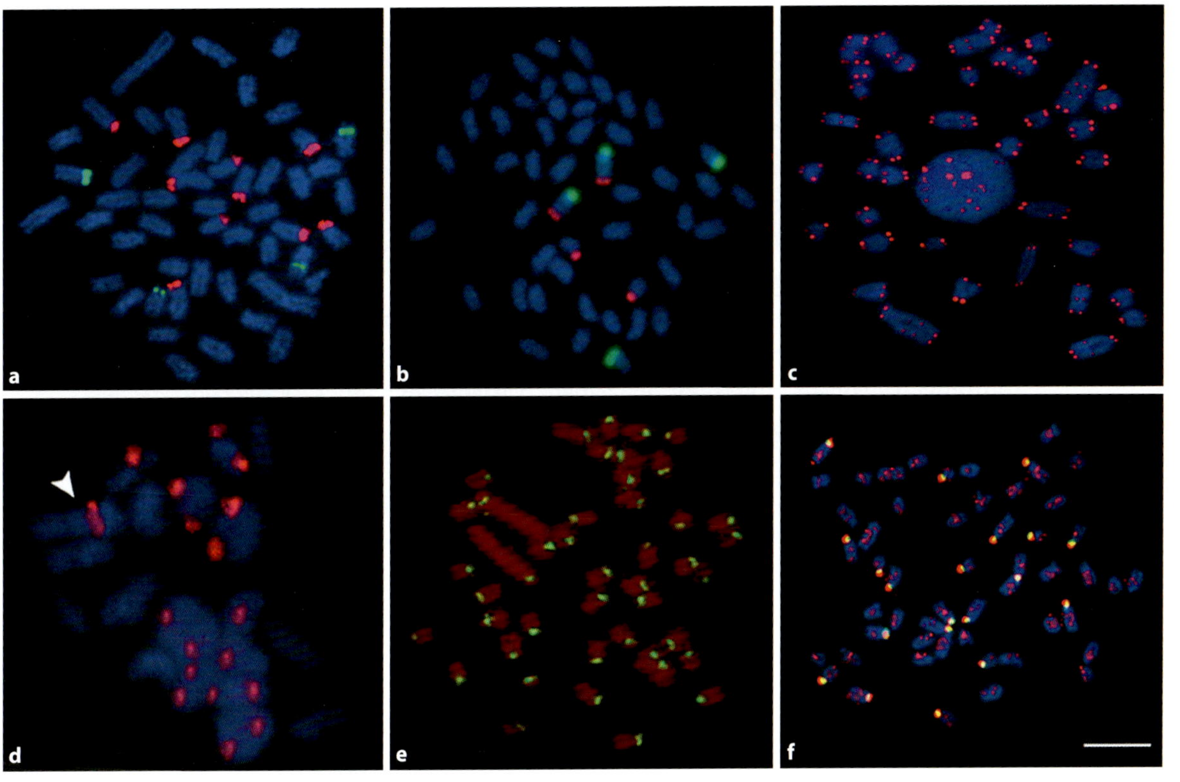

Fig. 1. Fluorescence in situ hybridization with various repetitive DNA probes on metaphase chromosomes of different fish species. **a** 5S rDNA (green) and 18S rDNA (red) in *Hyphessobrycon vinaceus*. **b** 5S rDNA (green) and H1 hisDNA (red) in *Bathygobius soporator*. **c** Telomeric DNA in *Oreochromis karongae* showing the presence of ITS on several chromosomes. **d** Satellite 5S*Hin*dIII-DNA in *Hoplias malabaricus* highlighting the sex trivalent during male meiosis (arrowhead). **e** The SATA satellite DNA family in *Oreochromis niloticus* has a centromeric distribution (courtesy of Cesar Martins, Universidade Estadual Paulista). **f** 5S rDNA (green) and the *Rex3* retroelement (red) in *Erythrinus erythrinus* showing their association on the chromosomes. Bar = 5 μm.

Satellite DNAs

Satellite DNA sequences (satDNAs) are tandemly arrayed and highly repeated sequences in which the size of a repeated unit varies from 100 to 1,000 nucleotides and which are present in the genome at copy numbers of 1,000 to >100,000. They are organized into large clusters located mainly in the centromeric and telomeric regions of chromosomes and are the main component of heterochromatin. The centromeres, which are the primary constriction of each eukaryote's metaphase chromosome, are essential for the correct segregation of chromosomes during cell division. Likewise, telomeres play a critical role in maintaining chromosomal stability and are involved in chromosome replication. Thus, the repetitive DNA sequences located in the centromeres and telomeres may be important for the correct function of the metaphase chromosome [1]. The molecular organization, chromosomal localization and possible functions of satDNAs have been studied in several animal groups, providing evidence that these sequences play an important role at both the nuclear and chromosomal levels of organization [12].

SatDNAs are dynamic regions of the genome that differentiated rapidly during evolution. Different species generally display high divergence between satDNA families as a result of concerted evolution mechanisms [13], leading to species-specific sequences. Studies of satDNAs have proven useful for addressing a number of questions related to fish genomes. These studies applied satDNAs toward the physical mapping of the genome, contributing to the development of genetic markers of significant importance to both the fundamental and applied biology of fish species. Additionally, considering their extremely variable dynamics, especially species-specific or chromosome-specific satDNAs provide useful information for microevolutionary studies [14].

The first descriptions of satDNA families in fish genomes date from the end of the last century, and since then, several studies have added important contributions concerning the origin and evolution of such sequences in fish genomes. These studies demonstrated that satDNAs are mainly located in the centromeric region of chromosomes and suggest that they might play an important role in the structure and function of fish centromeres (for a review, see [15]). For example, a satDNA family isolated from the genome of the Adriatic sturgeon, *Acipenser naccarii*, was preserved in the pericentromeric regions of the chromosomes of 7 species of the genus *Acipenser* [16]. Centromeric families were also isolated from the genome of the Nile tilapia *O. niloticus*; these same satDNA families were located in the centromeric regions of all chromosomes of the complement [17]. Important advances concerning the role of satDNAs in the origin and diversification of B and sex chromosomes will be discussed in the next sections.

Transposable Elements

Transposable elements represent another important class of repetitive DNA that is widely studied in the genome of many organisms. The majority of interspersed repetitive DNA is formed by TEs which have the ability to jump from one location within the genome to another. TEs include the 'copy and paste' retroelements, which transpose via an RNA intermediate, and the 'cut and paste' DNA transposons. However, this is only a very rough classification because the activity of TEs is very complex. In general, the TE distribution pattern in heterochromatin and euchromatin is extremely variable between different genomes. However, it seems that TEs tend to accumulate preferentially in the centromeric and in the telomeric heterochromatic regions of chromosomes. These distribution patterns can be correlated with a role of repeated sequences in controlling the structure and organization of these regions, besides preventing the occurrence of crossing-over in close vicinity, thus keeping integrity of these important architectural structures [2]. It is currently unknown whether TEs specifically target heterochromatic regions for transposition, or whether they merely accumulate at these locations due to the reduced recombination and silencing of these regions. However, it is becoming clear that TEs constitute part of the regulatory toolkit of the genome, with important roles in directing gene expression [4, 12].

Although fish genomes are more compact than those of mammals, a higher diversity of TEs is characteristic of fish genomes [18]. Given this, investigation of TEs in fish can contribute to the knowledge of their genomes, especially because they remain as gaps even within those genomes that have been reported to be completely sequenced. These gaps exist due to the difficulty of correctly identifying the position, array and repeat number of these repeated DNA elements. In addition, some specific TEs can also be applied as molecular markers to track the evolutionary history of particular clades [18].

The chromosomal location of various types of TEs in the genome of the pufferfish showed that these sequences are generally absent in gene-rich regions. The compact genomes of both pufferfishes, *T. rubripes* and *T. nigroviridis*, contain a smaller quantity of repeated sequences but a greater diversity of TE families than the much larger human and mouse genomes [18]. The TEs of the retrotransposon class are among the best studied within fish species. Among them is the *Rex* group, a retroelement characterized for the first time in the genome of the swordtail fish *Xiphophorus* [19]. These elements are present in the genome of different fish species and have undergone some retrotranspositions during their dispersion process, some of which have been identified as relatively recent events [18, 20].

The chromosomal distribution of TEs has been extensively studied in the Cichlidae species, and the in situ investigation of some retroelements in genomes of many African and South American species revealed that they are compartmentalized in centromeric heterochromatic regions, suggesting that these elements are part of the structure and organization of these heterochromatic areas [21]. However, in the genome of the Nile tilapia, *O. niloticus*, several other TEs aside from retroelements (such as the LINEs, CINEs and Tc1-like elements) have been mapped, all of them being dispersed throughout the whole genome [17, 22]. In some other fish species, despite a preferential localization to the centromeric region, TEs have a widely scattered distribution over all chromosomes, with intense hybridization signals in some specific regions [20, 23–25].

The presence of TEs in heterochromatic regions can be correlated with their role in the structure and organization of centromeres or with the reduced selective pressure acting on heterochromatic regions, which are poor in gene content. The centromeres of chromosomes of a large number of species also had TEs interspersed with satDNA sequences, suggesting an important role for these sequences in the expansion and stabilization of centromeres [26].

Despite the preferential distribution of TEs in the non-coding regions of fish genomes, they have a varied distribution both between and within chromosomes and are more often associated with sex chromosomes than autosomes, probably due to the increased concentration of heterochromatin in the first sex chromosomes [2]. In fact, TEs have a crucial role in the differentiation and evolution of sex chromosomes. Overall, the data now available indicate that TEs are important structural components of the heterochromatic regions and have played an important role in the evolutionary history of fish genomes.

Multigene Families

Multigene families represent a common structural element of eukaryotic genomes and are defined as a set of genes derived by duplication of an ancestral gene and displaying >50% similarity [27]. Multigene families are composed of hundreds to thousands of gene copies; examples include the rRNA and histone gene families. Eukaryote genomes contain multiple copies of rRNA genes, presumably because exceptionally high quantities of RNA transcripts are necessary [28]. These rRNA molecules are encoded by 2 distinct multigene families: the first one corresponds to the 45S rDNA that is organized in tandem arrays containing transcriptional units for the 18S, 5.8S and 28S rRNA, while the second ribosomal family transcribes the 5S rRNA.

The 2 rDNA classes (45S and 5S rDNA) have been extensively mapped in fish genomes. The organization of rDNA clusters in fish genomes indicates a non-association of the 2 classes, with most species having these sequences on different chromosomes, although in some species, the presence of clusters for these sequences near each other or far apart on the same chromosome was reported [15, 29]. For example, analyses conducted in 6 icefish species (Perciformes, Channichthyidae) showed that clusters of 45S and 5S ribosomal genes co-localize and likely compose the entire arm of a single pair of submetacentric chromosomes in all of these species [20].

In general, 5S rDNA occurs in interstitial regions of fish chromosomes, and this pattern could represent an ancestral condition or even some advantage for the genome organization of these sequences [29]. Additionally, genomes of fishes can also have 2 distinct 5S rDNA classes organized in different chromosomal regions or even on different chromosomes [29, 30]. In contrast, multiple 45S rDNA loci are a common feature in fish, with some species showing a very large number (10–15) [31, 32]. In fact, these results could be related to the chromosomal position of the different rDNA classes: while the interstitial location of 5S rDNA sites would preserve their conservative distribution, the telomeric position of 45S rDNA would promote chromosomal dispersion due to telomeric proximity within the interphase nucleus according to Rabl's model (reviewed in [33]).

These hypotheses are supported by analyses conducted in different fish species. In the genome of Cacho chub *Squalius pyrenaicus* (Cyprinidae), the 5S rDNA was consistently mapped to 3 chromosomes per haploid genome with apparently conserved locations on morphologically similar chromosomes; conversely, prominent intra- and interindividual variations of the 28S rDNA were detected with regard to number, size and location on the chromosomes, as well as syntenic sites with 5S rDNA [32]. However, great variation concerning the number and distribution of 5S rDNA clusters was also observed among some species such as the red wolf fish *Erythrinus erythrinus*, which possesses more than 20 sites [24], and the cichlid *Astatotilapia latifasciata*, with 15 clusters [34]. The variations observed for rDNA sites indicate a complex microevolutionary pattern that rules their organization in the genome. Ribosomal DNAs are able to spread through the genome, thus creating new rDNA loci, variant rDNA copies and even associations with other multigene families [35]. The genomic organization and evolution of rRNA genes seems to be governed by the combination of birth-and-death and concerted evolution mechanisms that generate the observed patterns of chromosomal distribution of rDNA clusters [36].

Another multigene family is the histone family which is known to be composed of moderately repeated genes [37]. The histone genes have an extraordinary organization in tandem arrays, are interspersed from each other with non-coding spacer sequences, and code for 5 histone proteins [37, 38]. The organization and arrangement of histone genes has undergone significant evolutionary change, but it is not yet clear how these different units originated and spread in the genome [38]. Histones are a class of basic proteins that associate with each other and with nuclear DNA to form the nucleosome, the fundamental unit of chromatin. The structure of histones, particularly the H3 and H4 histones, is generally highly conserved between diverse animal phyla and even between the animal and plant kingdoms, although remarkable variations have been reported in the DNA sequence for these proteins [39].

However, just a few studies have been conducted in fish to analyze the chromosomal distribution of such sequences. In a pioneer study in salmonid fish, a single pericentromerically situated locus for the major histone DNA (hisDNA) was described [40]. More recently, the chromosomal mapping of the H1 hisDNA was conducted in 3 Neotropical species of the genus *Astyanax*, showing that the major hisDNA clusters are also located close to the centromeric region on 2 chromosome pairs [41].

Microsatellites

Microsatellites, also known as simple sequence repeats, consist of very short motifs (1–6 nucleotides in length) repeated in tandem arrays. They are abundant in all eukaryotic genomes studied thus far and are found either between the coding regions of structural genes or between other repetitive sequences [42]. One of the most remarkable properties of these sequences is their ability to give rise to variants with a different number of repeats. In addition, microsatellite repeats may be organized in long stretches consisting of hundreds to several thousand tandem units, and they are associated with constitutive heterochromatin in many species [15, 42, 43].

The microsatellites are located in the heterochromatic regions (telomeres, centromeres and in the sex chromosomes) of fish genomes, where a significant fraction of repetitive DNA is localized. For example, the genomes of catfishes *Imparfinis schubarti, Steindachneridion scripta*, and *Rineloricaria latirostris* exhibit a remarkable accumulation of both $d(GA)_{15}$ and $d(A)_{30}$ microsatellites in telomeric regions [44]. Furthermore, a linkage map analysis of the genome of the zebrafish *D. rerio*, using 200 microsatellite markers of the CA and GT types, identified that the $(CA)_n$ and $(GT)_n$ repeats are more clustered in the centromeric and telomeric regions [45]. In situ chromosome hybridization detected $(GACA)_n$ sequence-enriched segments in the heterochromatic portion of the W and Y chromosomes of the guppy *Poecilia reticulata* [46]. In the genome of the wolf fish *Hoplias malabaricus*, 12 different microsatellite repeats showed strong hybridization signals at subtelomeric and heterochromatic regions of several autosomes, with a varied amount of signal on the sex chromosomes [43].

Telomeric Sequences

Telomeric sequences have been an important tool in the analysis of some evolutionary processes. These sequences, originally isolated from human repetitive DNA libraries, consist of short repeats that are rich in guanine – $(TTAGGG)_n$ – and are widely distributed and conserved in the genome of vertebrates. The complexes formed by these sequences with specific proteins are highly specialized structures at the ends of eukaryotic chromosomes, the telomeres, which perform several vital functions such as stabilizing chromosomes and allowing complete replication of chromosome ends [47]. Thus, $(TTAGGG)_n$ sequences are present at the telomeres of all vertebrates, and their study provides insight into the chromosomal rearrangements that have occurred during karyotype evolution of distinct organisms. The occurrence of interstitial telomeric sites (ITS) has been critical for detecting ancestral chromosomal fusions. However, many cases of tandem chromosome fusions or centric fusions may not have the expected ITS, probably due to loss or drastic reduction of the telomeric DNA during these rearrangements [48].

Although telomeric sequences are primarily distributed in the telomeres in many fish species [49], ITS have also been detected elsewhere in some cases [24, 50–52]. In the genomes of some Salmonidae species (*Salmo trutta, S. salar, Onchorhynchus kisutch* and *O. mykiss*), telomeric sequences were located in chromosomal regions corresponding to nucleolar organizing regions [53]. The presence of ITS in chromosome pairs of *S. salar* and the Nile tilapia, *O. niloticus* [50], was considered indicative of chromosomal fusion events that occurred during the karyotype differentiation of these species.

The comparative mapping of telomeric sequences in the chromosomes of tilapia *O. karongae* (2n = 38), which differs from the typical tilapia karyotype (2n = 44), has helped to clarify the chromosomal rearrangements that took place during the diversification of this karyotype. This work suggested that 3 separate chromosome fusions occurred in the development of the *O. karongae* karyotype [52].

In summary, the investigation of the chromosomal distribution of repetitive DNAs in fish genomes, although relatively recent, is a promising tool for the analysis of genomic organization and evolution of fish. Thus, the use of repetitive DNA, combined with other chromosomal analysis procedures, has provided useful contributions to the knowledge of the heterochromatic component of the genome and karyotype evolution for diverse fish species. Repetitive DNA studies have been particularly effective with regard to determining the origin, differentiation and evolution of particular classes of chromosomes such as the sex and B chromosomes, which will be further analyzed in detail.

The Repetitive DNA Fraction: A Primary Driving Force behind the Diversity of Fish

As described previously, an inherent feature of heterochromatin is its complex composition of various types of repetitive sequences, leading to changes in the amount and distribution of the heterochromatic genomic fraction of many fish species. Heterochromatic regions are known to harbor active genes and to have structural functions, such as centromeric activity and chromosome pairing [54]. Modifications in these regions might give rise to fertility barriers that promote evolutionary divergence and speciation. In fact, the correlation between repetitive sequences and chromosomal rearrangements has been extensively demonstrated [55].

It is obvious that chromosomal repatterning during evolution may alter the number and position of rDNA sites, but recent studies have shown that the dynamism of the rDNA clusters may be regarded as a strong indicator of significant intra-genomic processes. For example, the correlation between karyotype rearrangement and retrotransposon activity was demonstrated in the red wolf fish *E. erythrinus*. This species presents extensive intrapopulational chromosome diversity, with 4 karyomorphs differentiated by the number and morphology of chromosomes. Although having only two 5S rDNA sites is a common feature among *E. erythrinus*, one of its karyomorphs showed a surprising increase in the number of 5S rDNA sites, with 22 in females and 21 in males, and all of them were located in centromeric chromosome regions. Additionally, it was shown that all of these rDNA sequences co-localized with the retrotransposable element *Rex3* in the centromeric heterochromatin (fig. 1f). The synteny of both repetitive sequences strengthened the idea that the chromosomes have undergone rearrangements during evolution that were mediated by retrotransposon activity in which the insertion of the retrotransposable element *Rex3* into 5S rDNA sequences created a 5S rDNA-*Rex3* complex that moved and dispersed in the karyotype [24].

Preferential integration of retrotransposons into heterochromatic regions has also been observed in the genome of the black rockcod fish *Notothenia coriiceps*, which has the most derived karyotype among the marine perciform fishes analyzed thus far. This species exhibits a more compartmentalized retrotransposon distribution with accumulation in pericentromeric regions, suggesting a correlation between karyotype rearrangements and retrotransposon activity [20]. Similarly, correlations between the accumulation of repetitive elements, heterochromatin and chromosome rearrangements have been hypothesized to explain the karyotype differentiation that occurred among distinct species of the discus fishes of the *Symphysodon* genus. Meiotic analyses conducted in 2 species (*S. aequifasciatus* and

S. haraldi) revealed the presence of an intriguing meiotic chromosomal chain with up to 20 elements during the diplotene/diakinesis stages. Such chromosomal chains with high numbers of elements have not been observed in any other vertebrate. In these species the 5S rRNA gene co-localizes with constitutive heterochromatin and is flanked by some TEs [23]. The mobile elements may serve as substrates for DNA recombination due to their repetitive nature, especially when in the Rabl configuration [56]. Repetitive mobile elements, in addition to facilitating the chromosome rearrangements that caused the difference in position of the 5S rDNA, may have also caused the multiple translocations that produced this odd meiotic chain in *Symphysodon* species [23].

One of the first descriptions of a satDNA in Neotropical fishes was the isolation of the satellite called Hop in *H. malabaricus* [57]. This species shows a conspicuous karyotype diversification with 7 karyomorphs currently identified, suggesting the existence of distinct species [31]. Recently, the repetitive DNA class named 5S*Hin*dIII-DNA, which shares considerable identity with the Hop satellite DNA and 5S rDNA of teleosts, was isolated and characterized from the *H. malabaricus* genome [35]. The cloning and sequencing of this satellite family identified 350–360-bp DNA fragments with insertions, deletions and base substitutions among the clones. The main differences between the repeating units of the 5S rDNA and 5S*Hin*dIII-DNA family are the presence of an expanded microsatellite sequence TAAA in the 5S rDNA non-transcribed spacer and 2 small internal deletions in its transcriptional region. The transfer of 5S rDNA repeats to the centromeric position could have changed the selective pressure on these genes, allowing their multiplication and spread over the centromeres of several chromosomal pairs, as has been demonstrated for other centromeric satellite sequences [35]. The first copy of 5S*Hin*dIII-DNA may have been associated with other repetitive sequences located in the centromeric heterochromatin, facilitating its dispersion to other chromosomes due to the mechanisms of evolution. It is also likely that this DNA family has been favored during the evolutionary process due to a possible structural or functional role at the centromeres [58].

A comparative analysis between distinct *H. malabaricus* karyomorphs and other Erythrinidae species showed that this satellite DNA family is unique to the *H. malabaricus* genome [24, 58]. However, differences in the number and position of 5S*Hin*dIII-DNA sites were detected between distinct populations, suggesting that the 5S*Hin*dIII satellite DNA must have emerged during the divergence of *H. malabaricus* from the other groups of Erythrinidae and before the diversification of *H. malabaricus* karyomorphs, and that the 5S*Hin*dIII satellite DNA has accompanied the differentiation of these karyotype forms [31, 58]. Thus, this satellite DNA proves to be promising as a cytogenetic marker for chromosomal studies of the *H. malabaricus* species group.

Although it is most common for *H. malabaricus* to have eighteen 5S*Hin*dIII-DNA sites, an analysis of different allopatric populations of the same karyomorph revealed a variable number of chromosomes harboring such sequences, making the 5S*Hin*dIII-DNA sites good population markers for minor evolutionary divergences [31]. Due to its centromeric location, this satellite family may have propagated within the centromeric region of several chromosomes and been favored during evolution due to a possible role in centromere structure or function. Although evolutionary mechanisms have not caused major changes in the karyotypes of different populations of a specific karyomorph, these genomes are undergoing continuous evolution. The repetitive fraction of the genome seems to escape the selective pressure that acts on the non-repetitive segments, thus representing good evolutionary markers to detect recent events of evolution.

The repetitive DNA fraction also plays an important role during polyploidization and post-polyploidization changes in fishes. Polyploidy is a multiplication of entire chromosome sets and represents an uncommon phenomenon in higher vertebrates. However, it has appeared repeatedly during the evolution of several fish lineages, being a wide-spread phenomenon in the more ancestral lineages (such as elasmobranchs and lower teleosts as far as Protacathopterygii) and occurring independently and often repeatedly in many recent fishes (for a review see [59]).

It is noteworthy that there are some difficulties in compiling the known cases of polyploidy in fishes. Polyploidy is an ongoing process, where ancient events can be obscured in further karyotypic evolution, and more recent events can lead to multiple ploidy levels in one species [60]. A good example to illustrate this is represented by the order Acipenseriformes (represented by the popular sturgeons) in which all individuals are considered to be of polyploid origin, and the chromosomal distribution of repetitive DNA sequences allows a more detailed characterization of the ploidy relationships between these species. The karyotype of all sturgeons is characterized by a very high number of chromosomes, approximately half of which are microchromosomes. They can be divided into 2 groups: the first includes those species with 2n = 120 chromosomes, while the second includes those with a diploid number of approximately 240–260 chromosomes. The species with 120 chromosomes belong to the ancient tetraploids (paleotetraploids), but they have passed through significant diploidization that has resulted in their practically functional diploid state. At least 3 independent polyploidization events have taken place in sturgeon evolution, but, in fact, there seem to have been many more [61].

The isolation and chromosomal distribution of some repetitive DNAs provided further insights into ploidy levels in sturgeons. FISH with a 5S rDNA probe in the sturgeon genome reveals intense fluorescent signals in the middle regions of 4 and 8 small chromosomes in species with 120 and 240–260 chromosomes, respectively [16, 62]. A satellite DNA family (called HindIII) previously isolated from the genome of *A. naccarii* [63] was found in the centromeric region of 50 to 80 chromosomes in the species with 240–260 chromosomes, while just 8 to 10 chromosomes harbored this sequence in the species with 120 chromosomes [62]. These results suggest that the sturgeons with 120 chromosomes are diploid and those with 240–260 chromosomes are tetraploid. As polyploidization is an ongoing process, more investigation is needed to better understand the true extent and role of polyploidy in fishes.

The results discussed in this section represent examples of the use of repetitive sequences in revealing the evolutionary forces behind the large amount of diversity found in fish. Once again, it is evident that repetitive sequences are powerful tools for improving the comprehension of genome structure and evolution.

Structure and Evolution of Fish Sex Chromosomes

The processes working on sex chromosome evolution are still not completely understood. However, during this process, one general rule must be followed: the cessation or partial restriction of recombination between the sex chromosomes. The accumulation of repetitive DNA sequences is one of the first steps in the early stages of sex chromosome evolution [64]. Additionally, regions with suppressed or no recombination have the potential to accumulate these DNA sequences, and, therefore, the absence of recombination between the sex chromosomes favors their increase during evolution. In fact, a clear correlation between sex chromosomes and repetitive DNAs has been noted in several studies conducted in different organisms (for a review, see [65]), indicating that the accumulation of repeated DNAs and the gene loss are considered to be a convergent property of sex chromosome differentiation. It is well known that repetitive DNAs can also occur and accumulate in the autosomes of several organisms. However, sex chromosomes are preferred sites for these elements because repetitive DNA accumulation is favored by no or low recombination levels (fig. 2) [64].

Many species have heteromorphic sex chromosomes that manifest in karyotype differences between the sexes. For example, mammals have an XX/XY sex chromosome system (with XX females and XY males), while in birds and snakes a ZZ/ZW sex chromosome system (with ZW females and ZZ males) can be found. In both of these systems, the sex-specific chromosome (the Y or W) is usually smaller than the X or Z and is often degenerated or entirely heterochromatic. However, in contrast to higher vertebrates, where differentiated sex chromosomes and relatively stable sex-determining systems are found, most fish species lack heteromorphic sex chromosomes, although cases have been found with different sex chromosome determination systems [66]. This makes fishes, which are the oldest vertebrate group, a good model for analyzing the evolution of sex chromosome differentiation in vertebrates, as evolution can be followed from the absence of sex chromosomes to their presence through several stages.

The sex chromosome systems found within fish species show an amazing variety. ZZ/ZW and XX/XY systems are the most common, but several other systems (such as the XX/X0, 00/0W, $X_1X_1X_2X_2/X_1X_2Y$ and XX/XY_1Y_2 sex systems) have also been

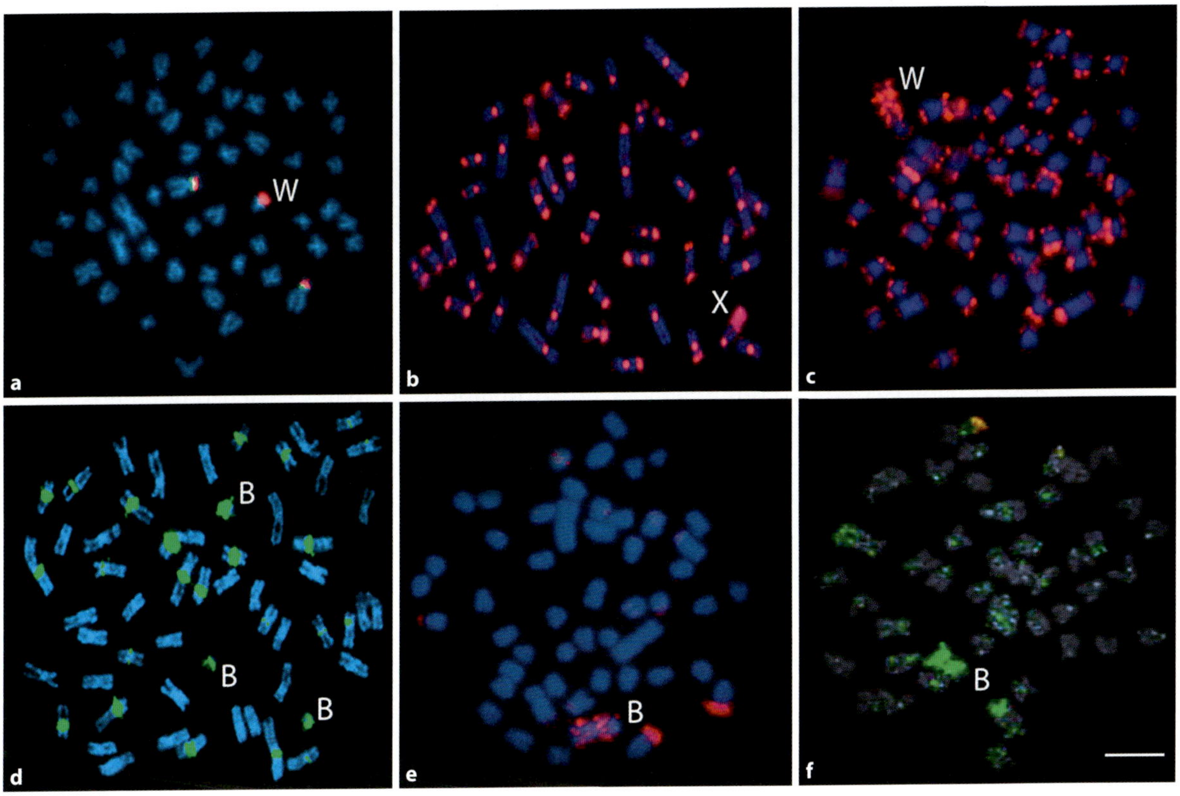

Fig. 2. Distribution of repetitive DNAs on the sex and B chromosomes of different fish species. **a** 5S rDNA (green) and 18S rDNA (red) in *Triportheus nematurus*, showing the unusual presence of the 18S rDNA on the W chromosome. **b** Cot-1 DNA fraction in *Hoplias malabaricus*, showing the accumulation of highly and moderately repetitive sequences on the X chromosome. **c** Simple sequence repeats $(GAA)_{10}$ in *Leporinus elongatus* are accumulated on the W chromosome. **d** SATH1 satellite DNA in *Prochilodus lineatus*, showing its presence on several B chromosomes. **e** As51 satellite DNA in *Astyanax scabripinnis*, highlighting its preferential accumulation on the macro B chromosome (courtesy of Marcelo R. Vicari, Universidade Estadual de Ponta Grossa, Brazil). **f** Retrotransposable *Rex1* element in *Astatotilapia latifasciata* highly accumulated on the B chromosome (courtesy of Cesar Martins, Universidade Estadual Paulista). Bar = 5 μm.

described, including systems with the presence of different sex chromosomes inside the same nominal species [67].

The medaka rice fish in the genus *Oryzias*, for example, show an extraordinarily high diversity in sex determination systems and sex chromosomes. In these species, both XY and ZW sex chromosome systems have been identified, but all of them have different origins nevertheless. These findings suggest that the frequent emerging of new sex chromosomes from autosomes during evolution in some fish species occurred possibly in association with the emergence and substitution of the master sex-determining gene [68]. Another well-studied species is the platyfish *X. maculatus* in which a variety of sex chromosomes coexist in a population [69]. In addition, along with well-differentiated heteromorphic sex chromosomes, others in the nascent state can also be found, as well as diverse sex chromosome systems between congeneric species or even populations of the same species [51].

As in other vertebrates, fish sex chromosomes can also be enriched in repetitive DNA sequences, as shown by the isolation and mapping of several sex-specific repetitive DNAs in this group [15, 65]. In some cases, even species that do not have differentiated sex chromosomes can reveal an initial stage of differentiation for these chromosomes, as evidenced by differential accumulation of repetitive DNAs on the morphologically undifferentiated sex pair. For example, in some Poeciliidae species in which sex chromosomes cannot be identified by conventional staining, they can be easily characterized by a differential heterochromatic block that is enriched in repetitive DNA sequences and located in the telomeric region of the Y and W chromosomes. The same heterochromatic region is present on the X and Z chromosomes as well, however it is smaller, suggesting that the accumulation of repetitive DNAs is related to the initial differentiation of the sex chromosomes [46]. A similar condition also occurs in the nascent XY system of the platyfish *X. maculatus* in which the expansion of a specific repeat (*XIR*) was one of the first molecular events associated with the divergence of the Y chromosome and recombinational isolation of the sex-determining locus [69].

Undoubtedly, among the most differentiated sex chromosomes in fish are the ZW chromosomes. However, in contrast to what is observed in higher vertebrates, the sex-specific W chromosome is larger than the Z chromosome, as a result of a large accumulation of repetitive DNAs. The *Leporinus* genus and Parodontidae family are examples of this situation.

The representatives of the genus *Leporinus* (Characiformes, Anastomidae) are characterized by a conservative karyotype in which all species show 2n = 54 chromosomes. While some species apparently lack differentiated sex chromosomes, others bear a clear heteromorphic ZZ/ZW sex system in which the W chromosome is always highly heterochromatic and enriched in repetitive sequences [70]. In one particular species, *L. elongatus*, 3 satellite DNA families have been isolated and used as probes for FISH analysis. One of them, named L5, was found on both Z and W chromosomes, whereas the second family, named L46, was specific to the W chromosome [71]. The third one, LeSpe I, was a sex-specific dispersed repetitive element showing distinct distribution patterns on 2 exclusive female chromosomes (W_1 and W_2). Therefore, it was suggested that instead of ZW sex chromosomes, this species has a multiple $Z_1Z_1Z_2Z_2/Z_1W_1Z_2W_2$ sex chromosome system [72].

The chromosomal distribution of 12 different microsatellite repeats was investigated in genomes of 4 ZW-*Leporinus* species. Because microsatellites are the most dynamic component of genomes and because non-recombining regions of the sex chromosomes give microsatellites the chance to expand, their contribution to sex chromosome differentiation is significant. In fact, the heterochromatinized W chromosomes of all species showed strong microsatellite accumulation. However, the distribution of microsatellite sequences on this chromosome differed greatly between species and with respect to the distinct microsatellites. Additionally, enrichment of some microsatellite sequences over the whole W chromosome and the accumulation of some repeats near the centromere was also detected (fig. 2c) [unpublished data].

It is possible that the accumulation of microsatellites plays an important role in sex chromosome evolution. This could be a consequence of their ability to adopt unusual DNA conformations, including hairpins, triplex, tetraplex, and 'sticky DNA'. The potential interactions of these structures can bring together distant regions of the same chromosome, thereby facilitating recombination-based processes such as gene conversion. Indeed, the role of gene conversion in the formation of large palindromes and the protection of genes located on the Y chromosome has been demonstrated in humans [73]. Similarly, non-B DNA conformations adopted by microsatellites can serve as breakpoints for large rearrangements, such as inversions, that are thought to play a key role in the evolution of sex chromosomes [43, 74].

The Neotropical Parodontidae fishes are characterized by 2n = 54 chromosomes. However, while some species apparently lack differentiated sex chromosomes, others bear a clear heteromorphic ZZ/ZW sex system. The species with differentiated sex chromosomes seem to have the same sex chromosome differentiation process: by increasing heterochromatic segments and accumulating repetitive DNAs in the W chromosome from an ancestral homomorphic pair. A satellite DNA family was isolated from the genome of *Parodon hilarii* using genomic DNA restriction [75]. This DNA fragment, called pPh2004, is a monomeric sequence of 200 bp and is 60% AT-rich. FISH analysis in *P. hilarii* using the pPh2004 probe revealed its presence on autosomes and in the terminal regions of the short and long arms of the Z and W chromosomes, respectively. Thus, a heterochromatic block in the short arms of the Z chromosome appears to have undergone an amplification in size from its ancestral homologue (primitive W), giving rise to the long arms of the current W chromosome. Therefore, pPh2004 was clearly involved in the differentiation of the W chromosome through an amplification process [75].

Variation in the amount of several types of repetitive DNA is associated with the genomic diversity and sex chromosome evolution of *H. malabaricus*. This species has different sex chromosome systems, as well as distinct evolutionary stages of sex chromosome differentiation found among its populations. For example, in some populations of this species, a well-differentiated XX/XY sex chromosome system can be found in which the X chromosome clearly differs from the Y by the accumulation of DNA repeats [76]. At least 15 distinct repetitive DNA classes (including satellites, TEs and microsatellite repeats) accumulated in the heterochromatic region of the X chromosome. This finding suggested that the X chromosome is the preferred site for the accumulation of repeats, representing an unusual example of an X chromosome accumulating more repetitive DNA than the Y in fish (fig. 2b) [76].

Transposable elements, which are expected to be abundant in chromosomal regions where recombination is reduced, can also accumulate on the sex chromosomes [1].

However, although several reports from different taxa refer to the accumulation of retroelements on the sex chromosomes, few studies correlate the association of TEs with the differentiation of fish sex chromosomes. The icefish *Chionodraco hamatus* [20] and the red wolf fish *E. erythrinus* represent 2 illustrative cases [24]. For both species, it has been proposed that a centric fusion between 2 non-homologous acrocentric chromosomes may have created the specific neo-Y chromosome and, consequently, the unpaired X_1 and X_2 chromosomes in the male karyotypes. The retrotroelement *Rex3* and the transposon Tc1-like, besides being part of heterochromatic regions, accumulate highly in the centromeric region of the neo-Y chromosomes of *C. hamatus* and *E. erythrinus*, suggesting their pre-existence before the occurrence of the fusion event and, therefore, their probable influence on the process of sex chromosome differentiation.

In summary, taking into account that the suppression of recombination between the sex chromosome pair is a prerequisite during the evolution of chromosomal sex systems and that the massive accumulation of repetitive sequences usually occurs in non-recombining regions, it is possible to predict a close relationship between differentiation of sex chromosomes and the accumulation of different kinds of repetitive DNAs, which can contribute to the physical differentiation of these chromosomes.

Origin and Maintenance of B Chromosomes

Supernumerary or B chromosomes are additional genetic elements found in the chromosome complement of approximately 15% of eukaryotic species. A number of hypotheses have been raised to explain the origin, frequency and evolution of B chromosomes, such as derivation from autosomes followed by gene silencing, heterochromatinization, and accumulation of repetitive DNAs. Moreover, B chromosomes may differ morphologically from the standard (A) chromosomes as they generally display a non-Mendelian inheritance pattern and accumulation mechanisms [77].

The occurrence of B chromosomes is not a rare event in fish genomes, with approximately 5% of the species cytogenetically analyzed having B chromosomes [78]. Almost all cases described are freshwater Neotropical fishes, with the order Characiformes showing the highest frequency, representing half of all of the species with B chromosomes [79]. The occurrence of B chromosomes in pufferfish *Sphoeroides spengleri* (Tetraodontiformes) represents the first report of such a chromosomal type in a marine species [80].

B chromosomes of fishes can vary greatly in size, ranging from microchromosomes, as in the curimbatá *Prochilodus lineatus* [81], *Poecilia formosa* [82] and *Cyphocharax spilotus* [83], and medium-sized chromosomes, as in *Rhamdia hilarii* [84] and *Parauchenipterus galeatus* [85], to macrochromosomes, as in *Astyanax scabripinnis* [86] and *Alburnus alburnus* [87]. The individual frequency of B chromosomes is also variable among the species, ranging from 3% in *Characidium cf. zebra* to 100% in *Moenkhausia sanctaefilomenae*, suggesting that B chromosomes may have different effects on their harboring species [78]. Overall, the differences concerning the frequency, morphology and size of the supernumerary chromosomes point to their variable origins in distinct fish species. Little is known about the supernumerary chromosomes among marine fishes, but their origins in freshwater and marine species do not seem to be phylogenetically related, instead representing independent evolutionary pathways [88].

Although B chromosomes are a highly heterogeneous class of chromosomes showing many particularities that cannot be extrapolated from one species to another, they do share some biological characteristics at different levels, from their basic structure to population dynamics [77]. Indeed, the molecular analysis of B chromosomes of different fish species has revealed that they are mostly composed of repetitive DNAs (frequently rDNA, satDNAs and mobile elements), which is consistent with their heterochromatic nature. Some of these repetitive DNAs are specific to Bs, whereas others are shared with the A chromosomes. In some cases, these sequences have provided a useful tool to ascertain whether the B chromosomes may have an intra- or interspecific origin (for a review, see [77]). For example, in *P. lineatus*, 2 satDNA families were isolated with monomeric units of 441 and 900 bp (named SATH1 and SATH2, respectively). Both families were located in the pericentromeric region of several chromosomes of the A complement, but only the SATH1 DNA was also located in the supernumerary microchromosomes, supporting the intraspecific origin of Bs in this species (fig. 2d) [81].

Similarly, several repetitive DNA elements have been mapped in the B macrochromosome of *A. scabripinnis*. In this species, the As51 satDNA, enriched in AT and with monomeric units of 51 bp, shows 58.8% similarity with a segment of the retrotransposon RT2 of the mosquito *Anopheles gambiae* and a lower similarity with the transposase gene of the transposon TN4430 of *Bacillus thuringiensis*, suggesting that this sequence might have arisen from a mobile element [89]. FISH demonstrated that this satellite DNA is mainly located in the distal heterochromatin of several acrocentric chromosomes and interstitially in the B chromosome (fig. 2e). The symmetric location of As51 in both arms of the metacentric B chromosome, along with their autopairing during meiosis, suggests that this B chromosome arose through misdivision and isochromosome formation from an A chromosome [89].

Many repetitive DNA sequences mapped to B chromosomes are known to be derived from TEs in distinct organisms such as the parasitoid wasp *Nasonia vitripennis* [90], the Australian daisy *Brachycome dichromosomatica* [91] and rye, *Secale cereale* [92]. As TEs represent a major component of the heterochromatin, they can also be frequently observed in association with the total or partial heterochromatic supernumerary chromosomes (fig. 2f). For example, in the fish species *A. alburnus*, which harbors a large supernumerary chromosome, comparable to the largest-sized A chromosome, the Gypsy/Ty3 retrotransposon is exclusively located on the B chromosome, suggesting the occurrence of a specific dispersion process for this retroelement during the evolution of the supernumerary chromosome [87]. Besides the detection of classical repeated sequences such as satDNA, transposons and 18S rRNA, genes were also detected in the B chromosomes of some species of the genus *Moenkhausia* [93], *Metynnis maculatus* [94] and *A. latifasciata* [34].

Because there is scarce data on the molecular organization of DNA sequences in B chromosomes, in general, and particularly in fishes, further studies are needed to add to the knowledge of the DNA composition of these chromosomes and to answer some questions that still remain: (1) What is their actual origin? (2) Why do some species harbor such large amounts of probably dispensable DNA? (3) Do B chromosomes in fish possess drive mechanisms? and (4) Might B chromosomes be advantageous to carriers in some circumstances, or are they most frequently parasitic or neutral, as in other organisms? Only answering these questions will elucidate the true nature of these enigmatic genomic elements.

Future Perspectives: Cytogenomics

Chromosome studies considerably advanced over the last 60 years, mostly based on microscopic analysis. The correct diploid chromosome number in humans was established in 1956 [95], the development of fluorescence in situ hybridization in the 1980s [96], chromosome painting and multicolor hybridization in the 1990s [97]. And now molecular cytogenetics has moved in the direction of cytogenomics which integrates genomics with chromosome data. The synergy of cytogenetics with molecular biology started in the 1980s with the application of DNA sequences as probes for cytogenetic mapping and is currently advancing due to the availability of hundreds of completely sequenced eukaryotic genomes in the last decade. The integration of genomic data with cytogenetic data broadens the perspective of cytogenomics to the study of karyotypes and chromosomes [98]. One of the most significant examples of cytogenomics can be illustrated by data obtained for a vertebrate model for genomic studies, the pufferfish *T. nigroviridis*. This species contains one of the most compact genomes in vertebrates, and molecular cytogenetics was applied to anchor genomic data to specific chromosomes of the species, allowing comparative analysis of other vertebrates and inferences about the ancestral bony vertebrate [99]. Furthermore, analysis of the *Tetraodon* and human genomes shows that whole-genome duplication occurred in the teleost fish lineage, subsequent to its divergence from mammals. Kohn and co-workers [100] applied so-called in silico (bioinformatics) cytogenetics to a large data set of human, chicken, zebrafish, medaka and pufferfish genes, advancing reconstruction of the ancestral vertebrate protokaryotype, which comprises 11 protochromosomes. Although some conflicts between data from in silico and cytogenetic analyses are apparent, increasing taxa sampling and the development of more sophisticated bioinformatic tools will allow for a better match between the cytogenetic and bioinformatic models [101].

Special attention must be paid to the relationship between the data obtained by cytogenetic mapping of genomes and the data provided by complete sequencing of genomes with respect to repetitive DNAs. Unfortunately the bioinformatic tools currently available do not allow the correct assembly and array of repeated DNA clusters. Large clusters of repeated DNAs clearly visible by cytogenetic techniques are not detected by genome sequencing, and dispersed single/low copy repeats detected by genome sequencing are not visualized by classical molecular cytogenetics. However, the development of the next-generation DNA sequencing technology promises improved contributions to fish cytogenomics. Individual microdissected or flow-sorted chromosomes can be completely sequenced and the data comparatively analyzed against chromosomal data. Similarly, whole genomes or chromosomes can be analyzed against microarray platforms, looking for the functional or structural patterns of specific chromosomes or genomes. Such approaches seem to be very efficient for investigating several important questions in the cytogenetic area related to sex and B chromosomes and structural polymorphisms, with special attention to repetitive DNA.

Acknowledgements

The authors are grateful to Drs. Christian Biémont and Juan Pedro Camacho for their helpful suggestions and critical reading of the manuscript. We are also grateful to Dr. César Martins for stimulating discussions, critical reading of the manuscript and substantial contributions during the preparation of the final topic. This work was supported by a grant from the Brazilian agencies FAPESP (Fundação de Amparo à Pesquisa do Estado de São Paulo) and CNPq (Conselho Nacional de Desenvolvimento Científico e Tecnológico).

References

1 Charlesworth B, Snlegowski P, Stephan W: The evolutionary dynamics of repetitive DNA in eukaryotes. Nature 1994;371:215–220.
2 Kidwell MG: Transposable elements and the evolution of genome size in eukaryotes. Genetica 2002; 115:49–63.
3 Biémont C: Within-species variation in genome size. Heredity 2008;101:297–298.
4 Biémont C, Vieira C: Junk DNA as an evolutionary force. Nature 2006;443:521–524.
5 Horvath JE, Bailey JA, Locke DP, Eichler EE: Lessons from the human genome: transitions between euchromatin and heterochromatin. Hum Mol Genet 2001;10:2215–2223.
6 Nelson JS: Fishes of the World, ed 4, New Jersey, Inc. Hoboken, 2006.
7 Venkatesh B: Evolution and diversity of fish genomes. Curr Opin Genet Dev 2003;13:588–592.
8 Gregory TR: Genome size evolution in animals; in Gregory TR (ed): The Evolution of the Genome. San Diego, Elsevier, 2005, pp 3–87.
9 Mayden RL, Tang KL, Conway KW, Freyhof J, Chamberlain S, et al: Phylogenetic relationships of *Danio* within the order Cypriniformes: a framework for comparative and evolutionary studies of a model species. J Exp Zool B Mol Dev Evol 2007;308:642–654.
10 Schwarzacher T: DNA, chromosomes, and in situ hybridization. Genome 2003;46:953–962.
11 Dimitri P, Junakovic N: Revising the selfish DNA hypothesis: New evidence on accumulation of transposable elements in heterochromatin. Trends Genet 1999;15:123–124.
12 Grewal SI, Jia S: Heterochromatin revisited. Nat Rev Genet 2007;8:35–46.

13 Arnheim N: Concerted evolution of multigene families; in Nei M, Koehn RK (eds): Evolution of Genes and Proteins. Sunderland, Sinauer Associates, 1983, pp 38–61.
14 Ugarkovic D, Plohl M: Variation in satellite DNA profiles – Causes and effects. EMBO J 2002;21:5955–5959.
15 Martins C: Chromosomes and repetitive DNAs: a contribution to the knowledge of fish genome; in Pisano E, Ozouf-Costaz C, Foresti F, Kapoor BG (eds): Fish Cytogenetics. Enfield, Science Publishers, 2007, pp 421–453.
16 Lanfredi M, Congiu L, Garrido-Ramos MA, de la Herrrán R, Leis M, et al: Chromosomal location and evolution of a satellite DNA family in seven sturgeon species. Chromosome Res 2001;9:47–52.
17 Ferreira IA, Martins C: Physical chromosome mapping of repetitive DNA sequences in Nile tilapia *Oreochromis niloticus*: Evidences for a differential distribution of repetitive elements in the sex chromosomes. Micron 2008;39:411–418.
18 Volff JN, Bouneau L, Ozouf-Costaz C, Fischer C: Diversity of retrotransposable elements in compact pufferfish genomes. Trends Genet 2003;19:674–678.
19 Volff JN, Korting C, Sweeney K, Schartl M: The non-LTR retrotransposon *Rex3* from the fish *Xiphophorus* is widespread among teleosts. Mol Biol Evol 1999;16:1427–1438.
20 Ozouf-Costaz C, Brandt J, Körting C, Pisavo E, Bonillo C, et al: Genome dynamics and chromosomal localization of the non-LTR retrotransposons *Rex1* and *Rex3* in Antarctic fish. Antarct Sci 2004;16:51–57.
21 Valente GT, Mazzuchelli J, Ferreira IA, Poletto A, Fantinatti BEA, Martins C: Cytogenetic mapping of the retroelements *Rex1, Rex3* and *Rex6* among cichlid fish: new insights on the chromosomal distribution of transposable elements. Cytogenet Genome Res 2011;133:34–42.
22 Oliveira C, Chew JSK, Porto-Foresti F, Dobson MJ, Wright JM: A LINE-like repetitive DNA sequence from the cichlid fish, *Oreochromis niloticus*: sequence analysis and chromosomal distribution. Chromosoma 1999;108:457–468.
23 Gross MC, Schneider CH, Valente GT, Porto JIR, Martins C, Feldberg E: Comparative cytogenetic analysis of the genus *Symphysodon* (Discus fishes, Cichlidae): chromosomal characteristics of retrotransposons and minor ribosomal DNA. Cytogenet Genome Res 2009;127:43–53.
24 Cioffi MB, Martins C, Bertollo LAC: Chromosomal spreading of associated transposable elements and ribosomal DNA in the fish *Erythrinus erythrinus*. Implications for genome change and karyoevolution in fish. BMC Evol Biol 2010;10:271.
25 Ferreira DC, Porto-Foresti F, Oliveira C, Foresti F: Transposable elements as a potential source for understanding fish genome. Mob Genet Elements 2011;1:112–117.
26 Hua-Van A, Le Rouzic A, Maisonhaute C, Capy P: Abundance, distribution and dynamics of retrotransposable elements and transposons: similarities and differences. Cytogenet Genome Res 2005; 110:426–440.
27 Martins C, Wasko AP: Organization and evolution of 5S ribosomal DNA in the fish genome; in Williams CR (ed): Focus on Genome Research. Hauppauge, Nova Science Publishers, 2004, pp 289–318.
28 Prokopowich CD, Gregory TR, Crease TJ: The correlation between rDNA copy number and genome size in eukaryotes. Genome 2003;46:48–50.
29 Martins C, Galetti PM Jr: Organization of 5S rDNA in species of the fish *Leporinus*: Two different genomic locations are characterized by distinct non transcribed spacers. Genome 2001;44:903–910.
30 Sajdak SL, Reed KM, Phillips RB: Intraindividual and interspecies variation in the 5S rDNA of coregonid fish. J Mol Evol 1998;46:680–688.
31 Cioffi MB, Martins C, Bertollo LAC: Comparative chromosome mapping of repetitive sequences. Implications for genomic evolution in the fish, *Hoplias malabaricus*. BMC Genetics 2009;10:34.
32 Gromicho M, Coelho M, Alves M, Collares-Pereira M: Cytogenetic analysis of *Anaecypris hispanica* and its relationship with the paternal ancestor of the diploid-polyploid *Squalius alburnoides* complex. Genome 2006;49:1621–1627.
33 Foster HA, Bridger JM: The genome and the nucleus: a marriage made by evolution. Genome organization and nuclear architecture. Chromosoma 2005;114:212–229.
34 Poleto AB, Ferreira IA, Martins C: The B chromosomes of the African cichlid fish *Haplochromis obliquidens* harbour 18S rRNA gene copies. BMC Genetics 2010;11:1.
35 Martins C, Ferreira IA, Oliveira C, Foresti F, Galetti PM Jr: A tandemly repetitive centromeric DNA sequence of the fish *Hoplias malabaricus* (Characiformes: Erythrinidae) is derived from 5S rDNA. Genetica 2006;127:133–141.
36 Pinhal D, Yoshimura TS, Araki CS, Martins C: The 5S rDNA family evolves through concerted and birth-and-death evolution in fish genomes: an example from freshwater stingrays. BMC Evol Biol 2011;11:151.
37 Kedes LH: Histone genes and histone messengers. Ann Rev Biochem 1979;48:837–870.

38 Nagoda N, Fukuda A, Nakashima Y, Matsuo Y: Molecular characterization and evolution of the repeating units of histone genes in *Drosophila americana:* coexistence of quartet and quintet units in a genome. Insect Mol Biol 2005;14:713–717.
39 Albig W, Warthorst U, Drabent B, Prats E, Cornudella L, Doenecke D: *Mytilus edulis* core histone genes are organized in two clusters devoid of linker histone genes. J Mol Evol 2003;56:597–606.
40 Pendás AM, Morán P, García-Vázquez E: Organization and chromosomal location of the major histone cluster in brown trout, Atlantic salmon and rainbow trout. Chromosoma 1994;103: 147–152.
41 Hashimoto DT, Ferguson-Smith MA, Rens W, Foresti F, Porto-Foresti F: Chromosome mapping of H1 histone and 5S rRNA gene clusters in three species of *Astyanax* (Teleostei, Characiformes). Cytogenet Genome Res 2011;134:64–71.
42 Tautz D, Renz M: Simple sequences are ubiquitous repetitive components of eukaryotic genomes. Nucleic Acids Res 1984;12:4127–4138.
43 Cioffi MB, Kejnovsky E, Bertollo LAC: The chromosomal distribution of microsatellite repeats in the wolf fish genome *Hoplias malabaricus*, focusing on the sex chromosomes. Cytogenet Genome Res 2011;132:289–296.
44 Vanzela ALL, Swarça AC, Dias AL, Stolf R, Ruas PM, et al: Differential distribution of $(GA)_9$+C microsatellite on chromosomes of some animal and plant species. Cytologia 2002;67:9–13.
45 Shimoda N, Knapik EW, Ziniti J, Sim C, Yamada E, et al: Zebrafish genetic map with 200 microsatellite markers. Genomics 1999;58:219–232.
46 Nanda I, Feichtinger W, Schmid M, Schröder JH, Zischler H, Epplen JC: Simple repetitive sequences are associated with differentiation of the sex chromosomes in the guppy fish. J Mol Evol 1990;30:456–462.
47 Blackburn EH: Telomeres: No end in sight. Cell 1994;77:621–623.
48 Schubert I, Schriever-Schwemmer G, Werner T, Adler ID: Telomeric signals in Robertsonian fusion and fission chromosomes: implications for the origin of pseudoaneuploidy. Cytogenet Cell Genet 1992;59:6–9.
49 Meyne J, Baker RJ, Hobart HH, Hsu TC, Ryder OA, et al: Distribution of nontelomeric sites of the (TTAGGG)n telomeric sequence in vertebrate chromosomes. Chromosoma 1990;99:3–10.
50 Chew JSK, Oliveira C, Wright JM, Dobson MJ: Molecular and cytogenetic analysis of the telomeric $(TTAGGG)_n$ repetitive sequences in the Nile tilapia, *Oreochromis niloticus* (Teleostei: Cichlidae). Chromosoma 2002;111:45–52.
51 Cioffi MB, Bertollo LAC: Initial steps in XY chromosome differentiation in *Hoplias malabaricus* and the origin of an X_1X_2Y sex chromosome system in this fish group. Heredity 2010;105:554–561.
52 Mota-Velasco JC, Ferreira IA, Cioffi MB, Ocalewicz K, Campos-Ramos R, et al: Characterization of the chromosome fusions in *Oreochromis karongae*. Chromosome Res 2010;18:575–586.
53 Abuín M, Martínez P, Sánchez L: Localization of the repetitive telomeric sequence (TTAGGG)n in four salmonid species. Genome 1996;39:1035–1038.
54 Dernburg AF, Sedat JW, Hawley RS: Direct evidence of a role for heterochromatin in meiotic chromosome segregation. Cell 1996;86:135–146.
55 Raskina O, Barber JC, Nevo E, Belyayev A: Repetitive DNA and chromosomal rearrangements: speciation-related events in plant genomes. Cytogenet Genome Res 2008;120:351–357.
56 Le Rouzic A, Capy P: The first steps of transposable elements invasion: parasitic strategy vs. genetic drift. Genetics 2005;169:1033–1043.
57 Haaf T, Schmid M, Steinlein C, Galetti PM Jr, Willard HF: Organization and molecular cytogenetics of a satellite DNA family from *Hoplias malabaricus* (Pisces, Erythrinidae). Chromosome Res 1993;1: 77–86.
58 Ferreira IA, Bertollo LAC, Martins C: Comparative chromosome mapping of 5S rDNA and 5S*Hin*dIII repetitive sequences in Erythrinidae fishes (Characiformes) with emphasis on the *Hoplias malabaricus*' species complex. Cytogenet Genome Res 2007;118:78–83.
59 Leggatt RA, Iwama GK: Occurrence of polyploidy in the fishes. Rev Fish Biol Fish 2003;13:237–246.
60 Taylor JS, Braasch I, Frickey T, Meyer A, Van de Peer Y: Genome duplication, a trait shared by 22,000 species of ray-finned fish. Genome Res 2003;13: 382–390.
61 Fontana F, Zane L, Pepe A, Congiu L: Polyploidy in Acipenseriformes: cytogenetic and molecular approaches; in Pisano E, Ozouf-Costaz C, Foresti F, Kapoor BG (eds): Fish Cytogenetics. Enfield, Science Publishers, 2007, pp 385–403.
62 Fontana F, Bruch RM, Binkowski FP, Lanfredi M, Chicca M, et al: Karyotype characterization of the lake sturgeon, *Acipenser fulvescens* (Rafinesque, 1817) by chromosome banding and fluorescent in situ hybridization. Genome 2004;47:742–746.
63 Garrido-Ramos MA, Soriguer MC, de la Herrán R, Jamilena M, Ruiz Rejón C, et al: Morphometric and genetic analysis as proof of the existence of two sturgeon species in the Guadalquivir river. Mar Biol 1997;129:33–39.
64 Charlesworth D, Charlesworth B, Marais G: Steps in the evolution of heteromorphic sex chromosomes. Heredity 2005;95:118–128.

65 Cioffi MB, Camacho JPM, Bertollo LAC: Repetitive DNAs and the differentiation of sex chromosomes in Neotropical fishes. Cytogenet Genome Res 2011; 132:188–194.
66 Dettai A, Bouneau L, Fischer C: FISH analysis of fish transposable elements: tracking down mobile DNA in teleost genomes; in Pisano E, Ozouf-Costaz C, Foresti F, Kapoor BG (eds): Fish Cytogenetics. Enfield, Science Publishers, 2007, pp 361–383.
67 Devlin RH, Nagahama T: Sex determination and sex differentiation in fish: an overview of genetic, physiological, and environmental influences. Aquaculture 2002;208:191–364.
68 Tanaka K, Takehana Y, Naruse K, Hamaguchi S, Sakaizumi M: Evidence for different origins of sex chromosomes in closely related medaka fishes: substitution of the master sex-determining gene. Genetics 2007;177:2075–2081.
69 Nanda I, Volff JN, Weis S, Körting C, Froschauer A, et al: Amplification of a long terminal repeat-like element on the Y chromosome of the platyfish, *Xiphophorus maculatus*. Chromosoma 2000;109: 173–180.
70 Molina WF, Schmid M, Galetti PM Jr: Heterochromatin and sex chromosomes in the Neotropical fish genus *Leporinus* (Characiformes, Anostomidae). Cytobios 1998;94:141–149.
71 Nakayama I, Foresti F, Tewari R, Schartl M, Chourrout D: Sex chromosome polymorphism and heterogametic males revealed by two cloned DNA probes in the ZW/ZZ fish *Leporinus elongatus*. Chromosoma 1994;103:31–39.
72 Parise-Maltempi PP, Martins C, Oliveira C, Foresti F: Identification of a new repetitive element in the sex chromosomes of *Leporinus elongatus* (Teleostei: Characiformes: Anostomidae): new insights into the sex chromosomes of *Leporinus*. Cytogenet Genome Res 2007;116:218–223.
73 Skaletsky H, Kuroda-Kawaguchi T, Minx PJ, Cordum HS, Hillier L, et al: The male-specific region of the human Y chromosome is a mosaic of discrete sequence classes. Nature 2003;423:825–837.
74 Kejnovsky E, Hobza R, Kubat Z, Cermak T, Vyskot B: The role of repetitive DNA in structure and evolution of sex chromosomes in plants. Heredity 2009;102:533–541.
75 Vicente VE, Bertollo LAC, Valentini SR, Moreira-Filho O: Origin and differentiation of a sex chromosome system in *Parodon hilarii* (Pisces, Parodontidae). Satellite DNA, G- and C-banding. Genetica 2003;119:115–120.
76 Cioffi MB, Martins C, Rebordinos L, Vicari MR, Bertollo LAC: Differentiation of the XX/XY sex chromosome system in the fish *Hoplias malabaricus*: Unusual accumulation of repetitive sequences on the X chromosome. Sex Dev 2010;4:176–185.
77 Camacho JPM: B chromosomes; in Gregory TR (ed): The Evolution of the Genome. San Diego, Elsevier, 2005, pp 223–286.
78 Oliveira C, Foresti F, Hilsdorf AWS: Genetics of Neotropical fish: from chromosomes to populations. Fish Physiol Biochem 2009;35:81–100.
79 Carvalho RA, Martins-Santos IC, Dias AL: B chromosomes: an update about their occurrence in freshwater Neotropical fishes (Teleostei). J Fish Biol 2008;72:1907–1932.
80 Alves AL, Porto-Foresti F, Oliveira C, Foresti F: Supernumerary chromosomes in the pufferfish *Sphoeroides spengleri* – first occurrence in marine Teleostean Tetraodontiformes fish. Genet Mol Biol 2008;31:243–245.
81 Jesus CM, Galetti PM Jr, Valentini SR, Moreira-Filho O: Molecular characterization and chromosomal localization of two families of satellite DNA in *Prochilodus lineatus* (Pisces, Prochilodontidae), a species with B chromosomes. Genética 2003;116: 1–8.
82 Schartl M, Nanda I, Schlupp I, Wilde B, Epplen JT, et al: Incorporation of subgenomic amounts of DNA as compensation for mutational load in a gynogenetic fish. Nature 1995;373:68–71.
83 Sampaio T, Gravena W, Gouveia J, Giuliano-Caetano L, Dias A: B microchromosomes in the family Curimatidae (Characiformes): mitotic and meiotic behavior. Comp Cytogenet 2011;5:301–313.
84 Fenocchio AS, Bertollo LAC: Supernumerary chromosome in a *Rhamdia hilarii* population (Pisces, Pimelodidae). Genetica 1990;81:193–198.
85 Lui RL, Blanco DR, Margarido VP, Moreira-Filho O: First description of B chromosomes in the family Auchenipteridae, *Parauchenipterus galeatus* (Siluriformes) of the São Francisco River basin (MG, Brazil). Micron 2009;40:552–559.
86 Maistro EL, Foresti F, Oliveira C, Almeida-Toledo LF: Occurrence of macro B chromosomes in *Astyanax scabripinnis paranae* (Pisces, Characiformes, Characidae). Genetica 1992;87:101–106.
87 Ziegler CG, Lamatsch DK, Steinlein C, Engel W, Schartl M, Schmid M: The giant B chromosome of the cyprinid fish *Alburnus alburnus* harbours a retrotransposon-derived repetitive DNA sequence. Chromosome Res 2003;11:23–35.
88 Alves AL, Martins-Santos IC: Cytogenetics studies in two populations of *Astyanax scabripinnis* with 2n = 48 chromosomes (Teleostei, Characidae). Cytologia 2002;67:117–122.

89 Mestriner CA, Galetti PM Jr, Valentini S, Ruiz IGR, Abel LDS, et al: Structural and functional evidence that a B chromosome in the characid fish *Astyanax scabripinnis* is an isochromosome. Heredity 2000; 85:1–9.

90 McAllister BF, Werren JH: Hybrid origin of a B chromosome (PSR) in the parasitic wasp *Nasonia vitripennis*. Chromosoma 1997;106:243–253.

91 Franks TK, Houben A, Leach CR, Timmis JN: The molecular organisation of a B chromosome tandem repeat sequence from *Brachycome dichromosomatica*. Chromosoma 1996;105:223–230.

92 Langdon T, Seago C, Jones RN, Ougham H, Thomas H, et al: De novo evolution of satellite DNA on the rye B chromosome. Genetics 2000;154:869–884.

93 Dantas ESO, Vicari MR, Souza IL, Moreira-Filho O, Bertollo LAC, Artoni RF: Cytotaxonomy and karyotype evolution in *Moenkhausia* Eigenmann, 1903 (Teleostei, Characidae). Nucleus 2007;50:509–522.

94 Baroni S, Lopes CE, Almeida-Toledo LF: Cytogenetic characterization of *Metynnis maculatus* (Teleostei; Characiformes): the description in Serrasalminae of a small B chromosome bearing inactive NOR-like sequences. Caryologia 2009;62:95–101.

95 Tjio JH, Levan A: The chromosome number of man. Hereditas 1956;42:1–6.

96 Pinkel D, Straume T, Gray JW: Cytogenetic analysis using quantitative, high-sensitivity, fluorescence hybridization. Proc Natl Acad Sci USA 1986;83: 2934–2938.

97 Speicher MR, Ballard SG, Ward DC: Karyotyping human chromosomes by combinatorial multi-fluor FISH. Nat Genet 1996;12:368–375.

98 Martins C, Cabral-de-Mello DC, Valente GT, Mazzuchelli J, Oliveira SG, Pinhal D: Animal Genomes under the Focus of Cytogenetics, ed 1. New York, Nova Science Publisher, 2011.

99 Jaillon O, Aury JM, Brunet F, Petit JL, Stange-Thomann N, et al: Genome duplication in the teleost fish *Tetraodon nigroviridis* reveals the early vertebrate proto-karyotype. Nature 2004;431:946–957.

100 Kohn M, Högel J, Vogel W, Minich P, Kehrer-Sawatzki H, et al: Reconstruction of a 450-My-old ancestral vertebrate protokaryotype. Trends Genet 2006;22:203–210.

101 Froenicke L, Caldés MG, Graphodatsky A, Müller S, Lyons LA, et al: Are molecular cytogenetics and bioinformatics suggesting diverging models of ancestral mammalian genomes? Genome Res 2006;16: 306–310.

Dr. Marcelo de Bello Cioffi
UFSCar – Universidade Federal de São Carlos
Departamento de Genética e Evolução
CP 676, 13565-905 São Carlos, SP (Brazil)
Tel. +55 16 3351 8431, E-Mail mbcioffi@ufscar.br

Author Index

Abbreviations

ALT	alternative lengthening of telomeres
APBs	ALT-associated PML bodies
APE	apurinic-apyrimidinic endonuclease
bp	base pairs
CENP	centromere protein
CBP	CREB binding protein
cDNA	complementary DNA
CSPs	chemosensory proteins
DDR	DNA damage response
DIRS	*Dictyostelium* intermediate repeat sequence
eccDNA	extrachromosomal circular DNA
EGF	epidermal growth factor
EN	endonuclease
ERV	endogenous retrovirus
ETS	external transcribed spacer
FAR	fatty acid reductase
FISH	fluorescence in situ hybridization
FXS	fragile X syndrome
FXTAS	fragile X-associated tremor/ataxia syndrome
GRs	gustatory receptors
HD	Huntington disease
HERV	human endogenous retrovirus
hisDNA	histone DNA
HOR	higher-order repeat
HP1	heterochromatin protein 1
HR	homologous recombination
HTT	*Het-A, TART, TAHRE*
H-type SNBPs	histone-type SNBPs
IAM	infinite allele model
IGS	intergenic spacer (= NTS + ETS)
IN or INT	integrase
IR	ionotropic receptor
ITS	(1) internal transcribed spacer (2) interstitial telomeric site
kb	kilobases
KO	knock out
LCR	low copy repeat

LF-SINEs	living fossil SINEs
LINE	long interspersed element
LTR	long terminal repeat
Mb	megabases
MBD	methyl-CpG binding domain
MHC	major histocompatibility complex
MIR	mammalian-wide interspersed repeat
MITE	miniature inverted-repeat transposable elements
mRNA	messenger RNA
MULEs	*Mutator*-like transposable elements
Mya	million years ago
Myr	million years
NOR	nucleolus organizer region
nt	nucleotide
NTS	non-transcribed spacer
OBPs	odorant-binding proteins
OR	olfactory receptor
ORC	origin of recognition complex
ORF	open reading frame
PCR	polymerase chain reaction
pg	picogram
piRNA	Piwi-interacting RNA
PLE	*Penelope*-like retrotransposon
PL-type SNBPs	protamine-like SNBPs
PML	promyelocytic leukemia
PNTR	perfect non-terminal repeats
Pol	polymerase
PR	proteinase
P-type SNBPs	protamine-type SNBPs
RAN translation	repeat-associated non-ATG translation
Rb	retinoblastoma
rDNA	ribosomal DNA, ribosomal RNA genes
RDRC	RNA-directed RNA polymerase complex
RH	RNase H
RISC	RNAi-induced silencing complex
RITS	RNA-induced transcriptional silencing complex
RLE	restriction-like endonuclease
RNAi	RNA interference
RNP	ribonucleoprotein
rRNA	ribosomal RNA
RT	reverse transcriptase
RTE	retroposon-like transposable element
satDNA	satellite DNA
SD	segmental duplication
SINE	short interspersed element
siRNA	small interfering RNA
SMM	stepwise mutation model
SNBPs	sperm nuclear basic proteins
snoRNA	small nucleolar RNA
SRP	signal recognition particle
SSR	simple sequence repeat

STR	short tandem repeat
TAS	telomere associated sequences
TE	transposable element
TERRA	telomeric repeat-containing RNA
TIR	terminal inverted repeat
TPM	two-phases model
TPRT	target-primed reverse transcription
TR	tandem repeats
TRD	telomere rapid deletion
tRNA	transfer RNA
T-SCE	telomere sister chromatid exchanges
TSD	target-site duplication
UTR	untranslated region
VNTR	variable number of tandem repeats
WGAC	whole-genome assembly comparison
WSSD	whole-genome shotgun sequence detection

Latin Species Names

Subject Index